Physics of
Electric Propulsion

Robert G. Jahn
Professor of Aerospace Sciences
Princeton University

Illustrations by
Woldemar von Jaskowsky

DOVER PUBLICATIONS, INC.
Mineola, New York

Copyright

Copyright © 1968 by McGraw-Hill, Inc.
Copyright © Renewed 1996 by Robert G. Jahn
All rights reserved.

Bibliographical Note

This Dover edition, first published in 2006, is an unabridged republication of the work originally published in 1968 by McGraw-Hill, Inc., New York.

International Standard Book Number

ISBN-13: 978-0-486-45040-7
ISBN-10: 0-486-45040-6

Manufactured in the United States by LSC Communications
45040609   2020
www.doverpublications.com

To **Kit** and the Kids:
Eric
Jill
Nina
Dawn

Preface

The purpose of this book is to provide students of aerospace science and engineering with the specific physical background and engineering concepts underlying the field of electric propulsion. It has been developed in connection with a graduate course in electric propulsion offered for the past several years at Princeton University, and thus tends toward tutorial presentation of a selection of topics appropriate to the needs and previous experience of typical graduate students in the aerospace sciences. The text addresses itself primarily to the fundamentals of those gas acceleration processes which may reasonably be implemented into useful spacecraft propulsion devices, i.e., *thrusters*. The specific topics developed, the examples cited, and the problems posed are selected mainly for their instructional value, rather than for their immediate relevance to currently popular thruster designs. Only in the bibliography is there an attempt to acknowledge more detailed studies and the broader extent of current research.

Part 1 provides a brief review of those aspects of electricity and magnetism and of ionized gas mechanics that underlie the various physical mechanisms of gas acceleration which are the focus of the text. Depending on the preparation of the individual student, some of these sections may be superfluous; others may need to be supplemented by the references. No similar review of elementary fluid dynamics is included, since the typical reader presumably is well versed in the continuum conservation laws from which each of the flows considered is developed, and since several excellent summaries of this topic exist.[1]

The broad division of the bulk of the text, Part 2, into the categories of electrothermal, electrostatic, and electromagnetic acceleration mechanisms conforms to the historical development of the field, and offers some conceptual organization for the new student. It is emphasized, however, that this division should not obscure the possible cooperation of two, or even of all three, classes of interaction in a single gas accelerating device—a cooperation which doubtless will characterize the more sophisticated electric thrusters of the future. Each chapter of Part 2 begins with the most elementary example of the type of acceleration mechanism to be considered, from which the ultimate potential and fundamental limitations of the class of device may be estimated. A succession of refinements and practical complications is then introduced eventually leading to satisfactory analytical representation of realistic engineering devices. Examples of such devices are then displayed, and their performance tabulated for comparison with the heretofore academic analysis. Thereby further refinements in both device and analysis may be discussed.

Throughout these chapters, periodic attempts are made to relate the particular accelerators under discussion to the all-important space power sources on the one hand, and to the equally critical space flight applications on the other. The book closes with an appendix which reviews these two adjacent areas of science and technology, but only in very brief and qualitative terms. Clearly, much more detailed analytic study of these topics must also be required if the student is to claim any breadth of preparation for creative contribution to the electric propulsion field.

The opportunity for such creative contribution is indeed great, for the field of electric propulsion is still very young. New concepts, promising but physically complex and technologically untried, continue to arise and undertake the engineering gauntlet of basic and applied research, development, and ground and space testing prerequisite to useful space application. Although several concepts have already survived this route and have emerged as practical space propulsion devices, and others are well along

[1] See, for example, H. W. Liepmann and A. Roshko, "Elements of Gasdynamics," Galcit Aeronautical Series, John Wiley & Sons, Inc., New York, 1957.

that road, superior ideas are doubtless yet to be proposed. In a field of such youthfulness and vigor, it seems unrewarding to dwell at any length on a survey of contemporary hardware, and little of this appears in this book. To the contrary, it is hoped that this presentation of the more basic elements of the field will foster a highly critical attitude of the student toward the existing technology, to the end that he will eventually generate those new ideas and more sophisticated implementations which will lead electric propulsion toward its maturity.

The list of those who have contributed substantially to the preparation of this book is very long, and even if space permitted, it would be impossible to express adequately my thanks and indebtedness to each by the conventional remarks. To begin with, there is the indispensable feedback from the many graduate students I have instructed who, by their honest challenging of points presented in lectures and notes, and by their invaluable corridor conversations after class, have helped to shape the tutorial format of the text. More particularly, the book reflects the efforts of R. L. Burton, who proofread the draft manuscript, A. C. Eckbreth, who evaluated and helped modify the problems, and N. A. Black, S. Baumgarth, K. E. Clark, W. R. Ellis, P. J. Turchi, P. J. Wilbur, and T. M. York, each of whom contributed some technical amplification or correction of the material presented.

During the preparation of the manuscript, I had the benefit of detailed critical reviews of various chapters by many eminent colleagues from other laboratories and institutions, each of whom is a recognized authority in his particular discipline, and could bring to bear an active research experience with the particular subject matter that I could not offer. I refer especially to Dr. A. T. Forrester, Dr. R. R. John, Mr. H. R. Kaufman, Professor J. L. Kerrebrock, Professor R. H. Lovberg, Professor W. R. Mickelsen, Dr. J. M. Sellen, and Dr. E. Stuhlinger. Other colleagues, too numerous to name, helped resolve particularly troublesome points, or contributed performance data or photographs of equipment otherwise inaccessible to me. Much of the research performed in the author's laboratory was supported by a NASA grant, with the encouragement of Mr. J. Lazar, Mr. J. P. Mullin, and their associates.

Special attention must be drawn to the technical illustrations throughout the book. These are the work of my research colleague, Mr. W. von Jaskowsky, who, beginning with my very sketchy idea for a particular figure, carried each through a series of tedious iterations to its final form. If the book gains any special appeal from its illustrations, it is due almost entirely to Waldo's efforts, supplemented by the pen work of Mr. J. Alcantara, Mr. T. Poli, and Mrs. K. Walter.

A similar priceless dedication to the formal preparation of the manuscript was displayed by my incomparable secretary, Miss Y. Pastor, who now knows far more of the mechanics of such things than I. The total tedium she has deflected from my desk during this three-and-a-half-year effort is staggering to contemplate.

Finally, I must attempt some hopelessly inadequate statement to the members of my family for the somewhat subtler, but no less essential, contribution they have made. To my wife who accepted this intense preoccupation with the same grace with which she accepted me originally, and to my children who accepted "the book" in their lives simply because it was mine, I can only mutter my thanks, and inscribe the dedication with all sincerity.

ROBERT G. JAHN

Contents

Preface   *vii*

List of Symbols   *xv*

PART 1   PHYSICAL BACKGROUND   1

CHAPTER 1   THE PROVINCE OF ELECTRIC PROPULSION   2
    1-1   High Impulse Space Missions   2
    1-2   Exhaust Velocity and Specific Impulse   4
    1-3   Definition of Electric Propulsion   6
    1-4   The Power Supply Penalty   7
    1-5   Historical Outline   9

CHAPTER 2   ELECTROMAGNETIC THEORY   12
    2-1   Electric Charges and Electrostatic Fields   12
    2-2   Currents and Magnetic Interactions   16
    2-3   Time-dependent Fields and Electromagnetic Wave Propagation   20
    2-4   Application to Ionized Gas Flows   23

CHAPTER 3 IONIZATION IN GASES   27
        3-1 Atomic Structure   27
        3-2 Ionization Processes   29
        3-3 Equilibrium Ionization of a Gas   34
        3-4 Ionization of a Gas Mixture   39
        3-5 Limits of Saha Calculations   42

CHAPTER 4 PARTICLE COLLISIONS IN AN IONIZED GAS   45
        4-1 Atomic Collision Nomenclature   46
        4-2 Electron-Atom Collisions   50
        4-3 Electron-Ion Collisions   55
        4-4 Electron-Electron and Ion-Ion Collisions   60
        4-5 Atom-Atom Collisions   61
        4-6 Ion-Atom Collisions   64
        4-7 Other Interactions   67

CHAPTER 5 ELECTRICAL CONDUCTIVITY OF AN IONIZED GAS   69
        5-1 Motion of a Charged Particle in Uniform Steady Electric and Magnetic Fields   70
        5-2 Effect of Collisions   74
        5-3 High-frequency Conductivity   79
        5-4 AC Conductivity in Steady Magnetic Fields   80
        5-5 Electromagnetic Wave Propagation in an Ionized Gas   83

PART 2 ELECTRICAL ACCELERATION OF GASES   89

CHAPTER 6 ELECTROTHERMAL ACCELERATION   90
        6-1 One-dimensional Model   90
        6-2 Enthalpy of High Temperature Gases—Frozen Flow Losses   93
        6-3 Resistojets   103
        6-4 Gaseous Discharges   110
        6-5 Arcjet Operation and Analysis   118
        6-6 Arcjet Thrusters   126
        6-7 Electrodeless-discharge Acceleration   134

CHAPTER 7 ELECTROSTATIC ACCELERATION   142
        7-1 One-dimensional Space-charge Flows   143
        7-2 Production of Positive Ions   148
        7-3 Design of the Accelerating Field   159
        7-4 Neutralization of the Ion Beam   165
        7-5 The Acceleration-Deceleration Concept   173
        7-6 Ion Thruster Design and Performance   177
        7-7 First Flight Test   184
        7-8 Heavy-particle Accelerators   188

CHAPTER 8 ELECTROMAGNETIC ACCELERATION—STEADY FLOW   196
        8-1 Classification of Electromagnetic Accelerators   197
        8-2 Magnetogasdynamic Channel Flow   198
        8-3 Ideal Steady-flow Acceleration   202

- 8-4 Thermal and Viscous Losses *214*
- 8-5 Field Geometry Considerations *216*
- 8-6 Tensor Conductivity—Hall Effect and Ion Slip *219*
- 8-7 Self-induced Fields *227*
- 8-8 Sources of the Conducting Gas *232*
- 8-9 The Magnetoplasmadynamic Arc *235*

CHAPTER 9 UNSTEADY ELECTROMAGNETIC ACCELERATION *257*
- 9-1 Motivation and Classification *257*
- 9-2 Pulsed Plasma Acceleration *259*
- 9-3 Circuit Analyses of Pulsed Accelerators *263*
- 9-4 Dynamical Models *269*
- 9-5 Dynamical Efficiency of Pulsed Acceleration *280*
- 9-6 Initiation of the Discharge *284*
- 9-7 Ejection of the Plasma *287*
- 9-8 Coaxial Guns and Pinch Accelerators *288*
- 9-9 Energy Storage and Switching *297*
- 9-10 Quasi-steady Acceleration *302*
- 9-11 Pulsed Inductive Acceleration *304*
- 9-12 Traveling Wave Acceleration *306*

APPENDIX SPACE POWER SUPPLIES AND LOW THRUST MISSION ANALYSIS *317*

Index *327*

List of Symbols

A	dimensional coefficient [12],[1] area [16], an atomic particle [30]
a	sound speed [206]
a_T	isothermal sound speed [203]
B	an atomic particle [32]
B	magnetic induction field [18]
$\mathfrak{B}$	magnetic stress tensor [244]
b	unit vector in direction of magnetic field [82]
C	electric capacitance [16]
c_0	propagation speed of electromagnetic waves in vacuum [23]
c_p	specific heat at constant pressure, per unit mass
D	electric displacement field [15]
d	width dimension [16], electrode gap [111]
dl	line element [17]
dP	collision probability [48]
ds	surface area element [16]
E	electric field [13]

[1] Page numbers indicate the first appearance of each symbol.

ε	energy of particle entering collision [53]; energy constant [229]
e	base of natural logarithms; an electron [30], internal energy per unit mass [94]
F	total partition function [35], mass flux [203]
F	force vector
$\mathbf{F}_g$	gravitational force on rocket [2]
f	internal partition functions [35], distribution function [170]
$\mathbf{f}_B$	magnetic body force density [197]
G, G'	dimensional coefficients [37, 38]
g	degeneracy [36]
g_0	gravitational acceleration at sea level [4]
H	magnetic field [19]
h	Planck's constant [30], enthalpy per unit mass [94], channel height [265]
I	total impulse of rocket [3], impurity particle [64]
I_s	specific impulse of rocket [4]
I_{ff}	free-free radiation intensity [59]
J	total current [17]
j	current density [16]
K_n	equilibrium constant [35]
k	propagation exponent [23], Boltzmann's constant [35], magnitude of wave propagation vector [83]
k	wave propagation vector [83]
L	self-inductance coefficient [21], channel length [210]
l	channel length [266]
ln	natural logarithm
M	mass of particle [35], Mach number [207]
M_{12}	mutual inductance coefficient [21]
m	mass; mass of electron [36]
m	magnetic dipole moment per unit volume [19]
N, n	number density of particles [13, 16]
n	unit vector normal to surface or line [245]
P	electric power [91], power input per unit volume [122], power input per unit length [204]
P	dielectric polarization field [13]
$\mathcal{P}$	perveance of ion accelerator [165]; momentum constant [229]
p	molecular polarizability [13], gas pressure [37]
p	particle momentum
p_m	molecular magnetic susceptibility [19]
Q	collision cross section [48], total electrostatic charge [16]
$\mathcal{Q}$	effective cross section [48]
q	electric charge [12], differential cross section [48]
R	ideal gas constant [203], electric resistance [263]

R_B	magnetic Reynolds number [228]
$\mathcal{R}$	aspect ratio of ion accelerator [164]
$\mathbf{r}$	position vector
r_B	gyro radius of charged particle in magnetic field [70]
T	absolute temperature; period of circuit oscillation [268]
$\mathbf{T}$	thrust of rocket [3]
t	time
U	shock speed [275]
$u, \mathbf{u}$	streaming velocity of propellant along accelerator channel [2]
V	electric potential [15]
V_c	sparking potential [111]
$\mathcal{V}$	electromotive force [20]
$\mathbf{v}$	velocity of rocket [3], velocity of particle [16]
$\dot{\mathbf{v}}$	acceleration of rocket [2]
v_p	phase velocity
v_s	slip velocity
W	work, or energy
W_0	energy stored in capacitor bank [265]
X	an unspecified particle [30]
Y	an unspecified particle [43]
Δx	space-charge neutralization gap [166]
Δv	characteristic velocity increment of a rocket [3] or of a given space mission [4]

SUBSCRIPTS

A	refers to atom [35]
a	related to accelerating grid [143], related to anode [243]
B	derived from magnetic field
b	bound state [60]
c	chamber conditions [91], related to constrictor [126], related to cathode [243]
d	related to dissociation [38]
E	derived from electric field
e	condition at exhaust or ejection from accelerator [2], refers to electron [35]
f	final condition [3], free state [59]
H	Hall component [201]
i	imaginary component [80], related to ionization [29]
i, j, k, l, m	running indices
L	condition at channel end [228]
m	magnetic quantity [19]
p	related to power plant [10]
r	real component; related to rotation of molecule [95]

r, θ, z	cylindrical components of vector
s	surface quantity [16]
t	total [14]
u	useful [10]
v	related to vibration of molecule [95]
x, y, z	cartesian components of vector
0	initial condition; sea-level condition [4], vacuum condition [13], amplitude [21], ground state [36], bias value [80]
1	denotes atomic species [94]
2	denotes diatomic molecule [38]
+	refers to positive ion [35]
⊥	perpendicular, or normal, component [16]
‖	parallel component [16]

SUPERSCRIPTS

.	single dot, directly over quantity denotes derivative with respect to time
⌢	directly over quantity denotes optimum or maximum value [8] or unit vector [12]
−	directly over quantity denotes mean value, value per unit volume of a particulate property [14], or value relative to a different reference [36]
∼	directly over symbol denotes high kinetic energy [30], a fictitious reference quantity [71], or a nondimensionalized quantity [79]
′	prime denotes effective or derived property [14, 18, 25] or dimensionless differentiation [270]
+	refers to positive ion [30]
−	refers to negative ion [34]
*	denotes an excited state [32], a complex quantity [80], or a nondimensional quantity [203]
**	denotes a doubly excited state [34]
0	denotes initial condition [40]
t	translational [35]
i	internal [35]
(β)	collision of class β
(p)	momentum transfer
(ε)	energy transfer

GREEK SYMBOLS

α	specific mass of power plant [7], degree of ionization [36], fraction of particles in a mixture [94], reference angle [224], scaling parameter [270], energy-division ratio [281]

β	class of collision [48], fraction of particles excited [95], magnetic interaction parameter [205], scaling parameter [270]
γ	ratio of specific heats [206]
Δ	beam half-width [160]
δ	phase angle [267]
δ_{ij}	Kronecker delta
ϵ	electric permittivity [15]
ϵ_0	electric permittivity of free space [13]
ε_d	dissociation energy [38]
ε_i	ionization potential [29]
ε_j	energy of jth electronic state [36]
ζ	a characteristic speed [208]
η	overall efficiency of conversion of electric power into thrust power [7, 290], a characteristic speed [207]
η_d	dynamic efficiency [281]
η_e	current-pulse efficiency [265]
η_f	frozen flow efficiency [93]
θ	scattering angle [48], various reference angles
κ	relative electric permittivity or dielectric constant [13], thermal conductivity [123]
κ_m	relative magnetic permeability [19]
λ	wavelength [23], mean free path [50]
λ_D	Debye length [57]
μ	magnetic permeability [19], scaling parameter [273]
μ_0	magnetic permeability of free space [18]
ν	frequency [30], collision frequency [50], pulse repetition rate [290]
ν_c	effective collision frequency [75]
ξ	frozen flow fraction [98], E/B [203]
π	$\equiv 3.1416$
ρ	mass density
σ	electrical conductivity (scalar) [17]
$\boldsymbol{\sigma}$	electrical conductivity (tensor) [82]
τ	pulse length [265], period of traveling wave [310]
ϕ	magnetic flux [20], polar angle [49], work function [149]
Ω	solid angle [49], Hall parameter [219]
ω	angular frequency [21]
ω_B	gyro frequency of charged particle in magnetic field [70]
ω_p	plasma frequency [79]

Physics of
Electric Propulsion

part one
Physical Background

1
The Province of Electric Propulsion

1-1 HIGH IMPULSE SPACE MISSIONS

The primary attraction of electric thrusters for the propulsion of spacecraft lies in their highly efficient utilization of propellant mass. The corresponding reduction in the propellant supply which must be contained and transported in the spacecraft permits the inclusion of a greater portion of useful payload and the achievement of space missions inaccessible to conventional chemical rockets. Rigorous demonstration of these potentialities involves detailed analyses of specific missions, but the essential concept may be illustrated by basic dynamical arguments.

The flight of a simple rocket in a gravitational field is described by the vector differential equation of motion [1],[1]

$$m\dot{\mathbf{v}} = \dot{m}\mathbf{u}_e + \mathbf{F}_g \qquad (1\text{-}1)$$

[1] Bracketed numbers indicate the References at the end of each chapter.

where $\dot{\mathbf{v}}$ = acceleration vector of rocket
$\dot{m}$ = rate of change of rocket mass by exhaust of propellant (a negative quantity)
$\mathbf{u}_e$ = exhaust velocity relative to rocket
$\mathbf{F}_g$ = local gravitational force

The first term on the right is commonly identified as the thrust of the rocket,

$$\mathbf{T} = \dot{m}\mathbf{u}_e \qquad (1\text{-}2)$$

and its integral over a complete mission is called the total impulse,

$$I = \int_{t_0}^{t_f} T\, dt \qquad (1\text{-}3)$$

For a mission of large total impulse requirement, it is apparent that the desired thrust should be achieved via high exhaust velocity rather than by excessive ejection of propellant mass, lest the craft be committed to an intolerably large initial propellant mass fraction. As a simple example, if the rocket operates at constant u_e in a region where the local gravitational field is negligible in comparison with the thrust, or if it exhausts its propellant over a negligibly short interval of time (impulsive thrust), the equation of motion integrates directly to the scalar form

$$\Delta\mathbf{v} = \mathbf{u}_e \ln \frac{m_0}{m_0 - \Delta m} \qquad (1\text{-}4)$$

where $\Delta\mathbf{v}$ is the magnitude of velocity increment achieved by the ejection of Δm of the initial mass m_0. By expending all its propellant mass in this way, the rocket can attain a maximum velocity increment

$$\Delta\mathbf{v} = \mathbf{u}_e \ln \frac{m_0}{m_f} \qquad (1\text{-}5)$$

where m_f includes the mass of the rocket casing, engine, tankage, etc., plus useful payload. Conversely, the fraction of the original rocket mass which can be accelerated through a given velocity increment Δv is a negative exponential in the ratio of that increment to the exhaust speed:

$$\frac{m_f}{m_0} = e^{-\Delta v / u_e} \qquad (1\text{-}6)$$

Clearly, it is necessary to provide u_e comparable with Δv if a significant fraction of the original mass is to be brought to the final velocity.

More complicated missions of practical interest, involving flight through planetary, lunar, or solar gravitational fields, with variable magnitude and direction thrust programs, staging, etc., can also be represented by characteristic velocity increments Δv, each of which satisfies relation (1-6) for the particular mission involved [2]. In general, long-range missions, such as interplanetary flights, or long-time missions, such as the maintenance of satellite position and orientation for several years, are characterized by correspondingly large Δv. For example, detailed analyses of certain interplanetary missions yield the characteristic velocity increments shown in Table 1-1.

Table 1-1 Characteristic velocity increments for planetary transfer missions

Mission	Δv, m/sec
Escape from earth surface (impulsive)	1.12×10^4
Escape from 300-mile orbit (impulsive)	3.15×10^3
Escape from 300-mile orbit (gentle spiral)	7.59×10^3
Earth orbit to Mars orbit and return†	1.4×10^4
Earth surface to Mars surface and return†	3.4×10^4
Earth orbit to Venus orbit and return†	1.6×10^4
Earth orbit to Mercury orbit and return†	3.1×10^4
Earth orbit to Jupiter orbit and return†	6.4×10^4
Earth orbit to Saturn orbit and return†	1.1×10^5

† Values are quoted for typical impulsive missions over minimum propellant semiellipse trajectories.

1-2 EXHAUST VELOCITY AND SPECIFIC IMPULSE

The propellant exhaust velocity u_e, which ideally should be comparable with the mission Δv, is determined by the detailed nature of the acceleration of the propellant gas within the rocket. It is directly related to another characteristic parameter of the rocket engine, the specific impulse, defined to be the ratio of thrust to the rate of use of propellant by sea-level weight:

$$I_s = \frac{\dot{m} u_e}{\dot{m} g_0} = \frac{u_e}{g_0} \quad \text{sec} \qquad (1\text{-}7)$$

where g_0 is the sea-level gravitational acceleration. In mks units, u_e and I_s are thus conveniently related by a factor of approximately 10. In conventional chemical rockets, the thrust is obtained by nozzled expansion of a propellant previously heated by its own chemical reaction. As such, the attainable exhaust velocity is limited by three factors: (1) the intrinsic energy available in the chemical reaction and convertible to chamber enthalpy of the gas, (2) the tolerable heat transfer to the combustion chamber or nozzle throat, and (3) the unrecoverable energy deposition in the internal modes of the gas, sometimes called "frozen flow losses," and radiation losses from the exhaust jet. The limiting performances of the more common chemical propellants are well known from practical experience, and are summarized in Table 1-2.

Table 1-2 Typical exhaust velocities to vacuum for chemical rockets

Propellant type	u_e, m/sec
Liquid monopropellants	1,700–2,900
Solid propellants	2,100–3,200
Liquid bipropellants (fuel and oxidizer)	2,900–4,500
"Exotic" bipropellants and tripropellants	4,000–6,000

The theoretically attainable exhaust velocities of the "exotic" reactions, such as beryllium-oxygen, hydrogen-ozone, etc., have been estimated by numerical calculations [3]. These higher performance chemical propellants tend to be difficult to handle and store. At the elevated chamber temperatures they produce, their combustion products may be extremely corrosive, and their attainable exhaust velocities may be limited by structural and material considerations rather than by their intrinsic reaction energies. Even at the values quoted, however, the best chemical rocket falls far short of the desired exhaust speed for planetary missions.

One concerted effort to overcome the basic limitations of chemical propulsion has been directed toward a nuclear-thermal rocket [4]. Conceptually, this device heats its propellant gas by passing it through the active elements of a nuclear reactor, thereby removing all dependence on the intrinsic chemical energy of the propellant. The thermal limitations on specific impulse here are thus primarily structural and material, and these are less severe for a given exhaust speed than for the exotic chemical rockets, say, since noncorrosive low-molecular-weight propellant gases may be used. A typical level of specific impulse for nuclear rockets is about 800 sec for a solid-core reactor running at 2000°K, using hydrogen as a propellant. Somewhat higher potentialities are forecast by those engaged in exploratory work on liquid- and gas-core reactors [5,6].

It should be noted that spacecraft of this class must carry their heating unit, i.e., the nuclear reactor, over the full thrust duration of the mission, and that the mass of this unit must logically be subtracted from the delivered payload when comparing such systems with chemical rockets. From this standpoint, although the solid-core nuclear rocket displays substantial advantages over high performance chemical units for many tasks, it still does not achieve the optimally high exhaust speeds for most deep space missions.

1-3 DEFINITION OF ELECTRIC PROPULSION

To obtain the exhaust velocities above 10,000 m/sec desirable for most planetary missions, it is evident that processes basically different from the simple heating of a propellant stream by chemical reaction or by solid-element heat transfer must be employed. If a thermal acceleration mechanism is to be retained at all, the heating must be accomplished by an external agent that has considerably higher heat release capability, and equally important, the propellant gas so heated must be constrained in some fashion away from the material walls of the heating chamber and nozzle. Even such sophisticated heaters will not provide the upper range of desired exhaust speeds, however. To achieve this domain, body forces must be directly applied to the propellant stream. Either process—high enthalpy heating of an insulated gas stream or direct acceleration of it by applied body forces—is most reasonably accomplished by electrical means. We are thus led to a definition of electric propulsion:

> *The acceleration of gases for propulsion by electrical heating and/or by electric and magnetic body forces.*

In more detail, this definition may be empirically subdivided into three distinct, but not necessarily isolated, concepts:

1. *Electrothermal propulsion*, wherein the propellant gas is heated electrically, then expanded in a suitable nozzle.
2. *Electrostatic propulsion*, wherein the propellant is accelerated by direct application of electric body forces to ionized particles.
3. *Electromagnetic propulsion*, wherein an ionized propellant stream is accelerated by interactions of external and internal magnetic fields with electric currents driven through the stream.

The exhaust velocities attainable by these mechanisms are found to be more than adequate to qualify for the large-velocity-increment space missions outlined above. Indeed, in thrusters[1] employing electrostatic

---

[1] In the technical literature of this and other space propulsion areas, the term *thruster*, or *thrustor*, is commonly used to denote any device which accelerates and ejects a propellant stream, thereby imparting thrust to the spacecraft on which it is mounted. The proper choice of the suffix is admittedly an academic point to this text, but since the literature displays disagreement on the subject, we have solicited editorial assistance in an attempt to resolve the issue. We are indebted to Mr. John Brinkley of the Princeton University Classics Department for the following statement:

THE PROVINCE OF ELECTRIC PROPULSION

and electromagnetic interactions, exhaust speeds higher than those needed for any interplanetary mission are readily attainable, implying possible extension of these concepts to flights beyond our solar system.

1-4 THE POWER SUPPLY PENALTY

Before blindly striving for extreme exhaust velocities, it is important to recognize that, like nuclear rockets, all electric thrusters require separate energy sources, in this case, electric power supplies. Thus, so far as space applications are concerned, enthusiasm for high specific impulses must be tempered by considerations of the power plant weight needed to drive these engines. For example, on a mission of given thrusting time Δt, performed at a given constant thrust level T, the total mass of propellant expended is indeed inversely proportional to the exhaust velocity or specific impulse:

$$\Delta m = \dot{m}\,\Delta t = \frac{T\,\Delta t}{u_e} = \frac{T\,\Delta t}{I_s g_0} \qquad (1\text{-}8)$$

The mass of the power supply needed, however, will probably scale monotonically with the power level involved, and hence directly with the specific impulse. Assuming a linear dependence and a constant conversion efficiency to thrust power η,

$$m_p = \alpha P = \frac{\alpha T u_e}{2\eta} = \frac{\alpha g_0 T I_s}{2\eta} \qquad (1\text{-}9)$$

In forming the agent-noun from the base *thrust*, established rules of English orthography require that the suffix be spelled *-er*. The applicable rules, representing general usage since at least the eighteenth century, may be summarized as follows. On all bases of non-Latin origin (e.g., *drive*, *push*) and on most of Latin derivation, such as nouns (e.g., *orbit*) and the present stems (the "first principal parts") of verbs (e.g., *design-*), the appropriate suffix is *-er;* only if the base is patently a Latin perfect-participial stem (the "fourth principal part") of a verb (e.g., *regulat-*, *compress-*) is the appropriate suffix *-or*. Apparent exceptions to the rule about *-or* occur in certain words borrowed from the Old French as technical terms of the law (e.g., *consignor*, *bailor*, *juror*); such words as *waiver*, also a legal technical term from Old French, do not in fact constitute a counter-exception even to this small category, having a quite different history. Clearly, as concerns new formations, *-er* is the "productive" suffix, and *-or* must be justified according to the strict specification above, which certainly does not apply to *thrust*. Occasionally, by the process of "false analogy," the motivation of which is often obscure, a spelling with *-or* is inappropriately used in a newly coined word, as happened a generation ago when some authorities used *propellor*, which has been almost universally replaced by the regular *propeller*. Note that pronunciation is not in question here, as both agent-suffixes are pronounced alike in Modern English, except in affected speech. Rather than risk affectation, we shall hereafter spell and pronounce the word *thruster*.

and it follows that there is an optimum I_s to maximize deliverable payload fraction on this mission (Fig. 1-1). This optimum value depends paramet-

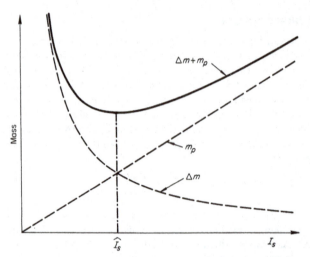

Fig. 1-1 Dependence of propellant mass Δm and power supply mass m_p on specific impulse I_s for a given constant-thrust mission.

rically on the specific power plant mass α, on the conversion efficiency η, and on the mission time Δt, but is independent of the thrust:

$$\hat{I}_s = \frac{1}{g_0}\left(\frac{2\eta\,\Delta t}{\alpha}\right)^{1/2} \tag{1-10}$$

Normally, it is more pertinent to specify a mission in terms of other combinations of parameters, say by the characteristic velocity increment requirement and by the available thrust of the rocket, from which more complicated expressions for $\hat{I}_s$ would develop (Prob. 1-3). More detailed analyses would include effects of imperfect conversion efficiency, variable thrust and specific impulse, and additional component mass contributions [6], but regardless of the details of the optimization procedure, the inescapable importance of power supply weight to the overall performance of an electric propulsion unit is emphasized, and three important points emerge: (1) for a given specific power plant mass, operation of a thruster at too high I_s will reduce payload ratio, just as will operation at too low I_s; (2) a premium is placed upon the efficiency with which the thruster converts electric power input to thrust power of the jet; and (3) the development of space power plants of low specific mass assumes importance comparable with the production of high specific impulse thrusters.

As one mitigating feature, it may well turn out that the power supply used to drive the thruster retains some utility at the destination, and as such may fairly be regarded as payload, at least in part. In other cases, the thruster may be used on a power-bearing satellite, such as a space station or a communications relay or transmitter. In these situations, the optimum I_s will increase.

The importance of high exhaust speed, tempered with considerations of power supply mass, is not confined to long-range interplanetary missions. Essentially similar arguments can be made with respect to certain near-earth tasks where minimization of propellant mass is also important, such as in the attitude control, station keeping, and orbit adjustment of long-lived earth satellites, or in cargo "ferry" missions between earth and lunar orbits. In the former, electric propulsion adds certain other advantages, such as a delicacy of control of thrust level over a wide range, and essentially unlimited restart capability not shared by chemical rockets. In the latter, reusability of the power plant and thruster for many missions projects the propellant conservation factor into a primary economic consideration.

1-5 HISTORICAL OUTLINE[1]

The physical concepts of electric propulsion have been recognized for many years. Robert H. Goddard, in 1906, expressed many of them informally [7], and Professor Herman Oberth, in 1929, included a chapter on electric propulsion in his classic book on rocketry and space travel [8]. For reasons outlined above, no serious research efforts were mounted until practical nuclear-fission power sources, solar panels, and other lightweight space electric power plants became a reasonable expectation. In fact, it was not until 1948 that the first serious feasibility studies of an integrated space flight system employing electric propulsion were undertaken [9]. Much of the early survey work in this country was done by Ernst Stuhlinger [10–12], whose initial publications had considerable effect on the subsequent development of the field.

Although physicists had long been familiar with the basic electrical acceleration mechanisms, actual electric propulsion experiments did not begin until the late 1950s, and then only on a small scale in a few government laboratories and independent companies. During this period, for example, the application of thermal arcjets to propulsion was first studied

---

[1] In his book "Ion Propulsion for Space Flight" [7], Stuhlinger presents a more detailed review of the historical development of the electric propulsion field, including certain elements of personal experience and insight only he could contribute.

[13]. Over the same period, in the electrostatic domain, experiments on ion production, beam neutralization, electrode design, and the construction of small ion thrusters were undertaken by several aircraft laboratories and private research corporations [14,15]. Electromagnetic acceleration experiments evolved from earlier work on T tubes, button guns, and other laboratory-scale plasma generators [16,17]. Although magnetohydrodynamic channel flows had been studied for some time, their application to propulsion was also delayed until this period, when techniques were developed for seeding gases to higher levels of electrical conductivity [18].

From these nuclei of early work the field blossomed abruptly around 1960 to become a major component of the space industry. Shortly thereafter, several models of each class of thruster were developed for space testing, and peripheral programs concerning the design and interpretation of these early tests, the performance of correlated ground tests in large vacuum tanks, the development of suitable power supplies and power conditioning equipment, and the detailed analyses of missions appropriate to electric propulsion occupied a steadily increasing number of research and development laboratories.

Logistical delays in the provision of suitable spacecraft and test capsules deferred the first successful space test of an electric thruster until July 20, 1964, at which time an electrostatic ion engine satisfactorily executed a prescribed program of thrust generation over a 25-min ballistic flight [19]. Other space tests followed shortly thereafter, and the electric thrusters thereby laid claim to a small niche in the space engine arsenal. The breadth of utility which they can ultimately achieve for the rapidly approaching era of extensive space flight will depend on man's cleverness in exploiting the physical phenomena on which they are based. It is to this phenomena that the following text is addressed.

PROBLEMS

1-1. Derive Eq. (1-1).

1-2. Verify the first three characteristic velocities quoted in Table 1-1. Explain why the values for gentle spiral ascent exceed those for impulsive ascent for a given earth orbit transfer.

1-3. For a mission characterized by a velocity increment Δv and a rocket of constant thrust level T, derive an expression for the optimum specific impulse in terms of power supply specific mass α and the initial total mass of the spacecraft.

1-4. If the mass delivered by a rocket, m_f, is separated into useful payload m_u and the power-producing system mass $m_p = \alpha P$, show that Eq. (1-6) takes the form

$$\frac{m_u}{m_0} = e^{-\Delta v/u_e} - \frac{\alpha u_e^2}{2\eta \, \Delta t}\left(1 - e^{-\Delta v/u_e}\right)$$

where Δt is the thrusting time, and η is the conversion efficiency. What is the optimum u_e for this mission?

REFERENCES

1. Goldstein, H.: "Classical Mechanics," chap. 1, Addison-Wesley Publishing Company, Inc., Reading, Mass., 1950.
2. Berman, A. I.: "The Physical Principles of Astronautics," pt. III, John Wiley & Sons, Inc., New York, 1961.
3. Dobbins, T. O.: Thermodynamics of Rocket Propulsion and Theoretical Evaluation of Some Prototype Propellant Combustions, *Wright Air Develop. Center Tech. Rept.* 59-757, December, 1959.
4. Bussard, R. W., and R. D. DeLauer: "Nuclear Rocket Propulsion," McGraw-Hill Book Company, New York, 1958.
5. Fox, R. H.: Nuclear Propulsion, *Astronaut. Aerospace Eng.*, vol. 1, no. 10, November, 1963.
6. Mickelsen, W. R.: Performance Parameters for Electric-propulsion Systems, *J. Spacecraft Rockets*, vol. 3, no. 2, p. 213, 1966.
7. Stuhlinger, E.: "Ion Propulsion for Space Flight," chap. 1, McGraw-Hill Book Company, New York, 1964.
8. Oberth, H.: "Man into Space," Harper & Row, Publishers, Incorporated, New York, 1957.
9. Shepherd, L. R., and A. V. Cleaver: The Atomic Rocket, *J. Brit. Interplanet. Soc.*, vol. 7, p. 185, 1948; vol. 8, pp. 23, 50, 1949.
10. Stuhlinger, E.: Possibilities of Electrical Space Ship Propulsion, *Ber. 5th Intern. Astronaut. Kongr.*, Innsbruck, p. 100, August, 1954.
11. Stuhlinger, E.: Electrical Propulsion Systems for Space Ships with Nuclear Power Source, *J. Astronaut.*, vol. 2, p. 149, 1955; vol. 3, pp. 11, 33, 1956.
12. Stuhlinger, E.: Flight Path of an Electrically Propelled Space Ship, *Jet Propulsion*, vol. 27, no. 4, p. 410, April, 1957.
13. Giannini, G. M.: The Plasma Jet and Its Application, *Office Sci. Res. Tech. Note* 57-520, 1957.
14. Speiser, R. C., C. R. Dulgeroff, and A. T. Forrester: Experimental Studies with Small Scale Ion Motors, 14th Annual Meeting of American Rocket Society, Washington, D.C., November, 1959, *ARS Preprint* 926.
15. Langmuir, D. B.: Low-thrust Flight: Constant Exhaust Velocity in Field-free Space, chap. 9 in H. S. Seifert (ed.), "Space Technology," John Wiley & Sons, Inc., New York, 1959.
16. Bostick, W.: Plasma Motors, *Advan. Astronaut. Sci.*, vol. 2, p. 2-1, American Rocket Society, Plenum Press, Inc., New York, 1957.
17. Clauser, M. U.: Magnetohydrodynamics, chap. 18 in H. S. Seifert (ed.), "Space Technology," John Wiley & Sons, Inc., New York, 1959.
18. Kantrowitz, A.: Introducing Magnetohydrodynamics, *Astronautics*, vol. 3, no. 10, October, 1958.
19. Cybulski, R. J., et al.: Results from SERT I Ion Rocket Flight Test, *NASA Tech. Note* D-2718, 1965.

2
Electromagnetic Theory

2-1 ELECTRIC CHARGES AND ELECTROSTATIC FIELDS

All the physical phenomena described as *electricity and magnetism* stem from the existence in nature of two types of discrete electric charge, positive and negative, as carried by protons and electrons, respectively, in exactly equal amount. Such charge is known to add linearly, so that heavier particles such as atoms, molecules, colloids, etc.—indeed, all macroscopic bodies, including spacecraft, satellites, and even planets—may carry a net charge of any integral multiple of the fundamental unit. Charged microscopic particles are called ions, and a substance containing a significant amount of them is said to be ionized.

In those situations where the electric charges are fixed (electrostatics), it is observed that like charges repel and unlike attract, with a force colinear with their coaxis, proportional to the product of their charge magnitudes and inversely proportional to the square of their separation (Coulomb's law):

$$\mathbf{F}_{ij} = A \frac{q_i q_j}{r_{ij}^2} \hat{\mathbf{r}}_{ij} \qquad (2\text{-}1)$$

ELECTROMAGNETIC THEORY

(Here and throughout, $\mathbf{r}_{ij} \equiv \mathbf{r}_i - \mathbf{r}_j$, and $\hat{\mathbf{r}}$ denotes a unit vector.) The proportionality constant A depends on the units employed; in the mks system it is most convenient to express it in the form $A = 1/4\pi\epsilon_0$, where $\epsilon_0 = 8.85 \times 10^{-12}$ farad/m, for q_i, q_j in coulombs, r_{ij} in meters.

In principle, the total electrostatic force on any charged body can be computed by vector summation of all such coulomb forces exerted by surrounding charges. It is more useful, however, to introduce an intermediary electric field E, defined as the force felt by a unit test charge at the particular position in the field:

$$\mathbf{F}_i = \mathbf{E}_i q_i \tag{2-2}$$

$\mathbf{E}_i$ is thus determined by the vector summation of the fields from the individual source charges:

$$\mathbf{E}_i = \frac{1}{4\pi\epsilon_0} \sum_j \frac{q_j}{r_{ij}^2} \hat{\mathbf{r}}_{ij} \tag{2-3}$$

The reaction of the charges involved in a particular problem to these mutual forces depends on the medium in which they reside: If this medium is a conductor, they will move about, thereby establishing currents which can exert additional forces on each other. If the medium is an insulator, the insulator molecules will respond to the electric field by polarizing, or by aligning themselves with the field if they are normally polar, thereby reducing the net local field intensity. In elementary theory this effect is represented by definition of a *dielectric constant* of the medium κ, in terms of which Coulomb's law becomes

$$\mathbf{F}_{ij} = \frac{1}{4\pi\kappa\epsilon_0} \frac{q_i q_j}{r_{ij}^2} \hat{\mathbf{r}}_{ij} \tag{2-4}$$

so that to retain the definition of E as a force per unit test charge, E also is reduced by the factor κ from relation (2-3).

In more detailed formulations, the response of the constituent molecules of the dielectric medium to the applied E field is described by a polarization field P, which is essentially a summation of the individual molecular dipoles induced by the external field. Specifically, P is related to the molecular number density N and to the individual molecular polarizability p in the form [1]

$$\mathbf{P} = \frac{3\epsilon_0 N p}{3\epsilon_0 - N p} \mathbf{E} \tag{2-5}$$

from which the dielectric parameter κ can then be expressed in terms of p (Clausius-Mosotti relation):

$$\frac{\kappa - 1}{\kappa + 2} = \frac{Np}{3\epsilon_0} \qquad (2\text{-}6)$$

For certain anisotropic media, **P** may not be parallel to **E**; hence p and κ are tensors, in general. In gases, however, p and κ are invariably scalar and, for all but very strong fields, independent of **E**. Since in a gas the molecules are relatively far apart, the polarization field is comparatively small, and hence κ is very nearly 1. In this case, Eq. (2-6) degenerates to the linear form (Gladstone-Dale relation)

$$\kappa - 1 = \frac{Np}{\epsilon_0} \qquad (2\text{-}7)$$

Typical values of κ for gases are shown in Table 2-1.

Table 2-1 Dielectric constants for various gases (1 atm and 20°C)

Gas	κ
Air	1.000536
Argon (A)	1.000517
Carbon dioxide (CO_2)	1.000922
Helium (He)	1.000065
Hydrogen (H_2)	1.000254
Nitrogen (N_2)	1.000548
Oxygen (O_2)	1.000495

SOURCE: D. E. Gray (ed.), "American Institute of Physics Handbook," McGraw-Hill Book Company, New York, 1957.

In situations involving more than just a few isolated source charges q_i, it is normally more convenient to introduce a charge density per unit volume $\bar{q}$ and to replace (2-3) by a divergence relation for **E** in terms of $\bar{q}$. In so doing it is important to recognize that the source of **E** is the total charge density, whatever its cause. In particular, the net charge density introduced by dielectric polarization must be included as a source of **E**:

$$\nabla \cdot \mathbf{E} = \frac{\bar{q}_t}{\epsilon_0} = \frac{\bar{q} + \bar{q}'}{\epsilon_0} \qquad (2\text{-}8)$$

where $\bar{q}$ now denotes the free charge density, and $\bar{q}'$ that net charge density corresponding to the local dielectric polarization,

$$-\bar{q}' = \nabla \cdot \mathbf{P} \tag{2-9}$$

Note that a uniform $\mathbf{P}$ implies zero net $\bar{q}'$. Comparison of (2-8) and (2-9) yields a divergence relation in $\bar{q}$ alone:

$$\nabla \cdot (\epsilon_0 \mathbf{E} + \mathbf{P}) = \bar{q} \tag{2-10}$$

(The arbitrary omission of ϵ_0 from the definition of $\mathbf{P}$ follows historical custom.) Since it is much easier to account for free charge sources, the vector sum $\epsilon_0 \mathbf{E} + \mathbf{P}$ is a useful combination. It is called the electric displacement field $\mathbf{D}$, and its sole source is the free charge density:

$$\nabla \cdot \mathbf{D} = \bar{q} \tag{2-11}$$

Again, $\mathbf{D}$, $\mathbf{E}$, and $\mathbf{P}$ need not necessarily be parallel. If we restrict ourselves to linear isotropic media, however,

$$\mathbf{D} = \kappa \epsilon_0 \mathbf{E} \equiv \epsilon \mathbf{E} \tag{2-12}$$

where κ and ϵ are scalars. If the medium is also uniform, then

$$\nabla \cdot \mathbf{E} = \frac{\bar{q}}{\kappa \epsilon_0} \tag{2-13}$$

and $\mathbf{E}$ is well defined both in terms of forces and source charges.

The electric potential at a point in space is defined by the work done on a unit charge in reaching the point from an arbitrary reference point (usually infinity), i.e., by the line integral of the electric field:

$$V_i - V_j = \int_i^j \mathbf{E} \cdot d\mathbf{l} \tag{2-14}$$

The concept has its greatest utility for conservative $\mathbf{E}$ fields, where the integral is independent of the path. Then

$$\mathbf{E} = -\nabla V \tag{2-15}$$

In free space, or in uniform dielectrics, it follows that

$$\nabla \cdot \mathbf{E} = -\nabla^2 V = \frac{\bar{q}}{\kappa \epsilon_0} \tag{2-16}$$

i.e., the potential function satisfies Poisson's equation. If the medium is charge-free, the potential satisfies Laplace's equation.

A very simple application of the source relations (2-11) and (2-13) and Gauss' and Stokes' theorems yields the boundary conditions on the normal component of **D** and the tangential component of **E** across a surface between two media:

$$(D_\perp)_1 - (D_\perp)_2 = \bar{q}_s$$
$$(E_\|)_1 = (E_\|)_2 \qquad (2\text{-}17)$$

where $\bar{q}_s$ is the free charge density on the surface.

The ability of a particular surface to store charge at a given potential may be expressed in terms of a capacitance C, defined as the ratio of the net charge maintained on the surface to the potential of the surface relative to infinity:

$$C = \frac{\int \bar{q}_s \, ds}{V} = \frac{Q_s}{V} \qquad (2\text{-}18)$$

As defined, C is a property of the surface geometry and of the surrounding medium only. For example, two uniformly charged plane-parallel surfaces separated by a dielectric of thickness d have a capacitance per unit area

$$\frac{C}{A} = \frac{\kappa \epsilon_0}{d} \qquad (2\text{-}19)$$

The inverse dependence of C on surface separation provides possibilities for compact electrostatic energy storage, provided dielectric separations of adequate high-voltage insulation can be employed.

2-2 CURRENTS AND MAGNETIC INTERACTIONS

In conducting media, such as metals and ionized gases, electric fields can force free charges into continuous motion, i.e., can drive currents. The current density is defined in terms of the flux of charged particles per unit area:

$$\mathbf{j} = \sum_i n_i q_i \mathbf{v}_i = \sum_i \bar{q}_i \mathbf{v}_i \qquad (2\text{-}20)$$

where n_i is the particle density of the specie of charge q_i, and $\mathbf{v}_i$ is its average velocity. If charge is to be conserved, the current density satisfies a continuity relation,

$$\nabla \cdot \mathbf{j} = -\frac{\partial}{\partial t} \sum_i n_i q_i = -\frac{\partial \bar{q}}{\partial t} \qquad (2\text{-}21)$$

If the electric field is the only significant force driving the charges, the

ELECTROMAGNETIC THEORY

current density will be directly related to **E** by an *Ohm's law:*

$$\mathbf{j} = \sigma \mathbf{E} \tag{2-22}$$

where σ is the electrical conductivity of the medium, defined in the same spirit as κ above, to represent on a macroscopic scale the integrated effect of all the atomic-scale charge motions induced by the applied **E** field. In general, σ, like κ, may be a complex space- and time-dependent tensor. In some cases, it may itself depend on **E**, making **j** nonlinear in **E**. Whereas this generality was not needed to describe the dielectric properties of gases, it frequently must be invoked when dealing with the conductivity of ionized gases, which can display complicated **j**(**E**) relations. These will be discussed in detail in Chap. 5. It should also be noted that (2-22) does not yet include net charge flux driven by magnetic fields or by gradients in specie partial pressure. Inclusion of such effects in a *generalized Ohm's law* is also discussed later and in Refs. 2 and 3.

Two currents are found to exert forces on each other, but in a some-

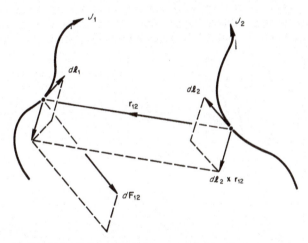

Fig. 2-1 Magnetic interaction of two current elements.

what different way from static charges (Fig. 2-1). The force felt by an element of current path $J_1\, dl_1$, in presence of another, $J_2\, dl_2$, is expressed by *Ampere's law:*

$$d\mathbf{F}_{12} = A' \frac{J_1 J_2}{r_{12}^2} [d\mathbf{l}_1 \times (d\mathbf{l}_2 \times \hat{\mathbf{r}}_{12})] \tag{2-23}$$

where $A' = \mu_0/4\pi$, and $\mu_0 = 4\pi \times 10^{-7}$ henry/m for J_1, J_2, in amperes (mks system). In analogy to electrostatics, it is convenient to introduce an intermediary field **B**, called the magnetic induction, which determines the force on a current element, as **E** does on a charge:

$$dB_{12} = \frac{\mu_0 J_2}{4\pi}\left(\frac{dl_2 \times \hat{r}_{12}}{r_{12}^2}\right) \quad (2\text{-}24)$$

in terms of which

$$dF_{12} = J_1 \, dl_1 \times dB_{12} \quad (2\text{-}25)$$

Note that, in general, $dF_{12} \neq dF_{21}$, an apparent contradiction to Newton's third law. Part of this difficulty is removed by noting that the total force between two complete circuits

$$\mathbf{F}_{12} = \frac{\mu_0}{4\pi} J_1 J_2 \oint_1 \oint_2 \frac{dl_1 \times (dl_2 \times \hat{r}_{12})}{r_{12}^2} = -\mathbf{F}_{21} \quad (2\text{-}26)$$

is symmetric, but the paradox may still exist for free charges and incomplete circuits carrying transient currents.

The definition of the magnetic induction field for a single source current (2-24) can be generalized to an integral form appropriate to a finite distribution of source-current density over a given volume [4]. With reference to Fig. 2-2, let **r** denote the position at which the value of **B** is

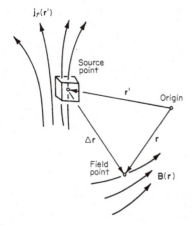

Fig. 2-2 Magnetic induction field from distributed current density.

desired, $\mathbf{r}'$ the coordinate of a typical source-current element $\mathbf{j}_t(\mathbf{r}') \, dV'$, and $\Delta \mathbf{r}$ the position of the field point $\mathbf{r}$ relative to the source element at $\mathbf{r}'$. Then

$$\mathbf{B}(\mathbf{r}) = \frac{\mu_0}{4\pi} \int \frac{\mathbf{j}_t(\mathbf{r}') \times \Delta \hat{\mathbf{r}}}{(\Delta r)^2} \, dV' \quad (2\text{-}27)$$

ELECTROMAGNETIC THEORY

Two important properties are derivable from this expression by vector differentiation in the **r** system (Prob. 2-2):

$$\nabla \cdot \mathbf{B} = 0 \qquad (2\text{-}28)$$

$$\nabla \times \mathbf{B} = \mu_0 \mathbf{j}_t \qquad (2\text{-}29)$$

The $\mathbf{j}_t$ appearing in (2-27) and (2-29) denotes the total current density from any source whatever, including that created in the medium by the applied **B** field. Much like the electrostatic situation, where material molecules polarize, and thereby introduce additional source-charge densities, these particles also possess internal currents which can be oriented to present effective interior macroscopic current densities. These are more conveniently described by a volume magnetic dipole moment **m**, defined by

$$\nabla \times \mathbf{m} = \mu_0 \mathbf{j}' = \mu_0(\mathbf{j}_t - \mathbf{j}) \qquad (2\text{-}30)$$

where $\mathbf{j}'$ is the interior current density, $\mathbf{j}_t$ the total current density, and hence $\mathbf{j}$ that current density actually driven by external electric fields.

A new field **H** is now definable in terms of these applied currents as sources:

$$\mathbf{H} = \frac{1}{\mu_0}(\mathbf{B} - \mathbf{m}) \qquad (2\text{-}31)$$

$$\nabla \times \mathbf{H} = \mathbf{j} \qquad (2\text{-}32)$$

Retaining the analogy to electrostatics, a bulk material property κ_m, called the relative magnetic permeability, may be defined to relate the interior dipole moment per unit volume **m** to the applied field (because of certain historical misconceptions, some of the analogy is distorted at this point):

$$\mathbf{m} = \frac{\kappa_m - 1}{\kappa_m}\mathbf{B} = (\kappa_m - 1)\mu_0 \mathbf{H} \qquad (2\text{-}33)$$

$$\mathbf{B} = \kappa_m \mu_0 \mathbf{H} \equiv \mu \mathbf{H} \qquad (2\text{-}34)$$

where, in the general case, **B**, **H**, and **m** are not necessarily colinear, and hence κ_m and μ are tensors. Again, **m** may be explicitly expressed in terms of the molecular number density and the individual molecular susceptibility p_m:

$$\mathbf{m} = \frac{3Np_m}{3 + 2Np_m}\mathbf{B} \qquad (2\text{-}35)$$

and thus, from the definition of κ_m,

$$p_m = \frac{3\mu_0}{N}\left(\frac{\kappa_m - 1}{\kappa_m + 2}\right) \qquad (2\text{-}36)$$

much like the Clausius-Mosotti relation (2-6).

Table 2-2 summarizes the distinctions which have been made up to this point between fields defined by the forces on charges and currents and fields defined by source charges and currents; that is, **E** and **B** are the true force fields and **D** and **H** are the simpler to compute from given sources.

Table 2-2 Sources of the electromagnetic fields and the forces they exert, mks units

Field	Source	Force exerted
E (volts/m; newtons/coul)	Total charge density q_t (coul)	$\bar{q}$**E** (newtons/m³)
D (coul/m²)	Applied charge density q (coul)	
B (webers/m²)	Total current density $\mathbf{j}_t$ (amp/m²)	**j** × **B** (newtons/m³)
H (amp-turns/m)	Applied current density **j** (amp/m²)	

2-3 TIME-DEPENDENT FIELDS AND ELECTROMAGNETIC WAVE PROPAGATION

To complete an outline of electrodynamics, it remains to consider the effects associated with time or space changes of the source currents. It is observed experimentally that a test loop in a changing magnetic field experiences an induced electric field along its length, proportional to the time rate of change of the flux of **B** through the loop (Faraday's law):

$$\mathcal{V} \equiv \oint_{\substack{\text{test}\\ \text{loop}}} \mathbf{E} \cdot d\mathbf{l} = -\frac{\partial}{\partial t} \oint_{\substack{\text{loop}\\ \text{area}}} \mathbf{B} \cdot d\mathbf{s} \qquad (2\text{-}37)$$

or
$$\mathcal{V} = -\dot{\phi} \qquad (2\text{-}38)$$

where $\phi = \oint \mathbf{B} \cdot d\mathbf{s}$ is called the magnetic flux, and $\mathcal{V}$ has the dimensions of potential difference and usually is called the electromotive force. The above result holds whether $\dot{\phi}$ is caused by a change in magnitude of the source of **B**, by motion of the source of **B**, or by motion of the test loop. In the first two cases, Stokes' law immediately permits the differential expression of this effect:

$$\nabla \times \mathbf{E} = -\dot{\mathbf{B}} \qquad (2\text{-}39)$$

In cases involving motion of the test loop, the same result is valid, but more subtle reasoning is required [4].

In the special case that the **B** field involved is generated by a current J_2 flowing in a separate circuit nearby, it follows from (2-32) and (2-39)

ELECTROMAGNETIC THEORY

that the electric field induced in the test loop will be proportional to the time rate of change of J_2 by a coefficient determined by the specific geometry of the two circuits and by the magnetic permeability κ_m of the intervening medium:

$$\mathcal{V}_1 = -M_{12}\dot{J}_2 \tag{2-40}$$

M_{12} is called the mutual inductance of the configuration.

The **B** field may also originate from a changing current in the test loop itself, J_1, in which case a self-induced electromotive force appears:

$$\mathcal{V}_1 = -L\dot{J}_1 \tag{2-41}$$

where L is called the self inductance of the loop. Relation (2-41) equivalently states that a voltage source $\mathcal{V}_1$ would be required to change the current through the loop at a rate $-\dot{J}_1$. In a circuit carrying a sinusoidal alternating current, for example, the voltage drop appearing across a purely inductive circuit element will be proportional to its self inductance, to the current amplitude, and to the frequency and will be 90° out of phase with the current:

$$\mathcal{V}_L = -L\frac{d}{dt}(J_0 e^{i\omega t}) = -i\omega LJ \tag{2-42}$$

Having recognized the ability of a changing **B** field to generate an **E**, one asks if the inverse process may occur. We already have a source relation for **H** [Eq. (2-32)], but in point of fact it needs to be amended. First, there is the obvious inconsistency that $\nabla \cdot (\nabla \times \mathbf{H})$ is identically zero yet $\nabla \cdot \mathbf{j}$ is not [Eq. (2-21)]. Second, it leads to various contradictions in dealing with transient currents. Consider the following paradox: Allow one plate of a capacitor in a simple charging circuit to be enclosed by a control surface S, around which is scribed a closed line element l, which divides the surface into two sections, S_1 and S_2, as shown in Fig. 2-3. When the switch is closed, a current $J(t)$ flows around the circuit to charge the two capacitor plates. From (2-32) and Stokes' law,

$$\int_{S_1} (\nabla \times \mathbf{H}) \cdot d\mathbf{s} = \oint_l \mathbf{H} \cdot d\mathbf{l} = J(t) \tag{2-43}$$

But l is also a boundary for S_2, and no current flows through that surface. Hence we should infer the contradiction

$$\int_{S_2} (\nabla \times \mathbf{H}) \cdot d\mathbf{s} = \oint_l \mathbf{H} \cdot d\mathbf{l} = 0 \tag{2-44}$$

Maxwell resolved this inconsistency by arguing that in the process of charging the capacitor in this way, an electric field is caused to increase

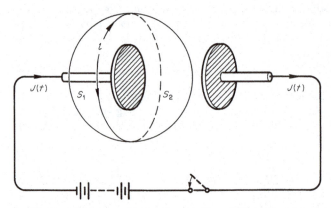

Fig. 2-3 Hypothetical experiment to illustrate Maxwell displacement current.

between the plates and that this time rate of change of **E** field is equivalent to a real current density for purposes of induction of an **H** field. Actually, to be dimensionally consistent, it is the time derivative of **D**, the "source" field, not **E**, that is needed:

$$\nabla \times \mathbf{H} = \mathbf{j} + \dot{\mathbf{D}} \qquad (2\text{-}45)$$

The sum $\mathbf{j} + \dot{\mathbf{D}}$ may be regarded as a generalized current density $\bar{\mathbf{j}}$, which is the true source of $\nabla \times \mathbf{H}$. Unlike $\mathbf{j}$, $\bar{\mathbf{j}}$ is divergenceless.

The identification of the *displacement current* $\dot{\mathbf{D}}$ clears the way for derivation of electromagnetic wave propagation phenomena. In free space, for example, relations (2-39), (2-32), (2-11), and (2-28), usually called Maxwell's equations, may be written

$$\nabla \times \mathbf{E} = -\mu_0 \dot{\mathbf{H}} \qquad (2\text{-}46)$$

$$\nabla \times \mathbf{H} = \epsilon_0 \dot{\mathbf{E}} \qquad (2\text{-}47)$$

$$\nabla \cdot \mathbf{E} = 0 \qquad (2\text{-}48)$$

$$\nabla \cdot \mathbf{H} = 0 \qquad (2\text{-}49)$$

Comparing the curl of (2-46) with the partial time derivative of (2-47) yields a wave equation for **E**:

$$\nabla^2 \mathbf{E} = \epsilon_0 \mu_0 \ddot{\mathbf{E}} \qquad (2\text{-}50)$$

Reversing the manipulation yields an identical wave equation for **H**. These relations imply that free space can support propagation of plane

ELECTROMAGNETIC THEORY

electromagnetic waves of the form

$$\mathbf{E} = \mathbf{E}_0 e^{i(\omega t \mp k_0 z)} \quad (2\text{-}51)$$

These waves propagate with a phase velocity

$$c_0 = (\epsilon_0 \mu_0)^{-1/2} \quad (2\text{-}52)$$

in the direction z with a wave number k_0, related to the angular frequency ω, or alternatively, to the wavelength λ_0:

$$k_0 = \frac{\omega}{c_0} = \frac{2\pi}{\lambda_0} \quad (2\text{-}53)$$

This derivation is readily extended to other media. For isotropic charge-free dielectrics, for example, the above relations are unchanged, except that the phase velocity, wavelength, and wave vector now are determined by the ϵ and μ appropriate to the medium. For isotropic media of finite conductivity, the wave equation for $\mathbf{E}$ acquires an additional term:

$$\nabla^2 E = \sigma\mu\dot{\mathbf{E}} + \epsilon\mu\ddot{\mathbf{E}} \quad (2\text{-}54)$$

and the formal solution (2-51) can be retained only by allowing the wave number k to be complex:

$$k = (\omega^2 \epsilon \mu - i\mu\sigma\omega)^{1/2} \quad (2\text{-}55)$$

The real part of k again defines the wavelength or phase velocity of the wave train; the imaginary part, when inserted in the wave function (2-51), provides a negative real exponent, indicating that the wave train attenuates with propagation in such a medium.

2-4 APPLICATION TO IONIZED GAS FLOWS

To summarize this brief review in the context of our problem of interest—the acceleration of a gas stream by electrical means—we note that, basically, there are but two modes of application of electromagnetic body forces to the gas, the interaction of an electric field with net free charge density in the gas,

$$\mathbf{F}_E = \bar{q}\mathbf{E} \quad (2\text{-}56)$$

and the interaction of a magnetic induction field with currents driven within the gas,

$$\mathbf{F}_B = \mathbf{j} \times \mathbf{B} \quad (2\text{-}57)$$

The space and time variations of the electromagnetic field vectors **E** and **B** are themselves related to the free charge density and current density patterns in the medium in accordance with the four Maxwell equations

$$\nabla \times \mathbf{H} = \mathbf{j} + \dot{\mathbf{D}} \qquad (2\text{-}58)$$

$$\nabla \times \mathbf{E} = -\dot{\mathbf{B}} \qquad (2\text{-}59)$$

$$\nabla \cdot \mathbf{D} = \bar{q} \qquad (2\text{-}60)$$

$$\nabla \cdot \mathbf{B} = 0 \qquad (2\text{-}61)$$

Actually, the last two statements are directly derivable from the first two, following the assumption of charge conservation and the definition of current density:

$$\nabla \cdot \mathbf{j} = -\frac{\partial \bar{q}}{\partial t} \qquad (2\text{-}62)$$

As such, (2-58) to (2-62) constitute seven independent (scalar) relations among sixteen quantities. The remaining relations must arise from the physical properties of the medium. For uniform gaseous media at rest we may use the simple forms

$$\mathbf{D} = \epsilon \mathbf{E} \qquad (2\text{-}63)$$

$$\mathbf{B} = \mu \mathbf{H} \qquad (2\text{-}64)$$

$$\mathbf{j} = \sigma \mathbf{E} \qquad (2\text{-}65)$$

where the constitutive parameters ϵ and μ are scalar quantities, close to the vacuum values, but σ may be a tensor with complex time-dependent elements, or may possibly be a nonlinear function.

If the gaseous medium is in motion relative to the frame of reference in which the fields are specified, as it must be in any interesting accelerator, Maxwell's equations remain valid, but certain modifications are required in the constitutive relations. These may be rigorously derived from considerations of relativistic invariance [5], or rationalized by empirical argument, in the form

$$\mathbf{D} + \frac{1}{c_0^2}(\mathbf{u} \times \mathbf{H}) = \epsilon(\mathbf{E} + \mathbf{u} \times \mathbf{B}) \qquad (2\text{-}66)$$

$$\mathbf{B} - \frac{1}{c_0^2}(\mathbf{u} \times \mathbf{E}) = \mu(\mathbf{H} - \mathbf{u} \times \mathbf{D}) \qquad (2\text{-}67)$$

$$\mathbf{j} - \bar{q}\mathbf{u} = \sigma(\mathbf{E} + \mathbf{u} \times \mathbf{B}) \qquad (2\text{-}68)$$

where **u** is the gas streaming velocity; $\bar{q}$ is again the net free charge density; ϵ, μ, and σ are the intrinsic properties of the medium, i.e., those an observer

ELECTROMAGNETIC THEORY

fixed in the medium would determine; and c_0 is the speed of light in a vacuum. For gaseous media, where $\epsilon\mu \approx \epsilon_0\mu_0 = 1/c_0^2$, (2-66) and (2-67) return to the form of (2-63) and (2-64), but Ohm's law (2-68) retains the motional term (Prob. 2-6). The physical interpretation of (2-68) is that current density will be driven through the gas by an effective electric field, $\mathbf{E'} = \mathbf{E} + \mathbf{u} \times \mathbf{B}$, composed of the applied field $\mathbf{E}$ and an induced field, $\mathbf{u} \times \mathbf{B}$. This induced field could have been derived from the discussion of Sec. 2-3. Indeed, the total effective field $\mathbf{E'}$ may be thought of as that actually seen by the moving medium, and is quite generally applicable in this sense. For example, an isolated particle of charge q moving with velocity $\mathbf{v}$, relative to a frame in which electric and magnetic fields $\mathbf{E}$ and $\mathbf{B}$ are measured, feels a force

$$\mathbf{F} = q\mathbf{E'} = q(\mathbf{E} + \mathbf{v} \times \mathbf{B}) \qquad (2\text{-}69)$$

Before applying the electromagnetic relations to a given gas acceleration problem, it is necessary to evaluate the conductivity function σ in terms of the atomic-scale properties and the thermodynamic state of the working fluid. The task of the next three chapters will be to examine the detailed constitution of ionized gaseous media, for the main purpose of evaluating this essential conductivity property.

PROBLEMS

2-1. Derive the boundary conditions on $\mathbf{E}$ and $\mathbf{D}$ [Eq. (2-17)].
2-2. Derive relations (2-28) and (2-29).
2-3. Derive the capacitance of a plane-parallel capacitor [Eq. (2-19)], a concentric-cylindrical capacitor, and a concentric-spherical capacitor. Show that the energy stored in any capacitor is $\frac{1}{2}CV^2$. Using the best tabulated values of dielectric breakdown strengths, estimate the maximum amount of electrostatic energy which could be stored in a capacitor of 1-m^3 volume.
2-4. Derive (2-54). Show it is equivalent to the form $\nabla^2\mathbf{E} + k^2[1 - i(\sigma/\epsilon\omega)]\mathbf{E} = 0$. Now suppose that σ is not uniform but has a finite gradient throughout the medium. Derive the appropriate wave equation. In the special case that $\nabla\sigma$ is everywhere parallel to the direction of propagation, what is the form of the permitted wave functions?
2-5. Derive (2-68) from arguments of induced electromotive force, starting from relation (2-37).
2-6. Show that relations (2-66) to (2-68) reduce to the form

$$\mathbf{D} = \epsilon\mathbf{E}$$
$$\mathbf{B} = \mu\mathbf{H}$$
$$\mathbf{j} - q\mathbf{u} = \sigma(\mathbf{E} + \mathbf{u} \times \mathbf{B})$$

to the order $(\epsilon\mu - \epsilon_0\mu_0)c_0^2 \ll 1$.

REFERENCES

1. Harnwell, G. P.: "Principles of Electricity and Electromagnetism," 2d ed., McGraw-Hill Book Company, New York, 1949.
2. Spitzer, L., Jr.: "Physics of Fully-ionized Gases," Interscience Publishers, Inc., New York, 1956.
3. Sutton, G. W., and A. Sherman: "Engineering Magnetohydrodynamics," McGraw-Hill Book Company, New York, 1965.
4. Panofsky, W. K. H., and M. Phillips: "Classical Electricity and Magnetism," Addison-Wesley Publishing Company, Inc., Reading, Mass., 1955.
5. Sommerfeld, A.: "Electrodynamics," Academic Press Inc., New York, 1952.
6. Stratton, J. A.: "Electromagnetic Theory," McGraw-Hill Book Company, New York, 1941.
7. Jackson, J. D.: "Classical Electrodynamics," John Wiley & Sons, Inc., New York, 1962.

3
Ionization in Gases

If significant electric and magnetic body forces are to be exerted on a body of gas, it must contain free electric charges. These may be provided by ionizing the gas thermally, chemically, electrically, or by irradiation or by combining separate sources of positive ions and electrons. Regardless of the means of preparation, the existence of free charges in a compressible fluid substantially complicates its thermodynamics, its particle kinetics, and its continuum dynamics. Several excellent texts consider various aspects of these topics in detail [1–18]; here we select only a few critical points for discussion. This chapter deals with atomic-scale ionization processes and the equilibrium ionization level they can establish. Chapter 4 outlines various types of particle collisions in an ionized gas, and Chap. 5 concentrates on the transport property of most interest for our purposes, the electrical conductivity. The continuum dynamics of ionized gases is fundamental to most of the acceleration concepts developed in later chapters, and will be discussed there in various specific contexts.

3-1 ATOMIC STRUCTURE

Without straying into detailed discussion of atomic physics, we must agree on a serviceable conceptual model of the atoms, molecules, and ions

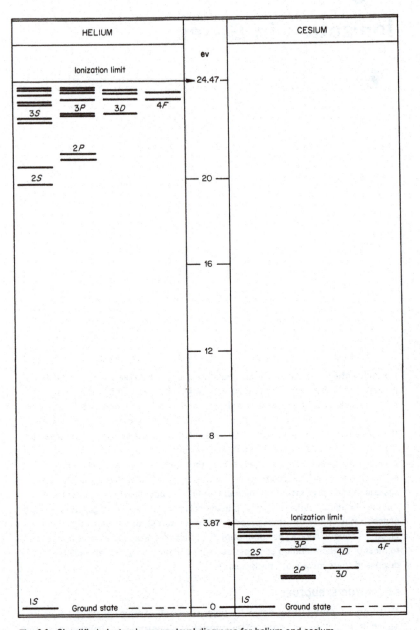

Fig. 3-1 Simplified electronic energy-level diagrams for helium and cesium.

which constitute the ionized gases of our interest. For our purposes, we shall regard an atom as composed of a very small compact nucleus of Z protons and Y neutrons ($Z \leq Y$; total dimension $\approx 10^{-12}$ cm), surrounded by a much larger region ($\approx 10^{-8}$ cm), wherein Z electrons arrange themselves in certain specific but diffuse configurations, depending on the internal energy of the atom and its internal angular momentum.

Bohr and Sommerfeld described these electronic configurations in terms of satellitelike orbits about the nucleus [19–21], an easily visualized concept which proved useful in the early interpretation of atomic spectra. A more sophisticated wave-mechanical, or probabilistic, formulation is required to treat the subtler features of atomic structure [19–23]. For our immediate purposes, we need only recognize that the atomic electron distribution has some minimum energy configuration, called the ground state, and that an atom can absorb energy to assume other configurations only in certain discrete amounts (quanta) and with certain discrete changes in internal angular momentum. Normally, such energy increases are manifested in an adjustment of only one of the bound electrons at a time. With the absorption of successive energy increments, that electron is driven progressively farther from the nucleus, until finally it detaches itself completely from the atomic system, leaving behind a configuration of net positive charge, i.e., a positive ion. The progression of these electronic energy levels to the ionization limit is conventionally represented by a so-called grotrian diagram (Fig. 3-1).

The total energy increment needed to remove the electron is called the ionization potential ε_i, and varies considerably from one gas to another. In general, the noble gases have high ionization potentials, and the alkali vapors, low (Table 3-1).

This ionization energy may be absorbed in a single atomic event, or the electron may be elevated by successive processes to the ionization limit. In general, the latter route is less likely because of the tendency for spontaneous radiative decay from one of the intermediate levels to interrupt the sequence of absorptions. A typical lifetime against spontaneous decay of an excited state is about 10^{-8} sec. There are special cases where stepwise ionization is important, but for the present we shall presume a single-increment process.

3-2 IONIZATION PROCESSES

In a gaseous environment, the energy increment needed for ionization of a constituent atom may be delivered in a variety of events (Fig. 3-2):

a. The atom may undergo an inelastic collision with another particle of sufficiently high relative kinetic energy. This may be symbolized

Table 3-1 Ionization potentials of various gases

Gas	ε_i, ev†
Argon (A)	15.75
Hydrogen, atomic (H)	13.59
Hydrogen, molecular (H$_2$)	15.42
Helium (He)	24.58
Nitrogen, atomic (N)	14.54
Nitrogen, molecular (N$_2$)	15.58
Neon (Ne)	21.56
Oxygen, atomic (O)	13.61
Oxygen, molecular (O$_2$)	12.20
Cesium vapor (Cs)	3.89
Potassium vapor (K)	4.34

† The unit of energy most commonly employed on the atomic scale is the electron volt, the energy a particle of one electronic charge acquires from falling through a potential difference of one volt. It is equal to 1.60×10^{-19} joule, or 1.60×10^{-12} erg.

SOURCE: A. Unsöld, "Physik der Sternatmosphaeren" Springer Verlag OHG, Berlin, 1955, and G. Herzberg, "Spectra of Diatomic Molecules," D. Van Nostrand Company, Inc., Princeton, N.J., 1957.

by the "reaction" notation

$$\tilde{X} + A \rightarrow A^+ + e + X \qquad (3\text{-}1)$$

where $\tilde{X}$ denotes a particle of high relative kinetic energy, and might be another atom, an ion, or an electron.

  b. The atom may absorb an electromagnetic photon of adequately high frequency:

$$h\nu + A \rightarrow A^+ + e \qquad (3\text{-}2)$$

where h = Planck's constant (6.62×10^{-34} joule-sec)
ν = frequency of radiation absorbed
$h\nu$ = corresponding energy increment delivered

  c. An electron may be forcibly extracted from the atom by a strong electric field (Prob. 3-1):

$$E(A) \rightarrow A^+ + e \qquad (3\text{-}3)$$

IONIZATION IN GASES

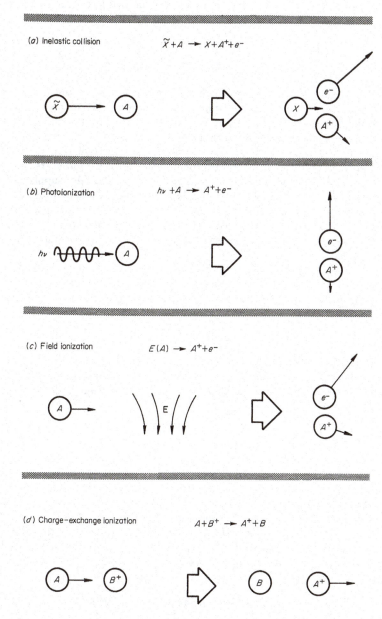

Fig. 3-2 Atomic-scale ionization processes.

d. In certain vulnerable atoms, an electron may be lost through a strong valence interaction with another atom:

$$A + B \to A^+ + B^- \tag{3-4}$$

or with an ion:
$$A + B^+ \to A^+ + B \tag{3-5}$$

Within each of these classes there are alternative ways to precipitate the ionization reaction. For example, the inelastic collision process (3-1) could be gas-kinetic, i.e., a purely thermal collision, or it could be with a charged particle accelerated in an external electric field, with a cosmic particle, or with a radioactive-decay product. The radiation source for (3-2) could be external to the gas (photoionization) or internal (resonance trapping). The electric field for (3-3) might be externally supplied as a steady field, a slowly alternating field, or a radio-frequency propagating electromagnetic wave train, or might arise from a localized net charge density within the body of gas. The charge-exchange reactions (3-4) and (3-5) can proceed in various ways, depending on the particular particles involved.

Ionization of molecules in the gas can occur through the same types of processes, with the additional possibilities that internal rotation and vibration modes associated with the molecular structure can interact with the electronic structure, for example,

$$AB^* \to AB^+ + e \tag{3-6}$$

where AB^* denotes a hyperenergetic molecular state which spontaneously surrenders its excitation to ionization. In a few rare exothermic chemical reactions, a molecular ion-electron pair is the normal product (chemi-ionization):

$$A + B \to AB^+ + e \tag{3-7}$$

The picture may be complicated further by the presence of material surfaces adjacent to the body of gas. Such solid surfaces may catalytically enhance some of the above gas-phase reactions or may themselves be prolific contributors of free electrons or positive ions via bombardment emission, photoelectric emission, field emission (cold cathode emission), charge-transfer trapping, or thermionic emission.

At the same time that the various ionization processes outlined above are acting to increase the free-electron and ion densities in the gas, a variety of recombination mechanisms compete to reduce these populations (Fig. 3-3):

a. Radiative recombination (free-bound transitions):

$$A^+ + e \to A + h\nu \tag{3-8}$$

IONIZATION IN GASES

(*a*) Radiative recombination: $A^+ + e^- \rightarrow A + h\nu$

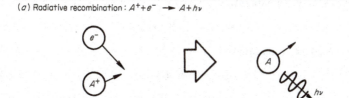

(*b*) Three-body recombination: $A^+ + e^- + X \rightarrow A + \tilde{X}$

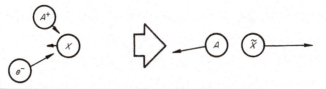

(*c*) Dissociative recombination: $(AB)^+ + e^- \rightarrow A + \tilde{B}$

(*d*) Dielectronic recombination: $A^+ + e^- \rightarrow A^{**}$

(*e*) Electron attachment – negative ion recombination:

$$A + e^- \rightarrow A^-$$

$$A^- + A^+ \rightarrow A + \tilde{A}$$

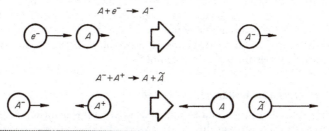

Fig. 3-3 Atomic-scale recombination processes.

b. Three-body recombination:

$$A^+ + e + X \to A + \tilde{X} \qquad (3\text{-}9)$$

c. Dissociative recombination:

$$AB^+ + e \to A + \tilde{B} \qquad (3\text{-}10)$$

d. Dielectronic recombination (two bound electrons excited):

$$A^+ + e \to A^{**} \qquad (3\text{-}11)$$

e. Electron attachment followed by negative-ion recombination:

$$\begin{aligned} A + e &\to A^- \\ A^- + A^+ &\to A + \tilde{A} \end{aligned} \qquad (3\text{-}12)$$

Again, this list is oversimplified. The product atom in (3-8) may be left in any of many excited states, whence it decays by radiation or subsequent collision. The third body in (3-9) may be another atom like A, an impurity particle, a wall, etc. Reaction (3-10) is not necessarily restricted to molecules; many atoms can form stable molecular ions. Here and in (3-11) the products may again be left in a variety of excited states. Alternative second steps are available in (3-12), involving excited states and/or radiation. As with the ionization reactions, the relative importance of the various mechanisms depends on the particular gas involved, its temperature, density, and surrounding circumstances of boundaries, applied fields, etc.

In principle, a complete statement and simultaneous solution of all ionization and recombination reactions and all related reactions involving the excited states and radiation fields, along with all appropriate initial and boundary conditions, should yield the ionization level of the gas and the net rate at which it is changing. Such a general analytical approach is clearly out of the question, however, not only because of the mathematical complexity, but because most of the individual "reaction rates" involved are poorly known. Only if one is willing to confine his attention to gases in thermodynamic equilibrium, and to abandon detailed representation of individual reactions in favor of statistical arguments, can ionization levels be accurately computed.

3-3 EQUILIBRIUM IONIZATION OF A GAS

Each of the ionization and recombination processes mentioned above can proceed in the reverse direction. Most of these reverse reactions are already included in the two lists, (3-1) to (3-12). In a body of gas in

IONIZATION IN GASES

complete thermodynamic equilibrium, every such process—indeed, every atomic excitation process, whether involving ionization or not—must be exactly balanced by its reverse process. This condition of detailed physical balancing provides the basis for application of statistical mechanics to calculation of the equilibrium composition of the gas.

The first published application of statistical methods to ionized gas problems was made by an Indian scholar, Megh Nad Saha, who was studying ionization phenomena in the solar chromosphere [24]. Actually, Saha phrased his development in terms of the "Nernst reaction isobar," a somewhat less direct method than is now commonly employed, but the essence of his approach remains valid; namely, for the case of a pure atomic gas, all the many ionization and recombination processes are to be represented by one overall reversible reaction,

$$A + \varepsilon_i \rightleftharpoons A^+ + e \tag{3-13}$$

where $+\varepsilon_i$ denotes the addition of the necessary ionization energy by any means whatever. This reaction is assigned the gross equilibrium constant

$$K_n = \frac{n_+ n_e}{n_A} \tag{3-14}$$

where n_+, n_e, n_A are the number densities of ions, electrons, and atoms, respectively. From the principles of statistical mechanics it develops that the above number density ratio in equilibrium is just given by the corresponding ratio of the sums of accessible energy states, appropriately weighted by degeneracies and by the Boltzmann factors in the relative energies of those states, i.e., by the corresponding ratios of total partition functions [25]:

$$K_n = \frac{n_+ n_e}{n_A} = \frac{F_+ F_e}{F_A} \tag{3-15}$$

For an atomic species, the total partition function F_A is the product of a translational and an internal part:

$$F_A = f_A{}^t \cdot f_A{}^i \tag{3-16}$$

where the translational part follows from "particle-in-a-box" quantum mechanics:

$$f_A{}^t = \frac{(2\pi M_A k T)^{3/2}}{h^3} \tag{3-17}$$

where M_A = mass of atom
k = Boltzmann's constant (1.38×10^{-23} joule/°K)
T = absolute temperature
h = Planck's constant

The internal part of the partition function is the sum of the available electronic energy levels, each weighted by its Boltzmann factor and degeneracy:

$$f_A{}^i = \sum_j g_j e^{-(\varepsilon_j/kT)} \qquad (3\text{-}18)$$

where the summation extends from $j = 0$, the ground state, to the highest bound level of the atom.

The total partition function for the atomic ions is constructed similarly, except that the internal energy levels are now differently spaced, and have a ground state of higher absolute energy than that of the neutral atoms:

$$F_+ = f_+{}^t \cdot f_+{}^i = \frac{(2\pi M_+ kT)^{3/2}}{h^3} \cdot \sum_j g_j^+ e^{-(\bar\varepsilon_j^+/kT)} \qquad (3\text{-}19)$$

The absolute ground state energy of the ion $\bar\varepsilon_0^+$ must differ from that of the atom ε_0 by the energy needed to make a ground state ion from a ground state atom, namely, by the ionization potential ε_i. Thus we may rescale each of the ionic levels relative to the ground state of the ion:

$$\varepsilon_j^+ = \bar\varepsilon_j^+ - \varepsilon_i \qquad (3\text{-}20)$$

(The ground state energy of the neutral atom is normally assigned the absolute value zero.)

The total partition function for the electron contains only the translational function and a factor of 2 for spin degeneracy, its only internal degree of freedom:

$$F_e = 2 \cdot f_e{}^t = \frac{2(2\pi mkT)^{3/2}}{h^3} \qquad (3\text{-}21)$$

Thus, to the order of approximation that $M_+ = M$, the equilibrium constant may be expressed

$$K_n = \frac{n_+ n_e}{n_A} = \frac{2(2\pi mkT)^{3/2}}{h^3} \frac{\sum_j g_j^+ e^{-(\varepsilon_j^+/kT)}}{\sum_j g_j e^{-\varepsilon_j/kT}} \; e^{-(\varepsilon_i/kT)} \qquad (3\text{-}22)$$

An alternative form of this relation follows from the definition of a degree of ionization α:

$$\alpha = \frac{n_+}{n_A + n_+} = \frac{n_+}{n_0} \qquad (3\text{-}23)$$

where n_0 is just the total number of atoms available for ionization; from the assumption of charge neutrality,

$$n_e = n_+ \qquad (3\text{-}24)$$

IONIZATION IN GASES

and from the assumption of a perfect gas type of equation of state,

$$p = (n_e + n_+ + n_A)kT = (1 + \alpha)n_0 kT \tag{3-25}$$

Insertion of these statements into (3-22) yields the celebrated *Saha equation*

$$\frac{\alpha^2}{1-\alpha^2} = \frac{2(2\pi m)^{3/2}(kT)^{5/2}}{ph^3}\left(\frac{f_+{}^i}{f_A{}^i}\right)e^{-(\epsilon_i/kT)} \tag{3-26}$$

The α thus expressed is asymptotic to 0 as $T \to 0$, and to 1 as $T \to \infty$, and has a reciprocal dependence on pressure (Fig. 3-4). At low tempera-

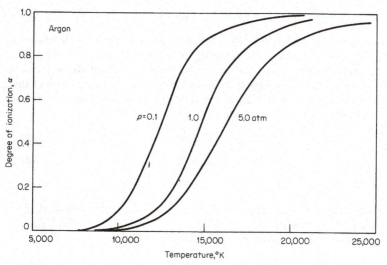

Fig. 3-4 Dependence of ionization of argon on temperature and pressure, from Saha equation. *(From K. S. Drellishak, C. K. Knopp, and A. B. Cambel, Partition Functions and Thermodynamic Properties of Argon Plasma, Phys. Fluids, ser. 6, vol. 9, p. 1280, 1963.)*

tures, the ratio of the internal partition functions is nearly constant since most of the atoms and ions are in their ground electronic states. The function is then dominated by the exponential factor, and may be approximated by the simpler relation

$$\alpha \approx G p^{-1/2} T^{5/4} e^{-(\epsilon_i/2kT)} \tag{3-27}$$

where G is nearly independent of T and depends otherwise only on the gas involved (Prob. 3-2). Since $\alpha \ll 1$ in this range, $p \approx n_0 kT$, and there is thus the simpler approximation

$$\alpha \approx G'n_0^{-1/2} T^{3/4} e^{-(\varepsilon_i/2kT)} \quad (3\text{-}28)$$

which directly displays the nonlinear dependence of the ion density on the total heavy-particle density in this regime:

$$n_+ = \alpha n_0 \sim \sqrt{n_0} \quad (3\text{-}29)$$

Considerable modification of the foregoing arguments is required if the gas is molecular, rather than monatomic. Beyond the inclusion of additional factors in the atomic and ionic partition function to account for the rotational and vibrational degrees of freedom,

$$f^i = f_{el} \cdot f_{rot} \cdot f_{vibr} \quad (3\text{-}30)$$

it is necessary to include many other reactions and species in the system of equilibrium equations. In general, molecular dissociation energies are comparable with or lower than the ionization energies of the molecule or its constituent atoms; hence the dissociation reactions must be included. For example, the ionization of the simplest bimolecular gas involves at least three interlocked reactions,

$$B_2 + \varepsilon_d \rightleftharpoons B + B \quad (3\text{-}31)$$

$$B_2 + \varepsilon_i \rightleftharpoons B_2^+ + e \quad (3\text{-}32)$$

$$B + \varepsilon_i' \rightleftharpoons B^+ + e \quad (3\text{-}33)$$

where ε_d, ε_i, and ε_i' are the dissociation energy, ionization potential of the molecule, and ionization potential of the atom, respectively. Each of these reactions has its own equilibrium constant,

$$K_n = \frac{(n_1)^2}{n_2} \quad (3\text{-}34)$$

$$K_n' = \frac{n_2^+ n_e}{n_2} \quad (3\text{-}35)$$

$$K_n'' = \frac{n_1^+ n_e}{n_1} \quad (3\text{-}36)$$

which are again expressible as ratios of the appropriate partition functions. These relations must be solved simultaneously with the conditions of conservation of species and conservation of charge:

$$n_2 + \tfrac{1}{2}n_1 + n_2^+ + \tfrac{1}{2}n^+ = n_2^0 \quad (3\text{-}37)$$

$$n_2^+ + n^+ = n_e \quad (3\text{-}38)$$

IONIZATION IN GASES

Similar complications enter the monatomic gas calculations if the second ionization potential is not very large or if extremely high temperatures are involved. In this event we need two ionization reactions,

$$A + \varepsilon_i \rightleftharpoons A^+ + e \qquad (3\text{-}39)$$

$$A^+ + \varepsilon_{ii} \rightleftharpoons A^{++} + e \qquad (3\text{-}40)$$

(where ε_{ii} is the second ionization potential), along with their respective equilibrium constants and the conservation conditions,

$$n_A + n_+ + n_{++} = n_0 \qquad (3\text{-}41)$$

$$n_+ + 2n_{++} = n_e \qquad (3\text{-}42)$$

3-4 IONIZATION OF A GAS MIXTURE

In many practical situations one must deal with the ionization of a mixture of gases of substantially different ionization potentials. For example, many practical problems involve air as the working fluid, where a myriad of atomic, molecular, and ionic species appear at elevated temperatures. To obtain a reasonable estimate of the equilibrium ionization of air, over twenty reaction equations must be solved simultaneously, with six or more conservation equations [26,27]. Somewhat less tedious problems of practical interest involve the ionization of a gas of relatively high ionization potential which contains an unavoidable trace of an easily ionized impurity, or to which such an easily ionized substance has purposely been added to increase its ionization level and thereby its electrical conductivity. Such *seeding* technique is common practice in certain magnetogasdynamic machinery, notably in electric generators and steady flow accelerators, in order to maintain adequate gas conductivity at tolerable temperature levels.

To illustrate this problem, consider the simplest example of the ionization of a mixture of two monatomic gases which are chemically inert to each other. Assume that each of the two gases, A and B, ionizes only singly:

$$A + \varepsilon_A \rightleftharpoons A^+ + e \qquad (3\text{-}43)$$

$$B + \varepsilon_B \rightleftharpoons B^+ + e \qquad (3\text{-}44)$$

where ε_A and ε_B are the respective ionization potentials. The essential complication in this calculation is the participation of free electrons in both reactions, thereby coupling the two processes in a statistical sense. As before, the equilibrium constants may be written in terms of the total partition functions:

$$K_A(T) = \frac{n_A^+ n_e}{n_A} = \frac{2(2\pi mkT)^{3/2}}{h^3} \left(\frac{f_A^{i+}}{f_A^i}\right) e^{-(\epsilon_A/kT)} \qquad (3\text{-}45)$$

$$K_B(T) = \frac{n_B^+ n_e}{n_B} = \frac{2(2\pi mkT)^{3/2}}{h^3} \left(\frac{f_B^{i+}}{f_B^i}\right) e^{-(\epsilon_B/kT)} \qquad (3\text{-}46)$$

to which we add the conservation requirements:

$$n_e = n_A^+ + n_B^+ \qquad (3\text{-}47)$$

$$n_A^+ + n_A = n_A^0 \qquad (3\text{-}48)$$

$$n_B^+ + n_B = n_B^0 \qquad (3\text{-}49)$$

These five relations determine the equilibrium number densities of the five species n_A, n_A^+, n_B, n_B^+, n_e in terms of n_A^0, n_B^0, and T. Note that $K_A(T)$ and $K_B(T)$ are identical with the values they would have in the case of a pure gas, since the difference in ground state energy of A and B would cancel in the ratio of the internal partition functions of the ion and atom.

The system (3-45) to (3-49) is equivalent to one cubic equation in n_e:

$$n_e = n_A^+ + n_B^+ = \frac{n_A^0 K_A}{n_e + K_A} + \frac{n_B^0 K_B}{n_e + K_B} \qquad (3\text{-}50)$$

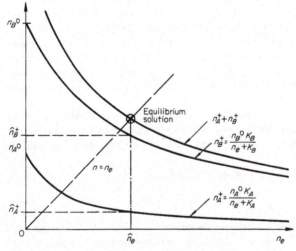

Fig. 3-5 Graphical solution of equilibrium ionization of binary gas mixture.

IONIZATION IN GASES

which can be displayed graphically to yield the equilibrium values of n_e, n_A^+, and n_B^+ at a given temperature. For example, if we plot $n_A^+ = n_A{}^0 K_A/(n_e + K_A)$ and $n_B^+ = n_B{}^0 K_B/(n_e + K_B)$ as functions of n_e (Fig. 3-5), the intersection of the sum curve, $n_A^+ + n_B^+$, with the 45° line, $n = n_e$, establishes the equilibrium value $\hat{n}_e$. The corresponding equilibrium values $\hat{n}_A^+$ and $\hat{n}_B^+$ lie on their respective curves at the same abscissa.

The special case where the temperature is so low that both species are only slightly ionized is of common occurrence, and displays the essence of the reaction coupling. To obtain reference values, let us imagine for the moment that each of the species A and B occupies identical but separate containers of this same volume, at the same particle density that each has in the mixture and at the same low temperature. The pure gas Saha relation (3-22) predicts respective ionization densities $\bar{n}_A^+$ and $\bar{n}_B^+$ for these pure gases:

$$\bar{n}_A^{+2} = \bar{n}_A K_A = (n_A{}^0 - \bar{n}_A^+) K_A \approx n_A{}^0 K_A \qquad (3\text{-}51)$$

$$\bar{n}_B^{+2} = \bar{n}_B K_B = (n_B{}^0 - \bar{n}_B^+) K_B \approx n_B{}^0 K_B \qquad (3\text{-}52)$$

and for the low-temperature mixture we have from (3-45) and (3-46)

$$n_A^+ n_e = n_A K_A = (n_A{}^0 - n_A^+) K_A \approx n_A{}^0 K_A \qquad (3\text{-}53)$$

$$n_B^+ n_e = n_B K_B = (n_B{}^0 - n_B^+) K_B \approx n_B{}^0 K_B \qquad (3\text{-}54)$$

Comparison of (3-51), (3-53) and (3-52), (3-54) yields

$$n_A^+ n_e \approx \bar{n}_A^{+2} \qquad (3\text{-}55)$$

$$n_B^+ n_e \approx \bar{n}_B^{+2} \qquad (3\text{-}56)$$

from which follow

$$\frac{n_A^+}{n_B^+} \approx \left(\frac{\bar{n}_A^+}{\bar{n}_B^+}\right)^2 \qquad (3\text{-}57)$$

and

$$n_e{}^2 = \bar{n}_A^{+2} + \bar{n}_B^{+2} < (\bar{n}_A^+ + \bar{n}_B^+)^2 \qquad (3\text{-}58)$$

The points made by relations (3-57) and (3-58) are that the total ionization of the mixture is less than the sum of that which its constituents would have as pure gases; indeed, that each constituent produces less ions than it would as a pure gas, and that the ratio of the relative contributions of the constituents is distorted in favor of the more prolific of the two (Prob. 3-3). The physical explanation is that the free electrons supplied by either specie are available for recombination with ions of the other, as well as with their own, and this situation preferentially discriminates against the net ionization level of the specie of higher ionization potential.

3-5. LIMITS OF SAHA CALCULATIONS

Before leaving the discussion of equilibrium ionization, it is perhaps worthwhile to emphasize the assumptions and corresponding precautions in application of the Saha type of statistical formulation outlined above.

1. Care should be taken that all accessible levels of ionization have been included. For example, the single ionization relation (3-26) clearly departs from reality as α approaches unity because of the onset of second ionization, etc.
2. It has been presumed that all particles involved in the system conform to maxwellian, or classical, statistics. In cases of extremely dense plasmas, more elaborate quantum statistics may be required.
3. Complete thermodynamic equilibrium has been presumed, including total containment of radiation. In many situations radiation losses are unavoidable, and depending on their intensity, the photon reactions such as (3-2) and (3-8) will be unbalanced and the ionization level will depart from its equilibrium value ([28]; Prob. 3-5).
4. The ratio of internal partition functions, for example, f_+^i/f_A^i, must be bounded. This may involve an artificial truncation of the electronic levels of an isolated atom at some level appropriate to the environment in which it exists in the composite gas. That is, as the outer electron is elevated to higher energy levels and wanders farther from its nucleus, a point is reached where disturbing influences from neighboring particles overcome its native attachment to its parent atom. At this point its bound levels effectively cease. Indeed, if these disturbing influences are severe, they may even lower the effective ionization potential perceptively, and this lower value would then be more appropriate for the exponential function in the Saha relations. This effective lowering of the ionization potential by the internal electric fields is characteristic of dense plasmas and can be observed spectroscopically.

The presumption of complete thermodynamic equilibrium clearly precludes formal application of the Saha relations to many interesting nonequilibrium ionized gas systems arising in nature and in practice, such as the radiating gas mentioned above. One particular type of nonequilibrium gas may be amenable to empirical application of a slightly modified Saha relation, however. In various low-density electric discharges it is common to find a gas composed of relatively low-temperature atoms and ions, $T_A \approx T_+$, and high-temperature free electrons, T_e. To the extent that the various ionization, excitation, recombination, and radiation processes are dominated by electron collisions with the heavy particles (rather than by heavy-particle collisions with themselves), it seems plausible to carry out the Saha calculation using the electron temperature T_e, particularly since the atomic and ionic temperatures will cancel in the ratio of their translational partition functions. Kerrebrock [29] has thus proposed the use of Saha relations based on T_e to describe the ionization level of nonequilibrium gases of this type, and various experimental evidence seems to support this attitude. It should be noted, however, that even these gases must conform to the other constraints outlined above if this approach is to be justified.

PROBLEMS

3-1. Estimate the electric field strength that would be needed to extract one electron from an argon atom initially in the ground state.

3-2. Evaluate the coefficients G and G' in Eqs. (3-27) and (3-28) for argon. Compute the degree of equilibrium ionization over the temperature range from zero to 3000°K at 1 atm pressure, and graph the result.

3-3. One cubic centimeter of a mildly ionized pure gas X contains 3×10^{10} free electrons in equilibrium at a given temperature. At the same temperature, but twice the pressure, 1 cm^3 of pure gas Y yields 1×10^{10} free electrons. If X and Y are combined in a 1-cm^3 volume at this temperature, what free-electron density will result? What portion of these will be contributed by X and Y, respectively?

3-4. Consider an equilibrium mixture of two gases, A and B, at a given temperature. A has so low an ionization potential and is present in so small a partial density that it is nearly completely (singly) ionized. B has so high an ionization potential that it is only slightly ionized. Derive expressions analogous to Eqs. (3-57) and (3-58) relating the relative contributions of free electrons from the two constituents in the mixture to the ionizations they would have achieved in the pure state at the same partial densities and temperature. Interpret your result.

3-5. Does a loss of radiation from a body of ionized gas tend to raise or lower its ionization level from that predicted by equilibrium theory? Discuss.

REFERENCES

1. Spitzer, L., Jr.: "Physics of Fully-ionized Gases," Interscience Publishers, Inc., New York, 1956.
2. Longmire, C. L.: "Elementary Plasma Physics," Interscience Publishers, Inc., New York, 1963.
3. Thompson, W. B.: "An Introduction to Plasma Physics," Pergamon Press, New York, 1962, and Addison-Wesley Publishing Company, Inc., Reading, Mass., 1962.
4. Allis, W. P.: Motions of Ions and Electrons, Handbuch der Physik, vol. 21, p. 383, Springer Verlag OHG, Berlin, 1956.
5. Gartenhaus, S.: "Elements of Plasma Physics," Holt, Rinehart and Winston, Inc., New York, 1964.
6. Uman, M. A.: "Introduction to Plasma Physics," McGraw-Hill Book Company, New York, 1964.

7. Chandrasekhar, S.: Plasma Physics, a course given at the University of Chicago, notes compiled by S. K. Trehan, The University of Chicago Press, Chicago, 1960.
8. Linhart, J. G.: "Plasma Physics," North Holland Publishing Company, Amsterdam, 1960.
9. Delcroix, J. L.: "Introduction to the Theory of Ionized Gases," transl. from the French by Melville Clarke, Jr., David J. BenDaniel, and Judith M. BenDaniel, Interscience Publishers, Inc., New York, 1960.
10. Cambel, A. B.: "Plasma Physics and Magnetofluidmechanics," McGraw-Hill Book Company, New York, 1963.
11. Rose, D. J., and S. M. Clark, Jr.: "Plasmas and Controlled Fusion," The M.I.T. Press, Cambridge, Mass., and John Wiley & Sons, Inc., New York, 1961.
12. Glasstone, S., and R. H. Lovberg: "Controlled Thermonuclear Reactions," D. Van Nostrand Company, Inc., Princeton, N.J., 1960.
13. Ferraro, V. C. A., and C. Plumpton: "An Introduction to Magneto-fluid Mechanics," Oxford University Press, Fair Lawn, N.J., 1961.
14. Pai, Shih-I: "Magnetogasdynamics and Plasma Dynamics," Prentice-Hall, Inc., Englewood Cliffs, N.J., 1962.
15. Cowling, T. G.: "Magnetohydrodynamics," Interscience Publishers, Inc., New York, 1957.
16. Vlasov, A. A.: "Many-particle Theory and Its Application to Plasma," Gordon and Breach, Science Publishers, Inc., New York, 1961 (originally published by State Publishing House for Technical-Theoretical Literature, Moscow, 1950).
17. Harris, L. P.: "Hydromagnetic Channel Flow," The M.I.T. Press, Cambridge, Mass., 1960.
18. Anderson, J. E.: "Magnetohydrodynamic Shock Waves," The M.I.T. Press, Cambridge, Mass., 1963.
19. Born, M.: "Atomic Physics," Hafner Publishing Company, Inc., New York, 1946.
20. Richtmyer, F. K., E. H. Kennard, and T. Lauritsen: "Introduction to Modern Physics," McGraw-Hill Book Company, New York, 1955.
21. Herzberg, G.: "Atomic Spectra and Atomic Structure," Dover Publications, Inc., New York, 1944.
22. Schiff, L. I.: "Quantum Mechanics," 2d ed., McGraw-Hill Book Company, New York, 1955.
23. Bohm, D.: "Quantum Theory," Prentice-Hall, Inc., Englewood Cliffs, N.J., 1951.
24. Saha, M. N.: Ionization in the Solar Chromosphere, *Phil. Mag.*, vol. 40, p. 472, 1920; Elements in the Sun, *ibid.*, p. 809.
25. Fowler, R., and E. A. Guggenheim: "Statistical Thermodynamics," chap. 5, Cambridge University Press, New York, 1952.
26. Gilmore, F. R.: Equilibrium Composition and Thermodynamic Properties of Air to 24,000°K, *Rand Corp. Res. Mem.* RM-1543, August, 1955.
27. Logan, J. G., Jr.: Thermodynamic Properties of Air at High Temperatures, *Cornell Aeron. Lab., Inc., Repts.*, no. AD-1052-A-1, A-2, A-3, 1956.
28. Griem, H. R.: "Plasma Spectroscopy," McGraw-Hill Book Company, New York, 1964.
29. Kerrebrock, J. L., and M. A. Hoffman: Nonequilibrium Ionization Due to Electron Heating, pts. I and II, *AIAA J.*, vol. 2, pp. 1072 and 1080, 1964.

4
Particle Collisions in an Ionized Gas

Virtually every gas property of engineering interest ultimately derives from the multitude of atomic-scale interactions, or collisions, which are incessantly in progress within the gas. The transport processes, such as mass diffusion, viscous friction, thermal conduction, conduction of electric current, etc., are accomplished by the migration of the atomic particles within the gas, or by transfer of their momenta and energy through the gas, and therefore are controlled primarily by the frequency and effectiveness of their collisions. The rate at which the gas can adjust to changes in its environment, and find a new equilibrium state after it has been disturbed from a previous one, is determined mainly by the efficiency of the collisional communication system among the atomic particles. As we have seen, the equilibrium state itself may be regarded as a balance between collisional events which produce a particular set of species products and other collisional events which reverse the reactions. Even the continuum thermodynamic properties, such as pressure, temperature, and entropy, are macroscopic manifestations of atomic collision processes, and kinetic theories of neutral gases have been carried to considerable elegance to establish the correlation between collisional mechanics and these thermodynamic properties [1,2].

4-1 ATOMIC COLLISION NOMENCLATURE

Important as atomic collision information may be for detailed understanding of ionized gas behavior, it is extremely difficult to accumulate and catalog. Even in the simplest ionized gas, composed only of electrons, single positive ions, and neutral atoms, there are six permutations of particle collision to be considered: electron-ion, electron-atom, electron-electron, etc. In all but the electron-electron collisions, the interaction may yield a variety of results. First, the event may be distinguished by one of several types of energy exchange (Fig. 4-1):

a. *Elastic*, wherein the total kinetic energy of the two particles is conserved and the particles retain their initial charges
b. *Inelastic*, wherein kinetic energy is transferred into some internal mode of one or both of the colliding particles, e.g., electronic excitation, vibrational excitation, etc., or into creation of a new particle system of higher potential energy, as in ionization, dissociation, etc.
c. *Superelastic*, wherein one or both of the colliding particles deliver some of their internal energy to the kinetic energy of the system, as in collisional deexcitation, or some potential energy is surrendered, as in the electron-ion recombination event
d. *Radiative*, wherein a charged particle radiates energy in the electromagnetic spectrum when accelerated through the potential field of the other particle
e. *Charge-reactive*, wherein electric charge is transferred from one particle to the other, in an otherwise elastic event

The foregoing list may encompass not only binary collisions, but also three-body collisions involving a "catalytic" particle which aids in the momentum conservation.

The detailed results of each of these classes of collision depend strongly on the relative kinetic energy of the colliding particles, on their impact parameter, and on their internal states of excitation. Strikingly different results are found for different atomic species colliding under otherwise identical circumstances. For example, elastic electron collisions on argon atoms are vastly different from those on helium at the same energy; electronic excitation collisions depend primarily on the specific atomic level structure of the specie excited; and so forth.

Much of this complexity in the atomic collision picture may be attributed to the essentially quantum-mechanical nature of the interactions. In many instances simple newtonian concepts and rigid-body mechanics are completely obscured by the wavelike behavior of the interacting particles. In particular, in any cases in which the *de Broglie*

PARTICLE COLLISIONS IN AN IONIZED GAS

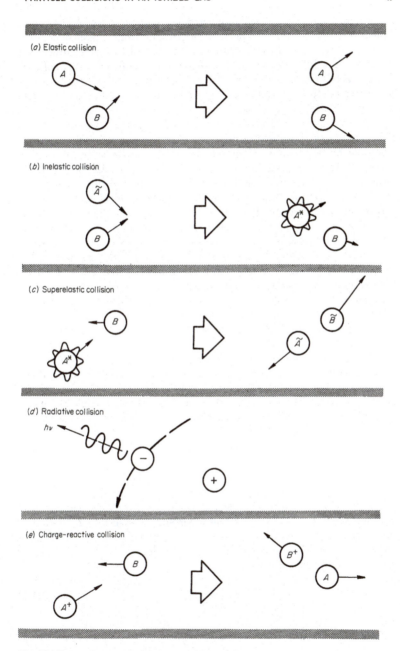

Fig. 4-1 Types of energy exchange in atomic collisions.

wavelength, $\lambda = h/mv$, of a colliding particle is comparable with the atomic dimensions of its collision partner, departure from classical collision behavior is to be expected. Quantum-theoretical formulation of these interactions is difficult, and little systematic tabulation of atomic collision effects has been achieved analytically. Only for relatively simple atomic structures are the necessary wave functions known well enough to justify detailed calculations of this sort.

Experimental data are also regrettably sparse. Atomic collision phenomena have been extensively studied, using atomic, ionic, and electronic beams scattered in target chambers at energies above 10 ev, but these techniques become more difficult, and the results less certain, at the lower energies corresponding to the ionized gas temperatures of engineering interest. Again, because of the quantum-mechanical effects at work, extrapolation of high energy particle beam data to very low energies may be misleading. Direct measurement of the desired ionized gas transport coefficients and reaction rates in ionized gases can be made only in a few relatively simple situations, and generalization of such results to less idealized environments is also risky. Vigorous theoretical and experimental research continues in the atomic collision field, but the present understanding of these important phenomena is incomplete, and will obstruct our later macroscopic calculations. Here we shall attempt only to introduce a bit of the nomenclature and physical concepts and to display a few typical results.

Any given collision may be represented by a cross section Q, defined as the effective cross-sectional area of one of the particles for a collision of a particular type with the other. Specifically, the probability that a particle of specie j will suffer a collision of type β with a particle of specie k in traveling a distance dx relative to a group of k particles is written

$$dP_{jk}^{(\beta)} = n_k Q_{jk}^{(\beta)} \, dx \tag{4-1}$$

where n_k is the number density of type-k particles in the vicinity of the interaction. As defined, $Q_{jk}^{(\beta)} \equiv Q_{kj}^{(\beta)}$. For example, we speak of a particular atom as having a certain cross section for elastic collision with another atom, or an ion as having a certain cross section for radiative interaction with an electron, etc. Occasionally, we deal with an effective cross section which represents a particle's vulnerability to all classes of collision by the other particle:

$$Q_{jk} = \sum_\beta Q_{jk}^{(\beta)} \tag{4-2}$$

We may also define a "differential" cross section $q(\theta)$ in terms of the angle

PARTICLE COLLISIONS IN AN IONIZED GAS 49

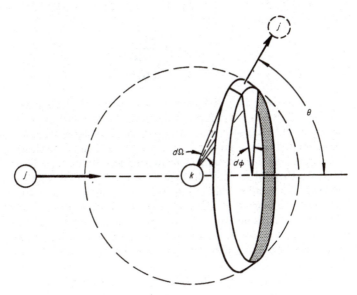

Fig. 4-2 Nomenclature for definition of differential cross section.

the scattered-particle trajectory makes with its incident direction (Fig. 4-2). The probability of the particle emerging into the solid angle $d\Omega = \sin\theta\, d\theta\, d\phi$ is defined

$$dP(\theta) = q(\theta) \sin\theta\, d\theta\, d\phi \tag{4-3}$$

so that the integral of $q(\theta)$ over the full solid angle yields the total cross section for the particular collision involved:

$$\int_0^{2\pi}\int_0^{\pi} q(\theta) \sin\theta\, d\theta\, d\phi = Q \tag{4-4}$$

The dimensions of a total cross section Q are clearly those of area, and the unit of square centimeters is most commonly employed, although square angstroms or square Bohr radii are also used. In view of typical electron-orbit dimensions $\approx 10^{-8}$ cm, we might regard $Q \approx 10^{-16}$ cm² as a standard size. Substantially larger values are indicative of long-range interactions or strong quantum-mechanical resonances. Substantially smaller values reflect a low efficiency of the collision for the process involved.

Closely related to the concept of cross section are those of mean free path and collision frequency:

$$\lambda_{jk} = \frac{1}{n_k Q_{jk}} \quad (4\text{-}5)$$

$$\nu_{jk} = n_k Q_{jk} \bar{v}_{jk} \quad (4\text{-}6)$$

where $\bar{v}_{jk}$ is essentially defined by (4-6) as an effective speed of the j particle relative to the group of k particles, for the particular process and environment involved. In the simple case of a collimated monoenergetic beam impinging on a static group of target particles, $\bar{v}_{jk}$ is clearly just the beam speed; in a multicomponent ionized gas, it will usually be close to the mean thermal speed of the faster specie of colliding particle, modified perhaps by a smaller relative streaming component. Actually, the collision frequency concept itself is complicated by the strong dependence of the cross sections on the relative kinetic energy of the particles, and is usually only applied in an average sense; e.g., the mean rate at which a swarm of j particles collide with a swarm of k particles might be expressed

$$n_j \bar{v}_{jk} = \iint n_j(\mathbf{v}_j) n_k(\mathbf{v}_k) Q_{jk}(|\mathbf{v}_j - \mathbf{v}_k|) |\mathbf{v}_j - \mathbf{v}_k| \, d\mathbf{v}_j \, d\mathbf{v}_k \quad (4\text{-}7)$$

where $n_j(\mathbf{v}_j) \, d\mathbf{v}_j$ is the number density of type-j particles having velocities in the differential range around $\mathbf{v}_j$, etc., and $\int d\mathbf{v}_j$ implies integration over the three-dimensional velocity space of $\mathbf{v}_j$ [2].

For the remainder of this chapter, we shall attempt only to construct a rudimentary catalog of the characteristics of various collision cross sections for the low energy range of our interest. For specific data, description of experimental techniques, and detailed theoretical analyses, the reader is referred to several extensive works devoted entirely to this subject, Refs. 3 to 10.

4-2 ELECTRON-ATOM COLLISIONS

Electrons are by far the most mobile particles in an ionized gas; hence their motion strongly influences the transport properties and the reaction rates. If the gas is only slightly ionized, the prevailing electron collisions are with neutral atoms; hence the electron-atom cross sections are major factors in determining these gas properties.

ELASTIC

In the low energy, or thermal, range, elastic electron-atom cross sections exhibit a severe energy dependence and an equally severe angular scattering profile and differ by several orders of magnitude from one atomic species to another (Figs. 4-3 and 4-4). Values may reach as high as 10^{-13} cm² for certain alkali metals which have a far-ranging valence electron and as low as 10^{-17} cm² for some parts of a noble gas profile where wave-mechanical resonance occurs (the *Ramsauer effect*).

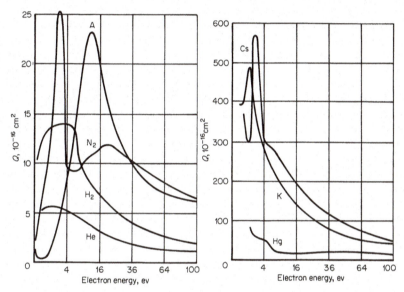

Fig. 4-3 Energy dependence of electron-atom elastic cross sections. (*From R. B. Brode, Rev. Mod. Phys., vol. 5, p. 257, 1933, and H. Margenau and F. P. Adler, Phys. Rev., vol. 79, p. 970, 1950.*)

INELASTIC

We include in this category both those inelastic collisions that leave the atom in an excited bound state,

$$\tilde{e} + A \rightarrow A^* + e \qquad (4\text{-}8)$$

and those that eject the bound electron completely, thereby ionizing the atom,

$$\tilde{e} + A \rightarrow A^+ + 2e \qquad (4\text{-}9)$$

where $\tilde{e}$ denotes an incident electron with kinetic energy at least as large as the excitation or ionization potential involved. In addition to their primary function of providing excited atoms and new ion-electron pairs, inelastic electron-atom collisions of these types may be the major factor

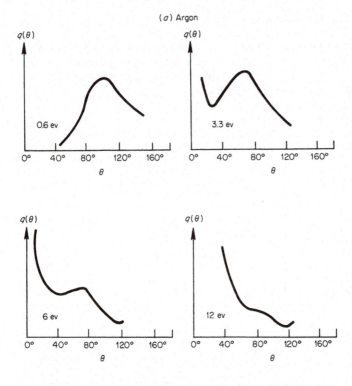

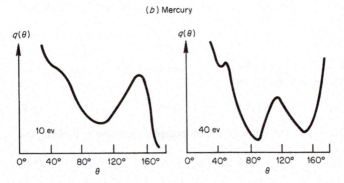

Fig. 4-4 Angular dependence of electron-atom elastic cross sections in argon and mercury. (*From E. C. Bullard and H. S. W. Massey, Proc. Roy. Soc., ser. A, vol. 130, p. 579, 1931, and E. C. Childs and H. S. W. Massey, ibid., vol. 141, p. 473, 1933.*)

in establishing the free-electron energy distribution in an ionized gas. Since most excitation and ionization potentials are at least a few electron volts, each inelastic event costs the electron component a significant portion of its mean thermal energy per particle. In fact, for gases of modest electron temperatures, say, $T_e < 10{,}000°K$, only the most energetic electrons in the distribution can accomplish these events at all, and, in so doing, revert abruptly to the low energy end of the distribution. The integrated effect of many such collisions is thus to suppress the electron temperature, and perhaps to distort the distribution toward the low energy end somewhat.

Electron-atom excitation cross sections have qualitatively the same energy dependence for most species and for most excitation levels (Fig. 4-5).

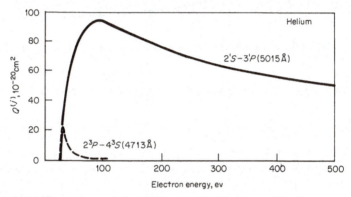

Fig. 4-5 Typical electron-atom excitation cross sections vs. energy. *(From L. Lees, Proc. Roy. Soc., ser. A, vol. 137, p. 173, 1932.)*

The magnitudes rise rapidly from zero at the threshold energy $\mathcal{E}_j$ to some maximum value at an incident energy $\hat{\mathcal{E}}$ which is a few times the particular transition energy, and then fall off with a high energy "tail" which is characteristic of the spectroscopic nature of the same transition. If the transition is "allowed" by the spectroscopic selection rules, the tail tends to fall as $\ln \mathcal{E}/\mathcal{E}$; if "forbidden," it falls more rapidly, as $1/\mathcal{E}$ or faster [3]. The maximum cross section is usually much greater for allowed transitions than for forbidden ones, but in either case depends strongly on the particular quantum jump involved. Values as high as $5 \times 10^{-18}\,cm^2$ are observed for certain strong resonance transitions. The angular dependence of the scattered electron trajectory tends to concentrate more around the forward direction than for elastic collisions, and this tendency increases with incident energy.

In general, the energy dependence of ionization cross sections is qualitatively the same as for the excitations (Fig. 4-6). Since all ioniza-

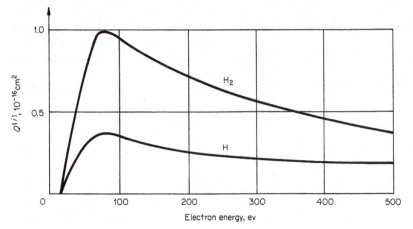

Fig. 4-6 Typical ionization cross sections vs. electron energy. *(From R. E. Fox, Westinghouse Elec. Corp., Res. Rept., 60-94439-4-R2, Aug. 15, 1956.)*

tion transitions are allowed, and since the end state is a continuum rather than a discrete level, the magnitudes are substantially larger. Peak values close to 10^{-15} cm² exist for some easily ionized species, and even the noble gases exceed 10^{-16} cm² in some cases.

In most engineering problems, the electron temperature of an ionized gas is so much lower than the ionization and excitation potentials involved that the complicated energy dependence of the cross sections can be replaced by a simple linear rise from threshold, for purposes of integration over the electron energy distribution, e.g., in a collision frequency calculation like Eq. (4-7) (Fig. 4-7).

SUPERELASTIC

Collisions wherein the electron acquires kinetic energy at the expense of the internal energy of an excited atom are known to occur:

$$e + A^* \rightarrow \bar{e} + A \qquad (4\text{-}10)$$

Experimental data on specific cross sections are very sparse, but mercury vapor has demonstrated values as high as 10^{-15} cm² for certain electronic transitions of the atom. Barring unexpectedly large values of Q, this type of collision will not seriously affect the electron energy distribution in

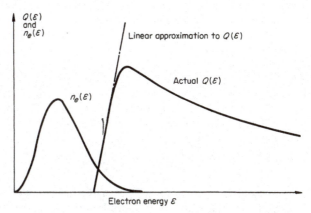

Fig. 4-7 Folding of inelastic cross section with electron energy distribution by linear approximation.

mildly ionized gases where the fraction of atoms in excited states is small.

RADIATIVE

In principle, the electron may undergo an electromagnetic interaction with the atomic potential field,

$$\bar{e} + A \rightarrow e + A + h\nu \tag{4-11}$$

but because the neutral atom field is extremely weak at all but very short range, the effective cross section is quite negligible, say, 10^{-22} cm² or less.

CHARGE-REACTIVE

Under certain circumstances, the incident electron may be trapped by the atom, thus creating a negative ion:

$$e + A \rightarrow A^- \tag{4-12}$$

This event is called electron attachment, and may require a third body to dispose of the net potential energy surrendered. It thus is more prevalent in certain polar molecules, such as O_2, NH_3, H_2S, SO_2, and N_2O, although it has been observed in atomic oxygen. Experimental data are incomplete, but as an example, O_2 displays a cross section $\approx 10^{-19}$ cm² with an irregular energy dependence.

4-3 ELECTRON-ION COLLISIONS

Two charged particles exert significant force on each other at much longer range than the usual atomic dimensions; hence the effective cross sections may be much larger than those involving neutral particles. It is typical for the electron-ion elastic collision frequency to exceed the electron-atom collision frequency in an ionized gas for ionization levels above a few tenths of one percent.

ELASTIC

Because the coulomb potential falls off only as $1/r$, a collision cross section is not well defined in terms of a simple deflection. A more useful concept here is the momentum-transfer cross section $Q^{(p)}$, defined in the average over many encounters by the relation

$$\frac{dp_x}{p_x} = n_+ Q^{(p)} \, dx \qquad (4\text{-}13)$$

where p_x and dp_x are the magnitude of the incident electron momentum and the differential change of magnitude of that component of momentum over the path length dx (Fig. 4-8). An energy-transfer cross section

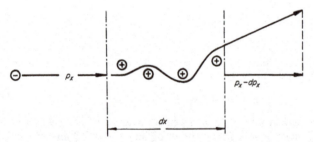

Fig. 4-8 Momentum transfer in electron-ion collisions.

$Q^{(\varepsilon)}$ could be defined similarly; it would be of the order of m/M_+ smaller than $Q^{(p)}$ (Prob. 4-1). Unlike most cross sections, the coulomb cross section can be accurately calculated theoretically. Using either classical central-force mechanics [11] or a parabolic separation of the Schrödinger wave equation [12], the following expression may be derived (Prob. 4-2):

$$Q^{(p)} = \frac{\pi e^4}{(4\pi\epsilon_0)^2 \mathcal{E}^2} \ln \frac{8\pi\epsilon_0 r_0 \mathcal{E}}{e^2} \qquad \text{mks} \qquad (4\text{-}14)$$

where e = electronic charge
ϵ_0 = vacuum dielectric constant
$\mathcal{E}$ = relative kinetic energy of particles before collision
r_0 = an empirical cutoff distance for effective range of ion's coulomb field in environment involved

For r_0 in an ionized gas, one might consider using a mean interionic spacing or a mean interatomic spacing, but in most cases the appropriate value can be shown to correspond to the Debye length,

$$\lambda_D = \left(\frac{\epsilon_0 k T_e}{n_e e^2}\right)^{1/2} \tag{4-15}$$

where T_e again denotes the temperature of the free-electron component of the gas, and is not necessarily the same as that for the ions or neutrals. The Debye length arises in several other situations as an index of the typical charge separation a plasma can sustain by virtue of the random thermal energy of its electrons (Prob. 4-3). Fortunately, the value of $Q^{(p)}$ is insensitive to the choice of r_0, since it appears in the logarithm and since the numerical values of the various reasonable possibilities are not widely different. When r_0 is identified with λ_D, it is common to replace the entire logarithmic factor with a single quantity, $\ln \Lambda$, which is well tabulated over the range of n and T_e of interest [(4-12)]. Considering the precision with which other associated calculations can be made, $\ln \Lambda \approx 10$ is usually an adequate approximation for ionized gases in the propulsion regime, and (4-14) may be written

$$Q^{(p)} \approx \frac{6.5 \times 10^{-13}}{\mathcal{E}^2} \quad \text{cm}^2 \text{ for } \mathcal{E} \text{ in electron volts} \tag{4-16}$$

The angular dependence of a coulomb scattering process is given by the familiar Rutherford relation

$$q(\theta) \propto \csc^4 \frac{\theta}{2} \tag{4-17}$$

Both (4-14) and (4-17) are clearly independent of the particular ionized specie involved.[1]

---

[1] It should be noted that the coulomb collision formulation yielding expressions (4-14) and (4-17) presumes an r^{-1} potential profile all the way to the origin; i.e., it neglects the "hard core" of the ion. For energies below about 10 ev this is an acceptable approximation, since the great bulk of the momentum transfer occurs far out on the potential profile, at much larger radii than the true ionic dimension. Indeed, the cumulative effect of many small deflections greatly exceeds the contribution of single wide-angle deflections in the determination of the effective collision frequency for momentum-transfer purposes. At very high energies, however, blind application of (4-14) or (4-15) yields a cross section smaller than that for electron-atom elastic collisions—clearly, an unreasonable result. The difficulty, of course, is that at these energies incident electrons with impact parameters comparable with the ionic dimension can reach and penetrate the bound electronic structure of the ion, wherein the simple coulomb interaction is totally invalid. Under these circumstances it is more appropriate to regard the ion as a hard sphere out to its effective atomic dimension, surrounded by a coulomb potential "skirt" at larger radii. The one specie not requiring this modification is the hydrogen atomic ion (proton) which has no bound electronic structure, and hence exhibits a "hard core" of only nuclear dimensions, $\approx 10^{-24}$ cm^2.

INELASTIC

Since the inelastic events require penetration of the incident electron into the bound electronic structure of the ion, cross sections for these processes may be expected to be similar to those for the electron-atom inelastic interactions. A slight increase in magnitude at low energies might be expected from the potential "funneling" of the incident electron toward the mass center of the ion. Also, since any given ion has a totally different electronic structure and ionization potential than its neutral atom, these differences will be reflected in the characteristics of the inelastic cross sections.

SUPERELASTIC

The foregoing remarks about inelastic collisions should apply to the collisional deexcitation processes as well, with the notable exception of electron-ion recombination, which, in the absence of radiation, requires a third body to conserve momentum and energy:

$$e + A^+ + X \to A + \tilde{X} \quad (4\text{-}18)$$

X here may denote another electron, ion, or atom or a material surface. An important special case of this reaction is the so-called *dissociative recombination*

$$e + A_2^+ \to \tilde{A} + \tilde{A} \quad (4\text{-}19)$$

where A_2^+ may be a short-lived, or *metastable*, concurrence of an atom and an ion. Representation of such reactions in terms of cross sections is less clear-cut than for the two-body reactions, and the literature more frequently quotes effective macroscopic recombination coefficients for maxwellian species at given temperature and pressure [6]. Experimentally, it is difficult to separate various recombination mechanisms and identify specific reaction rates, but a composite inverse energy dependence seems to prevail.

RADIATIVE

Deflection of an electron in the coulomb field of an ion may precipitate emission of a quantum of electromagnetic radiation, with a corresponding reduction in kinetic energy of the electron:

$$\tilde{e} + A^+ \to e + A^+ + h\nu \quad (4\text{-}20)$$

This radiation, called *bremsstrahlung*, may be associated classically with the acceleration of the electron in the ionic field, or equivalently, with the change in the electron-ion dipole moment. Spectroscopically, the event may be regarded as a *free-free* electronic transition between two energy states in the continuum. The cross section is usually expressed as a function of emitted photon frequency, and must be calculated quantum-mechanically [14]. In a relatively low temperature ionized gas, the radiative energy loss by this mechanism is unimportant, but the spectral continuum arising from these encounters may provide a useful diagnostic indication of the electron temperature and density within the gas. For example, if the electrons have a maxwellian distribution, and only single ions are involved, the intensity spectrum of the bremsstrahlung has the approximate form

$$I_{ff}(\nu) \propto \frac{n_e^2}{T_e^{1/2}} e^{-h\nu/kT_e} \qquad (4\text{-}21)$$

from which the electron temperature may be simply extracted (Fig. 4-9).

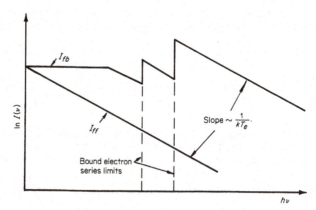

Fig. 4-9 Frequency spectrum of free-free and free-bound radiation. *(From W. Finkelnburg and Th. Peters, Handbuch der Physik, vol. 28, p. 96, Springer Verlag OHG, Berlin, 1957.)*

RADIATIVE RECOMBINATION

As the relative kinetic energy of the incident electron in (4-20) is decreased, there is an increased probability that emission of the photon will leave it with insufficient energy to escape the ion field. It thus will be captured by the ion and neutralize it—a so-called *free-bound* transition,

$$\bar{e} + A^+ \rightarrow A_j^* + h\nu \qquad (4\text{-}22)$$

where the photon now must dispose of the incident kinetic energy of the electron, plus the difference between the ionization potential and the energy of the particular bound electronic state in which the neutral atom is left, A_j^*:

$$h\nu = \varepsilon + (\varepsilon_i - \varepsilon_j) \quad (4\text{-}23)$$

For hydrogenlike atoms, a theoretical approximation for the cross section for radiative capture to the bound level j can be obtained in the form [13]

$$Q_{fj} \approx 2 \cdot 10^{-22} \frac{\varepsilon_i{}^2}{\varepsilon(\varepsilon + \varepsilon_i - \varepsilon_j)n_j{}^3}$$

$$\approx 2 \cdot 10^{-22} \frac{\varepsilon_i}{\varepsilon n_j} \quad \text{for } \varepsilon \ll \varepsilon_i - \varepsilon_j \quad (4\text{-}24)$$

where n_j is the appropriate principal quantum number. The indicated preference for recombination to the ground state is perhaps contrary to naive anticipation. Experimental verification of these individual cross sections is difficult because of their small size and because of other competing recombination processes. The overall inverse energy dependence of the total radiative recombination cross section, $Q_{fb} = \sum_j Q_{fj} \propto 1/\varepsilon$, is better established. The composite spectrum from this event is again indicative of electron density and temperature [15]:

$$I_{fb}(\nu) \propto \frac{n^2}{T_e{}^{3/2}} \sum_j \left[(\varepsilon_i - \varepsilon_j)^2 \frac{g_j}{n_j} \exp \frac{\varepsilon_i - \varepsilon_j - h\nu}{kT_e} \right] \quad (4\text{-}25)$$

where g_j is the quantum-mechanical degeneracy of the jth bound level. Note that $h\nu$ must exceed $\varepsilon_i - \varepsilon_j$ for each term in the sum, yielding abrupt intensity jumps at characteristic frequencies in this continuum spectrum (Fig. 4-9).

4-4 ELECTRON-ELECTRON AND ION-ION COLLISIONS

Because of their mass disparity, electrons in an ionized gas are energetically insulated to a certain extent from the heavy particles, and rely heavily on intraspecie collisions to establish their distribution of their energies in a given environment (Prob. 4-1). Ions are in better communication with other heavy particles, but because of the long range of ion-ion coulomb interactions, their intraspecie collisions also strongly influence the distribution of their energies.

ELASTIC

Although ion-ion and electron-electron collisions have the same coulomb potential fields as the electron-ion interactions, they differ from the latter in a few significant respects. First, because of the equal masses, the energy transfer is vastly more efficient, that is, $Q^{(p)} \approx Q^{(\varepsilon)}$, and the scattering angles of the two particles are equal in a center-of-mass frame of reference. Second, if the two particles are completely identical, both classical indistinguishability $[q(\theta) = q(\pi - \theta)]$ and quantum-mechanical exchange effects (Mott scattering) appear. With these exceptions, the energy and angular dependence of the cross section resemble the electron-ion scattering outlined above (Fig. 4-10).

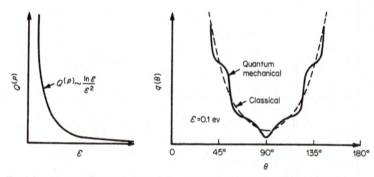

Fig. 4-10 Energy and angular dependence of elastic electron-electron collision cross section. *(From N. F. Mott, Collision between Two Electrons, Roy. Soc. London Proc., ser. A, vol. 126, p. 259, 1930.)*

INELASTIC

Only at energies above several electron volts can two heavy charged particles of like sign approach closely enough together to disturb either electronic structure, and even at these energies the interaction is highly inefficient (Sec. 4-5, under inelastic).

SUPERELASTIC

Same remarks as for inelastic.

RADIATIVE

Simultaneous deflection of two like charges involves no net charge acceleration or dipole moment change. Hence there is no bremsstrahlung to first order. Quadrupole radiation or relativistic effects can appear, but these are negligible in our regimes of interest.

4-5 ATOM-ATOM COLLISIONS

ELASTIC

The elastic collisions of neutral atoms have been extensively studied in the framework of the kinetic theory of gases. Detailed calculations of mean free paths, transport processes, rate processes, etc., are available for a variety of interatomic potential functions, ranging from simple rigid spheres to the van der Waals and Lennard-Jones models [1,2]. Typically, this potential combines a weak attraction of two dipoles, $\approx r^{-6}$, and a "hard-core" repulsion of much shorter range, say, $\approx r^{-12}$ (Fig. 4-11).

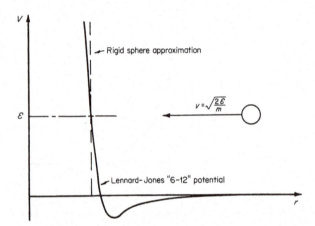

Fig. 4-11 Typical interatomic potential profiles for elastic scattering calculations.

Since the momenta of most heavy particles in a high temperature gas are large enough so that their associated de Broglie wavelengths, $\lambda = h/mv$, are small compared with atomic dimensions, quantum-mechanical effects are subdued. In most applications, a classical approach is completely adequate.

Although experimental determinations of individual cross sections are very difficult at low energies, macroscopic measurements of transport coefficients agree well with the classical theories. The effective cross sections deduced from these experiments lie in the range of 1 to 1,000 × 10^{-16} cm^2 for the common gases, depending on the particular electronic and molecular structure of the specie. Energy dependence is usually quite mild, except at very low energies (<0.1 ev), where the attractive portion of the interatomic potential may increase the cross section slightly, typically, $\propto \varepsilon^{-1/5}$. Recently, molecular beam techniques have advanced to the point where specific cross section vs. energy and angular scattering profiles can be traced for individual binary collisions of common gas atoms over the energy range below 1 ev [16,17]. The primary aim of these experiments is better understanding of the details of atomic structure and interactions, and although interesting quantum-mechanical resonances are being identified, it is doubtful that any major revision in the magnitudes of distribution-averaged cross sections for gas transport calculations will be required.

INELASTIC

The available atomic beam data on excitation or ionization by atom-atom collision is almost all at very high energies. Extrapolation back through the low energy range to threshold yields reasonable values, but little direct confirmation exists, except in swarm averages (Fig. 4-12). One point of view regards the colliding atoms simply as carriers which convey bound electrons into the particular interaction, suggesting the rule of thumb that, above threshold, an atom-atom collision is about as effective as an electron-atom collision, at the same relative velocity. As such we expect cross sections of the order of 10^{-20} cm^2 for excitations and 10^{-17} cm^2 for ionizations in the low energy range. The energy dependence again is probably linear just above threshold, but the magnitude of the slope may be strongly species-dependent. Recent shock tube measurements indicate slopes $\approx 10^{-19}$ cm^2/ev for the noble gases argon, krypton, and xenon [18,19].

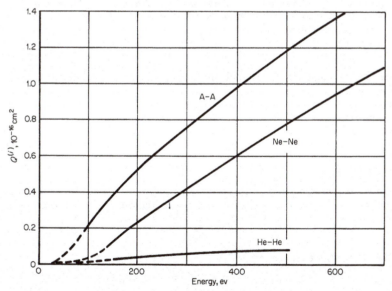

Fig. 4-12 Typical atom-atom ionization cross sections vs. energy. *(From A. von Engel, "Ionized Gases," chap. 3, p. 59, Oxford University Press, Fair Lawn, N.J., 1955.)*

SUPERELASTIC

Little is known about the collisional deexcitation process

$$A^* + A \to A + \tilde{A} \tag{4-26}$$

other than that, in a gas in complete equilibrium, it must just balance the reverse reaction, discussed in the previous section, and thus must have a correspondingly large cross section. It doubtless competes with the resonant exchange reaction,

$$A^* + A \to A + A^* \tag{4-27}$$

which clearly has no net effect on the population of excited states. Experimentally, it is much easier to acquire information about superelastic and exchange reactions with impurity particles in a gas by observing the quenching of radiation from the excited atom or the appearance of new radiation from the excited impurity:

$$A^* + I \to A + \tilde{I} \tag{4-28}$$

$$A^* + I \to A + I^* \to A + I + h\nu \tag{4-29}$$

Cross sections for the former reaction exceeding 10^{-14} cm² have been found for certain combinations of A^* and I. The latter reaction is favored by impurities having excited states of energy close to those of A^*, under which condition the cross section may have even larger values.

Three-body recombination,

$$A^+ + A + e \to A + \tilde{A} \tag{4-30}$$

can be regarded as a collision of this class, but has already been discussed in Sec. 4-2.

RADIATIVE

Because of the short range of atomic fields, no events of this class are expected or observed.

CHARGE-REACTIVE

Since this interaction requires formation of a negative ion, it can proceed only in certain electronegative gas mixtures, which are outside the scope of this outline.

4-6 ION-ATOM COLLISIONS

The interaction field between an ion and an atom is that between a charge and an induced dipole, that is, $\propto r^{-4}$, and hence slightly broader than that between two atoms, but still quite short range. The characteristics of elastic, inelastic, and superelastic collisions thus should be much like the corresponding atom-atom events. Experimentally, the preparation of a monoenergetic ion beam is simpler than for an atomic beam, and the cross-sectional data may be more reliable. Unfortunately, the energy range below 5 ev remains difficult, and little data yet exist.

ELASTIC

Ion beam experiments down to 5 ev show angular distributions of strong forward scattering preference, with total cross sections in the range 10^{-15} to 10^{-14} cm². The angular dependence becomes more isotropic, and the magnitude of the total cross section increases slightly as energy decreases in this range (Fig. 4-13).

INELASTIC

As in the atom-atom cases, the cross sections rise abruptly from zero at the threshold energy with a slope characteristic of the particular specie and transition, and then flatten and perhaps decay in the higher energy ranges beyond our interest.

SUPERELASTIC

Since the bound levels of an atom and its first ion are substantially different, resonant exchange reactions

$$(A^+)^* + A \to A^+ + A^*$$
$$A^+ + A^* \to (A^+)^* + A \quad (4\text{-}31)$$

are less likely to compete with the full deexcitations

$$(A^+)^* + A \to A^+ + \tilde{A}$$
$$A^* + A^+ \to A + \tilde{A}^+ \quad (4\text{-}32)$$

than in the atom-atom case; hence the latter must be regarded as important mechanisms. Beyond the detailed balancing arguments for a gas in equilibrium, little is known about the magnitude of these cross sections.

RADIATIVE

Because of the short-range interaction and large mass of the ion, bremsstrahlung from this source is negligible.

CHARGE-REACTIVE

Perhaps the most important of all ion-atom collisions is the charge-exchange event, wherein a fast ion and slow neutral change identity:

$$\tilde{A}^+ + A \to A^+ + \tilde{A} \quad (4\text{-}33)$$

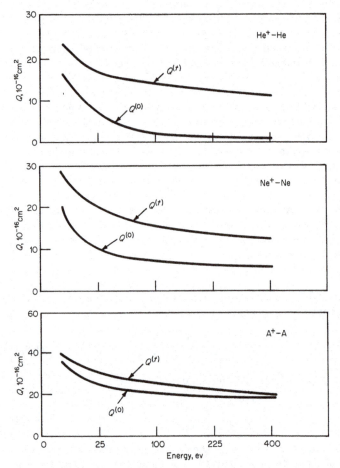

Fig. 4-13 Elastic scattering cross sections $Q^{(o)}$ and charge-transfer cross sections $Q^{(t)}$ for typical ion-atom collisions. (*From W. H. Cramer and J. H. Simons, He+ on He, J. Chem. Phys., vol. 26, p. 1272, 1957; W. H. Cramer, Ne+ on Ne, ibid., vol. 28, p. 688, 1958; A+ on A, ibid., vol. 30, p. 641, 1959.*)

This process, or its inverse, provides an effective coupling between the ionized- and neutral-gas components in many electromagnetic plasma accelerators. The cross section for this event is usually a maximum in the thermal energy range, where it can substantially exceed the elastic cross section, say 10^{-15} to 10^{-14} cm², and thereafter decreases gradually with energy (Fig. 4-13). Forward scattering strongly predominates, thereby providing a technique for the production of high energy atomic beams. Since the charge-exchange interaction has the same product particles as an elastic ion-atom collision, these two cross sections are frequently summed to one effective scattering cross section (Fig. 4-13).

4-7 OTHER INTERACTIONS

The above brief catalog is intended to give a flavor for the variety and complexity of collisional phenomena which can operate within a body of ionized gas, contributing to the thermodynamic, transport, and kinetic processes therein. It is by no means complete. Notable among its omissions are all the incident photon interactions, such as photoionization, photoexcitation, inverse bremsstrahlung, etc., which complement the radiative collisions; molecular interactions, which bring into play a variety of rotational and vibrational degrees of freedom that may participate in inelastic events; and most interactions involving negative ions. For details of these and other less common collisions, descriptions of experimental and theoretical techniques, and tabulations of cross-sectional values, the reader is referred to the References and to the bibliographies therein.

PROBLEMS

4-1. Consider a head-on elastic collision between two particles of masses m and M, respectively, where $m \ll M$. Compute the energy transferred from one to the other in terms of their initial velocities, v and V. Repeat for the case where the less massive of the two particles is deflected through an arbitrary angle θ from its initial direction. Repeat for the case where the incident velocity vectors are inclined to each other at an arbitrary angle ϕ.

4-2. Derive the momentum-transfer cross section for coulomb collisions [Eq. (4-14)].

4-3. Construct a hypothetical situation which clearly illustrates the significance of the Debye length as a useful plasma dimension.

REFERENCES

1. Jeans, J. H.: "The Dynamical Theory of Gases," 4th ed., Dover Publications, Inc., New York, 1925.
2. Chapman, S., and T. G. Cowling: "The Mathematical Theory of Non-uniform Gases," chap. 5, Cambridge University Press, New York, 1952.
3. Massey, H. S. W., and E. H. S. Burhop: "Electronic and Ionic Impact Phenomena," Oxford University Press, Fair Lawn, N.J., 1952.
4. Mott, N. F., and H. S. W. Massey: "The Theory of Atomic Collisions," 2d ed., Oxford University Press, Fair Lawn, N.J., 1949.
5. Bates, D. R.: "Atomic and Molecular Processes," Academic Press Inc., New York, 1962.
6. Brown, S. C.: "Basic Data of Plasma Physics," The M.I.T. Press, Cambridge, Mass., and John Wiley & Sons, Inc., New York, 1959.

7. Goldberger, M. L., and K. M. Watson: "Collision Theory," John Wiley & Sons, Inc., New York, 1964.
8. Drukarev, G. F.: "The Theory of Electron-Atom Collisions," Academic Press Inc., New York, 1965.
9. Gerjouy, E.: Low-energy Electron-Atom and Electron-Molecule Scattering Theory, circa 1964, *Phys. Today*, vol. 18, no. 5, pp. 24–30, May, 1965.
10. McDaniel, E. W.: "Collision Phenomena in Ionized Gases," John Wiley & Sons, Inc., New York, 1964.
11. Sutton, G. W., and A. Sherman: "Engineering Magnetohydrodynamics," McGraw-Hill Book Company, New York, 1965.
12. Schiff, L. I.: "Quantum Mechanics," 2d ed., McGraw-Hill Book Company, New York, 1955.
13. Spitzer, L., Jr.: "Physics of Fully-ionized Gases," Interscience Publishers, Inc., New York, 1956.
14. Heitler, W.: "The Quantum Theory of Radiation," 2d ed., Oxford University Press, Fair Lawn, N.J., 1950.
15. Griem, H. R.: "Plasma Spectroscopy," McGraw-Hill Book Company, New York, 1964.
16. Bates, D. R., and I. Esterman (eds.): "Advances in Atomic and Molecular Physics," vol. 1, Academic Press Inc., New York, 1965.
17. "Molecular Beams," in John Ross (ed.), Advances in Chemical Physics Series, vol. 10, Interscience Publishers, Inc., and John Wiley & Sons, Inc., New York, 1966.
18. Harwell, K. E., and R. G. Jahn: Initial Ionization Rates in Shock-heated Argon, Krypton and Xenon, *Phys. Fluids*, vol. 7, pp. 214–222, February, 1964.
19. Kelly, A. J.: Atom-Atom Ionization Mechanisms and Cross Sections in Noble Gases and Noble Gas Mixtures, Ph.D. thesis, California Institute of Technology, Pasadena, Calif., May, 1965; Atom-Atom Ionization Mechanisms in Argon, Krypton and Xenon, *J. Chem. Phys.*, vol. 45, p. 1723, 1966; Atom-Atom Ionization Mechanisms in Argon-Xenon Mixtures, *ibid.*, p. 1733.

5
Electrical Conductivity of an Ionized Gas

The extent to which magnetic body forces can be exerted on an ionized gas depends on the ability of that gas to conduct electric current. As we have seen, this current density may be related empirically to the applied electric and magnetic fields by a bulk electrical conductivity of the gas, σ [(2-68)]:

$$\mathbf{j} = \sigma(\mathbf{E} + \mathbf{u} \times \mathbf{B}) \tag{5-1}$$

but is more fundamentally expressed as a vector summation of all the individual charge motions induced by the applied fields [(2-20)]:

$$\mathbf{j} = \sum_i n_i q_i \bar{\mathbf{v}}_i \tag{5-2}$$

where n_i = number density of particles of type i
q_i = their charge
$\bar{\mathbf{v}}_i$ = their vector-averaged velocity, i.e., their migration, or swarm, velocity

The calculation of the conductivity function σ is thus essentially a calculation of the various $\bar{\mathbf{v}}_i$, and this is difficult, in general, because of the many internal and external conditions which influence them. For example, the mass and charge of the particles considered, their state of random thermal motion, the frequency and detailed characteristics of their collisions with themselves and with all other particles in the gas, the amplitude and frequency of the applied electric and magnetic fields, and the prevailing gasdynamic flow, all affect the migration velocity of each charged specie, and thereby the bulk property we call the electrical conductivity.

The purpose of this chapter is to develop a few particularly simple but instructive examples of electrical conductivity calculations, which, hopefully, will prepare the reader for the more sophisticated theories available in the referenced literature.

5-1 MOTION OF A CHARGED PARTICLE IN UNIFORM STEADY ELECTRIC AND MAGNETIC FIELDS

In gases of density or temperature so low that particle collisions occur infrequently on the time scale of interest, the migration velocities we seek are simply the free-particle motions. For example, an isolated charge q moving in a steady uniform magnetic induction field $\mathbf{B}$ feels a force normal to its velocity $\mathbf{v}$, normal to $\mathbf{B}$, and proportional to their vector product [(2-69)]:

$$\mathbf{F} = q\mathbf{v} \times \mathbf{B} \tag{5-3}$$

In the special case where $\mathbf{v}$ has only a component normal to $\mathbf{B}$, $\mathbf{v}_\perp$, the particle executes a circle of *gyro radius*

$$r_B = \frac{mv_\perp}{qB} \tag{5-4}$$

with angular velocity, or *gyro frequency*,

$$\omega_B = \frac{qB}{m} \tag{5-5}$$

If $\mathbf{v}$ also has a component parallel to $\mathbf{B}$, $\mathbf{v}_\parallel$, this is unaffected by the field, and the particle trajectory is a helix, with axis parallel to $\mathbf{B}$, radius given by (5-4), and pitch $2\pi v_\parallel/\omega_B$.

If an electric field is also present, the equation of motion involves other force components:

$$\mathbf{F} = q(\mathbf{E} + \mathbf{v} \times \mathbf{B}) = m\dot{\mathbf{v}} \tag{5-6}$$

The three scalar components of this equation can usually be integrated and combined to yield the particle trajectory for given initial velocity $\mathbf{v}_0$. However, the problem can be considerably simplified by the trick of sub-

ELECTRICAL CONDUCTIVITY OF AN IONIZED GAS

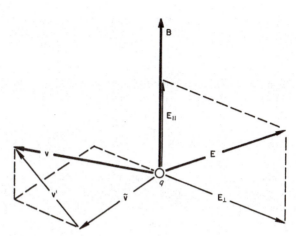

Fig. 5-1 Coordinate transformation for charged-particle trajectory calculation.

tracting out of **v** a fictitious velocity $\tilde{\mathbf{v}}$ perpendicular to both **E** and **B** (Fig. 5-1):

$$\mathbf{v} = \mathbf{v}' + \tilde{\mathbf{v}} \tag{5-7}$$

Breaking **E** into its components parallel and perpendicular to **B** and rewriting (5-6),

$$\mathbf{F} = q(\mathbf{E}_\parallel + \mathbf{E}_\perp + \mathbf{v}' \times \mathbf{B} + \tilde{\mathbf{v}} \times \mathbf{B}) = m\dot{\mathbf{v}}' \tag{5-8}$$

it follows from noting that the $\tilde{\mathbf{v}} \times \mathbf{B}$ is parallel to $\mathbf{E}_\perp$ that both can be eliminated by requiring[1]

$$\tilde{v} = \frac{E_\perp}{B} \tag{5-9}$$

The effect of the remaining components in the force equation can be readily visualized. $\mathbf{E}_\parallel$ produces a linear acceleration parallel to **B**. Superimposed on this is the helical motion brought about by $\mathbf{v}' \times \mathbf{B}$, as found above. All this is then convected by a constant *drift velocity* $\tilde{\mathbf{v}}$.

As an illustration of this technique, consider the case of $\mathbf{E}_\parallel = 0$, $\mathbf{v}_0 = 0$, that is, a charge starting from rest in uniform perpendicular **E** and **B** fields. Here the total force is normal to $\mathbf{v}'$ and to **B**:

$$\mathbf{F} = q(\mathbf{v}' \times \mathbf{B}) = m\dot{\mathbf{v}}' \tag{5-10}$$

[1] Throughout this discussion, the magnitude of a vector is denoted simply by italic type, that is, $\tilde{v} \equiv |\tilde{\mathbf{v}}|$, etc.

and the initial value of $\mathbf{v}'$ is simply the negative of the drift velocity $\tilde{\mathbf{v}}$:

$$\mathbf{v}_0' = \mathbf{v}_0 - \tilde{\mathbf{v}} = -\tilde{\mathbf{v}} = -\frac{\mathbf{E} \times \mathbf{B}}{B^2} \qquad (5\text{-}11)$$

The motion in a coordinate system moving with velocity $\tilde{\mathbf{v}}$ relative to the laboratory frame is a circle of radius r':

$$r' = \frac{mv_0'}{qB} = \frac{m\tilde{v}}{qB} = \frac{mE}{qB^2} \qquad (5\text{-}12)$$

which is transcribed with constant tangential speed, $\tilde{v} = E/B$, in the plane normal to $\mathbf{B}$. In the laboratory frame, this motion transforms into a cycloid in the same plane, advancing in the $\mathbf{E} \times \mathbf{B}$ direction with drift speed E/B (Fig. 5-2). Note that this drift is normal to $\mathbf{E}$ and independent of the charge sign.

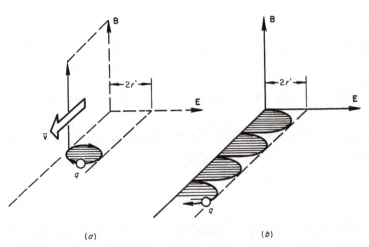

Fig. 5-2 Charged-particle drift from rest in crossed electric and magnetic fields. *(a)* Transformed system; *(b)* laboratory system.

If $\mathbf{v}_0 \neq 0$ but has some component parallel to $\mathbf{B}$, the motion in the transformed system will be a helix, instead of a circle, and in the laboratory frame, an inclined helix whose axis lies in the $\mathbf{B}, \mathbf{E} \times \mathbf{B}$ plane. If $\mathbf{v}_0$ also has a component normal to $\mathbf{B}$, this will contribute to the $\mathbf{v}' \times \mathbf{B}$ force, and hence change the radius of gyration and angular velocity, without affecting the drift velocity. The orbit then becomes a prolate cycloid (i.e., has loops) or curtate cycloid, depending on $v'_{0\perp} \gtrless E/B$. The cycloid always progresses along the positive $\mathbf{E} \times \mathbf{B}$ axis, but its position relative to that axis depends on the direction of $\mathbf{v}'_{0\perp}$. Note the very special case of $\mathbf{v}'_0 = 0$, that is, $\mathbf{v}_0 = \tilde{\mathbf{v}} = (\mathbf{E} \times \mathbf{B})/B^2$. A particle injected with this velocity moves in a straight line, undisturbed by the fields.

If, finally, we include a component of $\mathbf{E}$ parallel to $\mathbf{B}$, we find in the transformed frame a helix of quadratically increasing pitch:

$$v'_\| = \frac{qE_\|}{m} t + v'_{0\|} \tag{5-13}$$

In the laboratory system this transforms to a parabolically displaced helix, whose axis lies in the $\mathbf{B}$, $\mathbf{E} \times \mathbf{B}$ plane (Fig. 5-3).

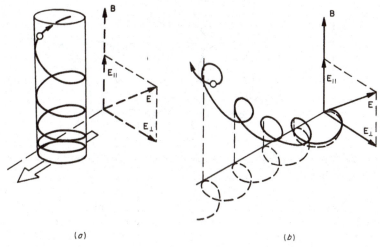

Fig. 5-3 Charged-particle drift in arbitrarily oriented electric and magnetic fields. *(a)* Transformed system; *(b)* laboratory system.

Admission of nonuniform electric and magnetic fields considerably complicates the calculation of charged-particle trajectories and yields additional possibilities for drift motions not included above [1]. Although such field gradient drifts are interesting physical processes (Prob. 5-2), they rarely need be explicitly included in gas accelerator analyses, and will not be pursued here.

Let us now attempt to express an electrical conductivity for a collisionless ionized gas by summing individual charged-particle motions for a particular combination of applied fields. Consider the simplest example of a uniform infinite body of such gas immersed in steady uniform perpendicular electric and magnetic fields. Note, first, that although we must now deal with a distribution of initial (thermal) velocities, these will be isotropic, and hence, while yielding a corresponding distribution of gyro radii and cycloid aspects, will not contribute to the migration velocity vector $\bar{v}_i$. Rather, this will be determined solely by the drift velocity $(\mathbf{E} \times \mathbf{B})/B^2$, which is the same for every particle. Indeed, it is the same, both in magnitude and direction, for the heavy, positively charged ions as it is for the small-mass, negatively charged electrons, and we are left with exactly equal electron and ion fluxes in the $\mathbf{E} \times \mathbf{B}$ direction, and hence no net current (Fig. 5-4).

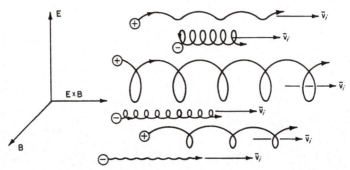

Fig. 5-4 Schematic representation of common migration velocities of thermal electrons and ions.

Current can be drawn by the addition of a component of $\mathbf{E}$ parallel to $\mathbf{B}$, which would drive components of electron and ion flux parallel to itself, but of different magnitude and sign. Now, however, we have the difficulty that in the absence of collisions, both the electron and ion migration speeds parallel to $\mathbf{B}$ increase linearly in time without limit, and thus the "conductivity" of the medium is again ill-defined. Both of these simple examples serve to illustrate the fundamental inadequacy of the concept of a scalar conductivity parameter for even the simplest ionized gas and to preface the important role of particle collisions, however infrequent, in establishing steady-state current conduction.

5-2 EFFECT OF COLLISIONS

When a charged particle accelerated in a steady electric field collides with another particle in the gas, it will recoil at some unspecified angle, with some unspecified residual speed, and then begin a new trajectory in the applied field, determined in part by the recoil parameters. To incorporate a series of such collisions into a calculation of the detailed motion of any given particle would be an unreasonable task. It is far simpler, and for many purposes quite adequate, to consider instead the mean motion of a large number, or swarm, of such particles, wherein the integrated effect of many such collisions may be represented heuristically as a damping agent on the organized migration of the swarm in response to the applied electric field. For example, one may define an effective collision frequency ν_c in terms of the rate at which the particle swarm loses its migration momentum via all such collisions. Thus, under a steady applied electric field, the motion of the swarm may be described by the relation

$$\frac{d}{dt} nm\bar{\mathbf{v}} = nq\mathbf{E} - \nu_c nm\bar{\mathbf{v}} \tag{5-14}$$

or more simply, by

$$\dot{\bar{\mathbf{v}}} = \frac{q}{m}\mathbf{E} - \nu_c\bar{\mathbf{v}} \tag{5-15}$$

which may be regarded as the equation of motion for a fictitious "average" particle in the swarm.

The collision frequency which has been introduced here is related to atomic-scale momentum-transfer cross sections like those discussed in Sec. 4-3 by the definition

$$\nu_c = \sum_j n_j Q_j^{(p)} \bar{v}_j \tag{5-16}$$

where $\bar{v}_j$ is the mean *scalar* speed of the charged particles relative to the various species j particles with which they can collide. This should by no means be identified with the *vector* average, or swarm migration velocity, $\bar{\mathbf{v}}$. Rather, v_j is normally determined primarily by the random thermal motions of the individual particles, possibly modified somewhat by the field-induced streaming. Despite this identification, the detailed evaluation of ν_c is usually difficult and depends heavily on the prevailing environment. For most ionized gases where the collisions may be regarded as discrete events, the indicated approach is via the Boltzmann equation [2]. In certain exotic plasmas, collective effects may dominate the binary collisions, and the concept of a collision frequency becomes more subtle [3]. For our purposes, we shall regard ν_c as an empirical parameter, defined by (5-14) and presumed independent of the swarm velocity $\bar{\mathbf{v}}$. It may be crudely estimated from atomic parameters via (5-16), or possibly may be experimentally measurable [4].

For constant ν_c, Eq. (5-15) for the swarm velocity has the solution

$$\bar{v} = \frac{q}{m\nu_c} E + Ce^{-\nu_c t} \qquad (5\text{-}17)$$

Thus, after the switching transient has died away, we achieve a steady-state current density,

$$j = nq\bar{v} = \frac{nq^2}{m\nu_c} E \qquad (5\text{-}18)$$

In the absence of other species of mobile charges, the coefficient of E can be identified as the conductivity of the medium and in this case is clearly a scalar quantity:

$$\sigma = \frac{nq^2}{m\nu_c} \qquad (5\text{-}19)$$

Collisions play a similar role in determining drift velocities when a magnetic field is also applied to the gas. In this case the ratio of the gyro frequency of the charged particle to the collision frequency distinguishes the response of the particle swarm to the applied E and B fields (Fig. 5-5). If $\omega_B/\nu_c \gg 1$, the charges execute many cycles of their cycloidal drift motion before they are disturbed by collision, and the major component of their current is in the $E \times B$ direction. If $\omega_B/\nu_c \ll 1$, the charges seldom complete one cycle of their drift motion before collision, and hence develop little cross-field motion. In this case, the primary component of current is parallel to E, with only a small off-diagonal element added to the scalar conduction described by Eq. (5-18). In intermediate cases, where $\omega_B/\nu_c \approx 1$, comparable components of current parallel and normal to E arise. Note that in all these examples, the effect of a collision implied by the definition of ν_c is to interrupt the "mean particle trajectory" and cause it to begin its motion again from rest in the applied fields.

The ratio ω_B/ν_c is popularly called the *Hall parameter*, and the $E \times B$ component of current, the *Hall current*, in honor of the discoverer of a similar *Hall effect* in metallic conduction in a magnetic field [5]. Clearly, the Hall parameter has different values for the various charged species in an ionized gas, depending primarily on their charge-to-mass ratio, and this disparity can considerably complicate the overall current conduction process. In most cases one is tempted to neglect ion currents entirely, because of the large inertia of these particles, but certain situations do not permit this. We have seen, for example, that collisionless cross-field ion drift is equal to that of free electrons. However, an ion takes far longer to complete one cycle in its drift motion and transcribes a far bigger orbit than does an electron. In some cases, then, we may find that

ELECTRICAL CONDUCTIVITY OF AN IONIZED GAS

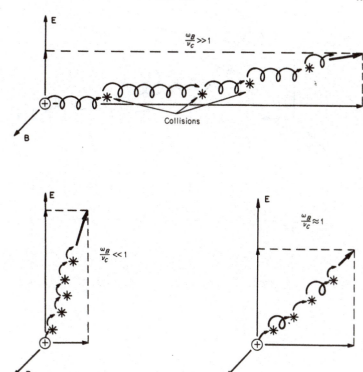

Fig. 5-5 Mean charged-particle motion in crossed electric and magnetic fields for various Hall parameters.

$\omega_B/\nu_c \gg 1$ for the electrons but $\omega_B/\nu_c \ll 1$ for the ions. In this environment the electrons will contribute a Hall current normal to **E**, while the ions supply a scalar conduction current parallel to **E** (Fig. 5-6a). This possibility is actually exploited in certain crossed-field accelerators, discussed in Sec. 8-6.

Even in the absence of particle collisions, a similar effect appears if the size of the gas container is intermediate to the ion and electron gyro radii. In this situation, the ions' tendency to drift is impeded, not by collisions with other particles, but by collisions with the walls of the container (Fig. 5-6b). Even in containers large compared with the ion gyro radius, this process may still occur near the negative electrode, forming, as it were, a type of ion conduction sheath there (Fig. 5-6c).

Quantitative expressions for the effect of collisions on the steady-field electrical conductivity may be derived directly in the same spirit as

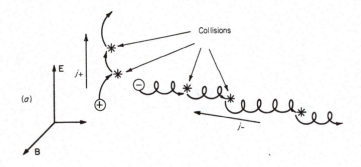
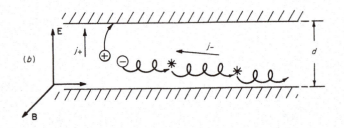
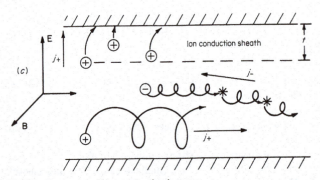

Fig. 5-6 Electron and ion conduction in a gas.

(a) $\left(\dfrac{\omega_B}{\nu_c}\right)_- \gg 1$ $\left(\dfrac{\omega_B}{\nu_c}\right)_+ \ll 1$

(b) $\left(\dfrac{\omega_B}{\nu_c}\right)_- \gg 1$ $\left(\dfrac{\omega_B}{\nu_c}\right)_+ \gg 1$ $r_B^+ > d$

(c) $\left(\dfrac{\omega_B}{\nu_c}\right)_- \gg 1$ $\left(\dfrac{\omega_B}{\nu_c}\right)_+ \gg 1$ $d > r_B^+ > t$

ELECTRICAL CONDUCTIVITY OF AN IONIZED GAS

(5-19) (Prob. 5-3), but these will also appear as special cases of the oscillating-field results to follow.

5-3 HIGH-FREQUENCY CONDUCTIVITY

Because of the inherent linearity of the electromagnetic field relations, the response of an ionized gas to arbitrary time-varying electric fields can be profitably studied on the basis of simple sinusoidal oscillations. Consider, for example, the application of an alternating electric field of frequency ω to a collisionless swarm of particles of charge q. An equation of motion may again be written in terms of a mean migration velocity:

$$\dot{\bar{v}} = \frac{q}{m}\mathbf{E} = \frac{q}{m}\mathbf{E}_0 e^{i\omega t} \tag{5-20}$$

This has a steady-state solution for the current carried by the swarm,

$$\mathbf{j} = nq\bar{v} = -i\frac{nq^2}{m\omega}\mathbf{E}_0 e^{i\omega t} \tag{5-21}$$

which implies a conductivity function that is a pure imaginary frequency-dependent scalar quantity. The negative imaginary coefficient is indicative of a 90° phase difference between the current and the driving field, arising from the inertial response of the particles in inverse proportion to their mass and to the applied frequency. The apparent singularity at the limit of zero frequency again reflects the inability to achieve a steady-state dc solution in the absence of any damping agent, such as collisions (Prob. 5-4).

The particular group of quantities

$$\omega_p = \left(\frac{nq^2}{\epsilon_0 m}\right)^{1/2} \tag{5-22}$$

arises in many problems of this type and has certain physical significance as a natural resonance of the charge swarm. It is called the *plasma frequency*, and its ratio to the applied frequency is a useful dimensionless parameter. In terms of ω_p we can also define the dimensionless conductivity $\bar{\sigma}$:

$$\bar{\sigma} = \frac{\sigma}{\epsilon_0 \omega} = -i\left(\frac{\omega_p}{\omega}\right)^2 \tag{5-23}$$

Again we should note that each of the properties thus defined, σ, ω_p, $\bar{\sigma}$, refers only to the single specie of charged particles considered, and each will have different values for other species, depending on the charge and mass. In this collisionless regime, the free-electron component typically overwhelms all others, and expressions (5-21) to (5-23), evaluated for electrons, adequately describe the gas as a whole.

The effect of collisions on ac conduction can be explored by the same heuristic representation in terms of a damping force, linear in the effective collision frequency. The equation of motion for the swarm of charged particles will be written

$$\dot{\bar{v}} = \frac{q}{m} \mathbf{E}_0 e^{i\omega t} - \nu_c \bar{v} \tag{5-24}$$

which yields a drift velocity

$$\bar{v} = \frac{q}{m(\nu_c + i\omega)} \mathbf{E}_0 e^{i\omega t} + \mathbf{C} e^{-\nu_c t} \tag{5-25}$$

After the transient has died out, the current becomes

$$\mathbf{j} = \frac{nq^2}{m(\nu_c + i\omega)} \mathbf{E}_0 e^{i\omega t} \tag{5-26}$$

and the conductivity is thus a complex scalar, $\sigma^* = \sigma_r + i\sigma_i$, indicative of current components in phase and out of phase with the applied field:

$$\sigma_r = \frac{nq^2}{m}\left(\frac{\nu_c}{\omega^2 + \nu_c^2}\right) = \epsilon_0 \omega \frac{\nu_c}{\omega}\left[\frac{(\omega_p/\omega)^2}{1 + (\nu_c/\omega)^2}\right]$$
$$\sigma_i = -\frac{nq^2}{m}\left(\frac{\omega}{\omega^2 + \nu_c^2}\right) = -\epsilon_0 \omega \left[\frac{(\omega_p/\omega)^2}{1 + (\nu_c/\omega)^2}\right] \tag{5-27}$$

The dc limit is now well behaved, and agrees with the earlier calculation (5-19). The high-frequency limit $\nu_c/\omega \to 0$ converges to the collisionless result (5-21).

5-4 AC CONDUCTIVITY IN STEADY MAGNETIC FIELDS

Addition of a steady magnetic induction field $\mathbf{B}_0$ to the ac conduction problem has the effect of destroying the isotropy of the medium and of permitting conduction normal to the applied electric field; i.e., the conductivity function becomes a tensor whose elements depend on the relative magnitudes of the four frequencies ω, ω_p, ν_c, and ω_B.

If $\mathbf{B}_0$ is spatially uniform as well as steady, the equation of motion of the swarm of charged particles may be written

$$\dot{\bar{v}} = \frac{q}{m}(\mathbf{E} + \bar{v} \times \mathbf{B}_0) - \nu_c \bar{v} \tag{5-28}$$

The $\bar{v} \times \mathbf{B}_0$ term generates off-diagonal elements in the conductivity tensor and needs to be reduced before $\bar{v}$ can be extracted. To accomplish this

ELECTRICAL CONDUCTIVITY OF AN IONIZED GAS

we first accumulate four relations:[1] the first time derivative of (5-28),

$$\ddot{\bar{v}} = \frac{q}{m}(\dot{E} + \dot{\bar{v}} \times B_0) - \nu_c \dot{\bar{v}} \qquad (5\text{-}29)$$

the cross product of (5-28) with B_0,

$$\dot{\bar{v}} \times B_0 = \frac{q}{m}(E \times B_0) + \frac{q}{m}[(\bar{v} \times B_0) \times B_0] - \nu_c(\bar{v} \times B_0) \qquad (5\text{-}30)$$

the scalar product of (5-28) with B_0,

$$\dot{\bar{v}} \cdot B_0 = \frac{q}{m}(E \cdot B_0) - \nu_c(\bar{v} \cdot B_0) \qquad (5\text{-}31)$$

and the vector identity,

$$(\bar{v} \times B_0) \times B_0 = (\bar{v} \cdot B_0)B_0 - B_0^2 \bar{v} \qquad (5\text{-}32)$$

Substituting (5-29), (5-28), and (5-32) into (5-30) yields an expression free of cross products of $\bar{v}$:

$$\ddot{\bar{v}} + 2\nu_c \dot{\bar{v}} + \left[\left(\frac{qB_0}{m}\right)^2 + \nu_c^2\right]\bar{v} = \frac{q}{m}\nu_c E + \frac{q}{m}\dot{E}$$
$$+ \left(\frac{q}{m}\right)^2 (E \times B_0) + \left(\frac{q}{m}\right)^2 (\bar{v} \cdot B_0)B_0 \qquad (5\text{-}33)$$

Assuming that $\bar{v}$, like E, will be harmonic in time, we can replace all derivatives with appropriate powers of $i\omega$:

$$\left[(\nu_c + i\omega)^2 + \left(\frac{qB_0}{m}\right)^2\right]\bar{v} = \frac{q}{m}(\nu_c + i\omega)E$$
$$+ \left(\frac{q}{m}\right)^2 (E \times B_0) + \left(\frac{q}{m}\right)^2 (\bar{v} \cdot B_0)B_0 \qquad (5\text{-}34)$$

Rather than attempting to factor $\bar{v}$ from the last term, we evaluate it by again invoking the harmonic time dependence in (5-31):

$$\bar{v} \cdot B_0 = \frac{q}{m}\left(\frac{E \cdot B_0}{\nu_c + i\omega}\right) \qquad (5\text{-}35)$$

Substitution of this into (5-34) finally yields the desired expression for the swarm velocity:

$$\bar{v} = \frac{(q/m)(\nu_c + i\omega)E + (q/m)^2(E \times B_0) + (q/m)^3(\nu_c + i\omega)^{-1}(E \cdot B_0)B_0}{(\nu_c + i\omega)^2 + (qB_0/m)^2} \qquad (5\text{-}36)$$

[1] The author is indebted to F. A. Albini for a portion of the following derivation.

At this point we can identify three distinct components of the charge motion. The first term in the numerator lies along **E** and corresponds to the scalar conduction in the absence of $\mathbf{B}_0$. The second term represents the $\mathbf{E} \times \mathbf{B}_0$ drift. The third term lies along $\mathbf{B}_0$ and is proportional to the component of **E** parallel to $\mathbf{B}_0$; this component of the current is undisturbed by the presence of $\mathbf{B}_0$.

After assembling the gyro frequency and plasma frequency and suitably nondimensionalizing, the current density may be written

$$\mathbf{j} = \epsilon_0 \omega \left(\frac{\omega_p}{\omega}\right)^2 \frac{[(\nu_c/\omega) + i]\mathbf{E} + (\omega_B/\omega)(\mathbf{E} \times \mathbf{b}) + (\omega_B/\omega)^2[(\nu_c/\omega) + i]^{-1}(\mathbf{E} \cdot \mathbf{b})\mathbf{b}}{[(\nu_c/\omega) + i]^2 + (\omega_B/\omega)^2} \quad (5\text{-}37)$$

where **b** denotes a unit vector along $\mathbf{B}_0$.

The "conductivity" of this swarm, as anticipated, has become a tensor of rank 2. To display it, let $\mathbf{B}_0$ define the z axis of a cartesian coordinate system. The dimensionless conductivity may then be written

$$\tilde{\sigma} \equiv \frac{1}{\epsilon_0 \omega} \sigma = \left(\frac{\omega_p}{\omega}\right)^2 \begin{Bmatrix} A_{xx} & A_{xy} & A_{xz} \\ A_{yx} & A_{yy} & A_{yz} \\ A_{zx} & A_{zy} & A_{zz} \end{Bmatrix} \quad (5\text{-}38)$$

where

$$A_{xx} = A_{yy} = \frac{\nu_c/\omega + i}{[(\nu_c/\omega) + i]^2 + (\omega_B/\omega)^2} \quad (5\text{-}39)$$

$$A_{zz} = \frac{1}{(\nu_c/\omega) + i} \quad (5\text{-}40)$$

$$A_{xy} = -A_{yx} = \frac{\omega_B/\omega}{[(\nu_c/\omega) + i]^2 + (\omega_B/\omega)^2} \quad (5\text{-}41)$$

$$A_{xz} = A_{zx} = A_{yz} = A_{zy} = 0 \quad (5\text{-}42)$$

Limiting cases can readily be extracted. For example, we can recover the scalar conductivity results for $B \to 0$:

$$A_{xx} = A_{yy} = A_{zz} \to \frac{1}{(\nu_c/\omega) + i} \quad (5\text{-}43)$$

$$A_{xy} = A_{yx} = 0 \quad (5\text{-}44)$$

or the collisionless case, $\nu_c/\omega \to 0$:

$$A_{xx} = A_{yy} \to \frac{i}{(\omega_B/\omega)^2 - 1} \quad (5\text{-}45)$$

$$A_{zz} \to -i \quad (5\text{-}46)$$

$$A_{xy} = -A_{yx} \to \frac{\omega_B/\omega}{(\omega_B/\omega)^2 - 1} \quad (5\text{-}47)$$

ELECTRICAL CONDUCTIVITY OF AN IONIZED GAS

To extract a steady-field (dc) limit, we must first rewrite the tensor to remove ω from the denominators:

$$\frac{1}{\epsilon_0} \boldsymbol{\sigma} = \omega_p^2 \begin{Bmatrix} b_{xx} & b_{xy} & 0 \\ b_{yx} & b_{yy} & 0 \\ 0 & 0 & b_{zz} \end{Bmatrix} \tag{5-48}$$

where
$$b_{xx} = b_{yy} = \frac{\nu_c + i\omega}{\omega_B^2 + (\nu_c + i\omega)^2} \to \frac{\nu_c}{\omega_B^2 + \nu_c^2} \tag{5-49}$$

$$b_{zz} = \frac{1}{\nu_c + i\omega} \to \frac{1}{\nu_c} \tag{5-50}$$

$$b_{xy} = -b_{yx} = \frac{\omega_B}{\omega_B^2 + (\nu_c + i\omega)^2} \to \frac{\omega_B}{\omega_B^2 + \nu_c^2} \tag{5-51}$$

(a result to be compared with Prob. 5-4).

It should be noted that the conductivity tensor for a swarm of negatively charged particles has opposite signs preceding its off-diagonal elements σ_{xy} and σ_{yx} if ω_B is defined simply as the absolute value of qB/m. Strictly, ω_B is a *spinor* quantity, requiring a sign to distinguish between clockwise and counterclockwise rotations. Also, it should be noted that the relative magnitude of the off-diagonal elements of the tensor to the diagonal elements depends on the mass-to-charge ratio of the particle. For these reasons, and because of the wide disparity in ν_c for electrons and ions, the electron conduction tensor can be qualitatively different from the ion conduction tensor in an ionized gas.

5-5 ELECTROMAGNETIC WAVE PROPAGATION IN AN IONIZED GAS

One of the most dramatic manifestations of the peculiar electrical conductivity of an ionized gas is in its effect on the propagation of electromagnetic waves through such a medium. The wave equation for conducting material derived earlier [(2-54)],

$$\nabla^2 \mathbf{E} = \sigma\mu \frac{\partial \mathbf{E}}{\partial t} + \epsilon\mu \frac{\partial^2 \mathbf{E}}{\partial t^2} \tag{5-52}$$

and its solution,

$$\mathbf{E} = \mathbf{E}_0 e^{i(\omega t \mp \mathbf{k}\cdot\mathbf{r})} \tag{5-53}$$

where $\mathbf{k}$ is a complex wave vector of size

$$k = (\omega^2 \epsilon\mu - i\mu\sigma\omega)^{1/2} \tag{5-54}$$

remain applicable to propagation in a body of ionized gas, but are further complicated by the various irregular forms of σ to be inserted. For exam-

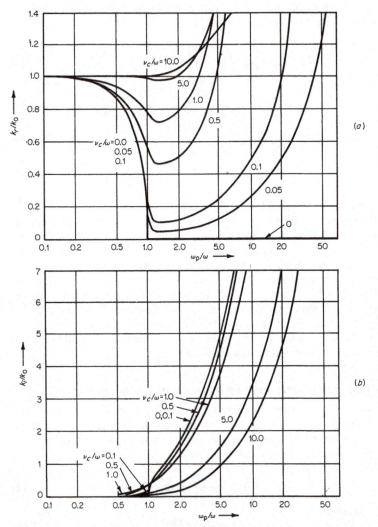

Fig. 5-7 Variation of propagation exponent with electron density and collision frequency. (a) Real part k_r; (b) imaginary part k_i.

ple, in the scalar case ($B_0 = 0$), substitution of (5-27) into (5-54) yields, for $k \equiv k_r - ik_i$,

$$k_r = \frac{k_0}{\sqrt{2}} \left\{ (1 - P) + \left[(1 - P)^2 + P^2 \left(\frac{\nu_c}{\omega}\right)^2 \right]^{\frac{1}{2}} \right\}^{\frac{1}{2}} \quad (5\text{-}55)$$

$$k_i = \frac{k_0}{\sqrt{2}} \left\{ -(1 - P) + \left[(1 - P)^2 + P^2 \left(\frac{\nu_c}{\omega}\right)^2 \right]^{\frac{1}{2}} \right\}^{\frac{1}{2}} \quad (5\text{-}56)$$

where
$$P \equiv \frac{(\omega_p/\omega)^2}{1 + (\nu_c/\omega)^2} \quad (5\text{-}57)$$

and $k_0 = \epsilon\mu\omega^2 = 2\pi/\lambda_0$ would be the propagation exponent for the same medium devoid of its free charge.

Both the wavelength and the damping rate of the waveform are seen to be implicit functions of the free electron density and collision frequency. Thus these two properties will determine the ability of a given wave train to penetrate the medium, the phase change it will undergo therein, and the reflection coefficients at the surfaces of the medium. Conversely, judicious probing of a gas sample by electromagnetic waves can provide information about $k_r(n,\nu_c)$ and $k_i(n,\nu_c)$, and thereby about n and ν_c, separately. Figure 5-7 displays the range of sensitivity of k_r and k_i to n and ν_c.

Application of a biasing magnetic field to the medium introduces an anisotropy in its electromagnetic wave propagation corresponding to that produced in its conductivity function and yields further idiosyncrasies in the wave patterns and further possibilities for diagnostics of the gas. Development of these formulas is beyond the scope of this chapter, but is available in Ref. 16.

The heuristic formulations of the electrical conductivity of an ionized gas presented above are intended primarily to display as simply as possible the various atomic-scale effects which influence this bulk property. The relations developed can be quite serviceable when properly interpreted for those ionized gases where discrete two-body collisions control the charged-particle migrations and for applied fields not so strong as to disturb sensibly the random thermal distribution functions. More sophisticated treatments capable of incorporating collective plasma effects and variable collision frequencies are available in the literature, to which the reader is now referred [6–20].

PROBLEMS

5-1. Set up the scalar component equations of motion for a charged particle in crossed electric and magnetic fields and find the trajectories for the following cases:

    a. $v_0 = 0$

b. $v'_{0\perp} < \dfrac{E}{B}$

c. $v'_{0\perp} > \dfrac{E}{B}$

d. $\mathbf{v}_0 = \dfrac{\mathbf{E} \times \mathbf{B}}{B^2}$

5-2. A magnetic field in the z direction, B, has a gradient in its magnitude in the x direction, ∇B. A charged particle starts in the xy plane with speed v_0. Show that to the approximation $\nabla B/B \ll 1$, the particle trajectory drifts in the y direction with a speed

$$v_d = r_B v_0 \dfrac{\nabla B}{B}$$

5-3. Show that the collision frequency defined by Eq. (5-16) is the proper coefficient of the damping term in the heuristic equation of motion (5-15) for the average charged particle.

5-4. Derive the conductivity tensor for an ionized gas in crossed steady $\mathbf{E}$ and $\mathbf{B}$ fields in terms of the ion and electron Hall parameters $(\omega_B/\nu_c)_+$ and $(\omega_B/\nu_c)_-$, directly from the heuristic equations for the migration velocities, without reference to the oscillating-field formulation of the text. Describe the direction and magnitude of the ion current, electron current, and total current vectors for example values of the Hall parameters.

5-5. Construct a total conductivity tensor for an ionized gas in a steady $\mathbf{B}$ field and sinusoidally time-varying $\mathbf{E}$ field. Estimate the relative importance of the electron and ion contributions to each element.

5-6. Compute the effective collision frequency ν_c for the electrons in a weakly ionized gas, presuming that they have a maxwellian velocity distribution, that electron-neutral collisions are dominant, and that the momentum-transfer cross section is a constant, Q_0. What is the temperature dependence of the dc scalar conductivity of such a gas? Repeat the calculation for the case of a momentum-transfer cross section rising linearly with relative particle velocity.

5-7. Compute the collision frequency for the electrons in a fully ionized gas, i.e., one containing only electrons and single ions, presuming that the coulomb encounters can be properly treated as two-body events of short duration. What is the temperature dependence of the dc scalar conductivity of this gas?

REFERENCES

1. Spitzer, L., Jr.: "Physics of Fully-ionized Gases," Interscience Publishers, Inc., New York, 1956.
2. Margenau, H.: Conduction and Dispersion of Ionized Gases at High Frequencies, *Phys. Rev.*, vol. 69, pp. 509–513, 1946; Conductivity of Plasmas to Microwaves, *ibid.*, vol. 109, pp. 6–9, 1958; with E. A. Desloge and S. W. Matthysse, Conductivity of Plasmas to Microwaves, *ibid.*, vol. 112, pp. 1437–1440, 1958.
3. Dawson, J., and C. Oberman: High-frequency Conductivity and the Emission and Absorption Coefficients of a Fully Ionized Plasma, *Phys. Fluids*, vol. 5, no. 5, pp. 517–524, May, 1962.
4. Jahn, R. G.: Microwave Probing of Ionized Gas Flows, *Phys. Fluids*, vol. 5, no. 6, pp. 678–686, June, 1962.

5. Hall, E. H.: On a New Action of the Magnet on Electric Currents, *Am. J. Math.*, pt. 2, vol. 1-2, p. 287, 1879.
6. Landshoff, R.: Transport Phenomena in a Completely Ionized Gas in Presence of a Magnetic Field, *Phys. Rev.*, vol. 76, p. 904, 1949; vol. 82, p. 442, 1951.
7. Cohen, R. S., L. Spitzer, and P. Routly: The Electrical Conductivity of an Ionized Gas, *Phys. Rev.*, vol. 80, p. 230, 1950.
8. Spitzer, L., and R. Harm: Transport Phenomena in a Completely Ionized Gas, *Phys. Rev.*, vol. 89, p. 977, 1953.
9. Bernstein, I. B., and S. K. Trehan: Plasma Oscillations, I, *Nucl. Fusion*, vol. 1, p. 3, 1960.
10. Oberman, C., A. Ron, and J. Dawson: High-frequency Conductivity of a Fully Ionized Plasma, *Phys. Fluids*, vol. 5, p. 1514, December, 1962.
11. Perel, V. I., and G. M. Eliashberg: Absorption of Electromagnetic Waves in a Plasma, *Soviet Phys. JETP*, vol. 14, p. 633, 1962.
12. Robinson, B. B., and I. B. Bernstein: A Variational Description of Transport Phenomena in a Plasma, *Ann. Phys.*, vol. 18, p. 110, 1962.
13. Dawson, J., and C. Oberman: Ion Correlations on High-frequency Plasma Conductivity, *Phys. Fluids*, vol. 6, p. 394, March, 1963.
14. Dubois, D. F., V. Gilinsky, and M. G. Kivelson: Propagation of Electromagnetic Waves in Plasmas, *Phys. Rev.*, vol. 129, p. 2376, 1963.
15. Oberman, C., and A. Ron: High-frequency Conductivity of Quantum Plasma in a Magnetic Field, *Phys. Rev.*, vol. 130, p. 1291, 1963.
16. Oberman, C., and F. Shure: High-frequency Plasma Conductivity in a Magnetic Field, *Phys. Fluids*, vol. 6, p. 834, June, 1963.
17. Ron, A., and N. Tzoar: Absorption of Electromagnetic Waves in Quantum and Classical Plasmas, *Phys. Rev. Letters*, vol. 10, p. 45, 1963.
18. Shkarofsky, I. P., I. B. Bernstein, and B. B. Robinson: Condensed Presentation of Transport Coefficients in a Fully Ionized Plasma, *Phys. Fluids*, vol. 6, p. 40, January, 1963.
19. Berk, H.: Frequency and Wavelength Dependent Electrical Transport Equation for a Plasma Model, *Phys. Fluids*, vol. 7, p. 257, February, 1964.
20. DeWolf, D. A.: Frequency Dependence of the Resistivity of a Fully Ionized Plasma, *Proc. IEEE*, vol. 52, p. 197, 1964.

part two
Electrical Acceleration of Gases

6
Electrothermal Acceleration

Electrothermal propulsion comprises all techniques whereby a propellant gas is heated electrically and then expanded through a nozzle to convert its thermal energy to a jet of directed kinetic energy. In this chapter we shall consider three substantially different electrical means for heating the propellant flow: (1) by passing it over an electrically heated solid surface, the so-called *resistojet;* (2) by passing it through an arc discharge, usually termed an *arcjet;* and (3) by high-frequency excitation, which is referred to by a variety of trade names, depending on the particular implementation. In each case we shall find that the attainable exhaust velocity is determined primarily by the maximum temperature that the chamber and nozzle surfaces can tolerate and by the gas-kinetic and thermodynamic properties of the propellant gas.

6-1 ONE-DIMENSIONAL MODEL

The gross performance of accelerators of this class can be forecast by reference to a general conceptual model such as that shown in Fig. 6-1.

ELECTROTHERMAL ACCELERATION

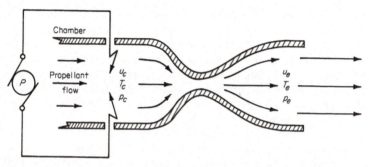

Fig. 6-1 Conceptual model of electrothermal thruster.

Electric power from an external supply of given capacity P is delivered in some manner, such as through a solid resistance element, a high current arc, or an electrodeless discharge, to heat a propellant stream in a suitable chamber to some maximum temperature T_c set by the thermal limitations of the material walls in the prevailing flow. The electrically heated gas is then allowed to expand through a supersonic nozzle to a low pressure determined by the nozzle area ratio and by the chamber pressure p_c. Under the crude assumptions of one-dimensional adiabatic constant specific heat expansion through the nozzle, the attainable exhaust speed u_e may be written from a simple energy balance:

$$\tfrac{1}{2}u_e^2 = \tfrac{1}{2}u_c^2 + c_p(T_c - T_e) \approx c_p T_c \tag{6-1}$$

where the flow speed in the chamber, u_c, and the exit temperature T_e are usually negligible in a first approximation. The constant-pressure specific heat of the propellant gas per unit mass c_p is seen to be a particularly critical quantity, since it defines the stagnation enthalpy which can be imparted to the gas at a given temperature, and thereby limits the attainable exhaust speed. As a first guess, hydrogen seems the most attractive propellant from this standpoint, since its molecular degrees of freedom and its low molecular weight give it a very high c_p in the temperature range of interest.

As one example, suppose the particular heating process employed limits T_c to the softening point of the surface material, as will be the case in all resistojet devices. Arbitrarily taking 3000°K as an upper limit for refractory materials in this application, at which temperature hydrogen has a specific heat of about 2×10^4 joules/(kg)(°K)(at 1 atm pressure), we estimate an exhaust speed of about 10^4 m/sec, corresponding to a specific impulse in the useful range of 1,000 sec.

The thrust that a device of this type can achieve clearly depends on the mass flow which can be heated and expanded, and thus on the size of the device and on the chamber pressure level. Actually, a more fundamental limit is set by the available electric power supply which drives the unit. For example, perfect conversion of a 30-kw source into kinetic energy of the 10^4-m/sec exhaust beam limits the mass flow to $\dot{m} = 2P/u_e^2 = 6 \times 10^{-4}$ kg/sec, with the corresponding thrust of $T = \dot{m}u_e = 6$ newtons (≈ 1.3 lb).

The example chosen has considerable practical interest. If this combination of specific impulse and thrust can be approached in a real device, it can perform a variety of modest space missions, such as attitude control and station keeping of satellites, better than small chemical rockets, cold gas jets, or the inherently large-power nuclear rockets. Indeed, considerably lower power levels, even down to 1 watt or less, can be shown to have logistical advantages for certain missions of this class [1].

It follows from Eq. (6-1) that further increase in the specific impulse of this rocket requires a quadratic increase in mean chamber temperature, a somewhat discouraging route from the materials standpoint. However, if the heating is accomplished by a gaseous discharge, suitably channeled by the propellant flow itself, extremely high temperatures can be sustained in the center of the chamber without damage to the walls. In this way, it is possible to extend electrothermal propulsion into the 2,000-sec range, where many more space applications become reasonable.

All practical electrothermal thrusters depart from the ideal model used above in several important respects. First, the flow is far from one-dimensional. Depending on the particular electrical heating mechanism employed, substantial temperature and density gradients are set up in the chamber, and vestiges of these may persist into the nozzle and out into the exhaust jet. Superimposed on these gradients are the viscous and thermal boundary layers developed in the high-speed high-temperature flow down the nozzle. In the simple surface heaters, or resistojets, these two-dimensional effects can usually be handled in semiempirical fashion and will manifest themselves in relatively small nozzle inefficiencies, or jet profile losses, which detract perhaps 10 percent from the ideal exhaust speed. In electrothermal accelerators in which the electric energy is deposited directly in the body of the gas flow, as in the arcjets and high-frequency discharge devices, however, the temperature and density gradients are inherently first-order effects, and must be so included in any rigorous analyses of the devices. Indeed, the severe gas-property gradients are responsible for the higher performance of these accelerators.

The second practical departure from ideal performance involves radiant heat loss from the thruster body or jet. In a gross sense this may be regarded as a loss of some fraction of the electric input power to thermal

ELECTROTHERMAL ACCELERATION

radiation, but the actual conversion may occur in several ways. The heater element itself may dissipate some of its energy input to surrounding elements of the thruster, which in turn radiate to the space vacuum; the hot propellant stream may radiate or conduct heat to cooler nozzle walls, or viscous dissipation in adjacent boundary layers may heat the walls, and they in turn then radiate to space; finally, some of the radiant energy in the hot gas flow may escape axially out the exhaust nozzle. Again, in resistojet devices, reduction of such losses to tolerably small levels can usually be handled by intuitive or semiempirical procedures of generous insulation, baffling, and reentrant gas flow passages. In arcjets and other high-temperature accelerators, however, radiation processes can play a major role in the overall electric energy conversion sequence, and thus must be handled more precisely.

The most serious departure of electrothermal flows from the ideal model above, however, arises from the strong temperature dependence of the specific heats of real propellant gases and the inability of these gases to maintain internal energy equilibration during their rapid expansion through the nozzle. This problem can seriously impair the performance of any accelerator of this class, and has much to say about the choice of propellant for a given impulse range.

The nonideality of an electrothermal accelerator can be cataloged in a slightly different way by defining the partial efficiency with which electric energy is delivered from the source to heat the gas stream, η_h; the aerodynamic, or nozzle, efficiency with which the stream follows a one-dimensional adiabatic route through the expansion process, η_a; and the efficiency with which it converts internal energy in the propellant stream to directed kinetic energy, η_f. As such, $1 - \eta_h$ accounts for losses in the heating process; $1 - \eta_a$ covers the nozzle viscous and profile losses; and $1 - \eta_f$ describes the unrecovered internal energy in the exhaust jet, often called *frozen flow* losses. By any classification, these losses determine the utility of a given electrothermal accelerator to function as a space thruster. In view of the premium on minimization of power supply weight discussed in Chap. 1, it is essential that the overall thruster efficiency, $\eta = \eta_h \eta_a \eta_f$, be kept high if the complete electrothermal propulsion system is to retain its advantage over chemical rockets. Thus the major problem of electrothermal thruster development is really the reduction of these losses. Much of the balance of this chapter is addressed to the basic electrical, atomic, and gasdynamic processes which determine them.

6-2 ENTHALPY OF HIGH-TEMPERATURE GASES—FROZEN FLOW LOSSES

To allow for temperature dependence of the specific heat, the one-dimensional energy equation (6-1) may better be expressed in terms of an enthalpy function [2]:

$$\tfrac{1}{2} u_e{}^2 = \tfrac{1}{2} u_c{}^2 + (h_c - h_e) \qquad (6\text{-}2)$$

where h, the enthalpy per unit mass, or specific enthalpy, is the sum of the internal energy e and a "flow-work" quotient of pressure and mass density,

$$h = e + \frac{p}{\rho} \qquad (6\text{-}3)$$

and yields the specific heat at constant pressure by partial differentiation:

$$c_p = \frac{\partial h}{\partial T} \qquad (6\text{-}4)$$

The specific enthalpy, like the internal energy it contains, embodies contributions from a variety of internal molecular degrees of freedom, in addition to the random translational modes. To illustrate possible contributions, consider a unit mass of diatomic gas initially at a low temperature where the only type of particle present is the molecule, in amount N_0, numerically equal to the reciprocal of the mass of the molecule. This gas is now heated to an elevated temperature T, where it contains many species of molecular, atomic, and ionic particles, in a variety of internal energy states. To simplify the illustration, assume that at the temperature of interest, we need include only the following species:

1. Neutral molecules, in amount $\alpha_2 N_0$
2. Neutral atoms, in amount $\alpha_1 N_0$
3. Molecular single ions, in amount $\alpha_2^+ N_0$
4. Atomic single ions, in amount $\alpha_1^+ N_0$
5. Free electrons, in amount $\alpha_e N_0$

Constraints on the α coefficients are provided by requirements on conservation of atomic particles:

$$\alpha_2 + \alpha_2^+ + \tfrac{1}{2}\alpha_1 + \tfrac{1}{2}\alpha_1^+ = 1 \qquad (6\text{-}5)$$

and conservation of electric charge:

$$\alpha_2^+ + \alpha_1^+ = \alpha_e \qquad (6\text{-}6)$$

We shall assume that the high-temperature gas is quasi-perfect, i.e., that its equation of state follows from the partial pressures of perfect-gas components:

$$\frac{p}{\rho} = (\alpha_2 + \alpha_1 + \alpha_2^+ + \alpha_1^+ + \alpha_e)N_0kT = (1 + \tfrac{1}{2}\alpha_1 + \tfrac{1}{2}\alpha_1^+ + \alpha_e)N_0kT$$
$$= (1 + \tfrac{1}{2}\alpha_1 + \alpha_2^+ + \tfrac{3}{2}\alpha_1^+)N_0kT \equiv \alpha_p N_0kT \qquad (6\text{-}7)$$

where k again is Boltzmann's constant, and α_p is the indicated factor modifying the usual ideal-gas relation.

The internal energy of the high-temperature gas can be constructed from the contributions of its constituent species. The internal energy of a particular specie, in turn, can be computed from classical energy-partition relations provided by statistical thermodynamics [3]. We shall summarize the results in a form which lends itself to the available tabulations of molecular data [4,5].

1. *Neutral molecules*

$$e_2 = \alpha_2 N_0 \left(\tfrac{3}{2}kT + \beta_r kT + \beta_v kT + \sum_j \beta_j \varepsilon_j\right) \qquad (6\text{-}8)$$

        trans-   rota-   vibra-   electronic
        lation   tion   tion   excitation

where β_r = effective fraction of molecules with rotation excited[1]
       β_v = effective fraction of molecules with vibration excited
       β_j = effective fraction of molecules in jth excited electronic state
       ε_j = energy of jth electronic state, above ground state

2. *Neutral atoms*

$$e_1 = \alpha_1 N_0 \left(\tfrac{3}{2}kT + \sum_k \beta_k \varepsilon_k\right) \qquad (6\text{-}9)$$

where β_k = effective fraction of atoms in kth excited electronic state
       ε_k = energy of kth electronic state

3. *Molecular ions*

$$e_2^+ = \alpha_2^+ N_0 \left(\tfrac{3}{2}kT + \beta_r^+ kT + \beta_v^+ kT + \sum_l \beta_l \varepsilon_l\right) \qquad (6\text{-}10)$$

4. *Atomic ions*

$$e_1^+ = \alpha_1^+ N_0 \left(\tfrac{3}{2}kT + \sum_m \beta_m \varepsilon_m\right) \qquad (6\text{-}11)$$

5. *Electrons*

$$e_e = \alpha_e N_0 (\tfrac{3}{2}kT) \qquad (6\text{-}12)$$

[1] Strictly, rotational and vibrational energies are quantized in discrete levels, and summations like that for electronic excitation would be more rigorous. However, since these levels are closely and regularly spaced and since the full equipartition values are well defined statistically, the effective excitation parameters β_r and β_v are normally quite serviceable and far more convenient.

In addition, we must include the energy absorbed in dissociation and ionization:

$$e_d = N_0 \frac{\alpha_1 + \alpha_1^+}{2} \varepsilon_d \qquad (6\text{-}13)$$

$$e_i = N_0(\alpha_2^+ \varepsilon_i + \alpha_1^+ \varepsilon_i') \qquad (6\text{-}14)$$

where ε_d = dissociation energy of molecules
ε_i = ionization potential of molecules
ε_i' = ionization potential of atoms

The enthalpy of the unit mass mixture thus becomes

$$h = \frac{p}{\rho} + \sum e = \alpha_2 N_0 \left[(\tfrac{5}{2} + \beta_r + \beta_v)kT + \sum_j \beta_j \varepsilon_j \right]$$
$$+ \alpha_1 N_0 \left[\tfrac{5}{2}kT + \sum_k \beta_k \varepsilon_k + \tfrac{1}{2}\varepsilon_d \right]$$
$$+ \alpha_2^+ N_0 \left[(\tfrac{5}{2} + \beta_r^+ + \beta_v^+)kT + \left(\sum_l \beta_l \varepsilon_l + \varepsilon_i\right) \right]$$
$$+ \alpha_1^+ N_0 \left[\tfrac{5}{2}kT + \sum_m \beta_m \varepsilon_m + \tfrac{1}{2}\varepsilon_d + \varepsilon_i' \right] + \tfrac{5}{2}\alpha_e N_0 kT \qquad (6\text{-}15)$$

Alternative forms follow from substitution of the conservation relations (6-5) and (6-6) for α_2 and α_e. Note that h is linear in N_0, and thus inversely proportional to the molecular weight of the parent specie, indicating an a priori desirability of light gases for electrothermal propellants.

To evaluate the enthalpy of a given gas at any temperature, we thus need to know all its α's and β's. In equilibrium, these can be expressed as functions of temperature and pressure by statistical arguments based on the partition functions, much like the earlier derivation of the Saha equation (Sec. 3-3). So long as the gas remains in equilibrium, the enthalpy is a well-defined function of T and p, which can be tabulated or graphed in some convenient form (Fig. 6-2).

Unfortunately, in high speed flows, the local gasdynamic conditions may change so rapidly that atomic-scale collisions become insufficiently frequent to maintain equilibrium adjustments of all the α's and β's, and some may lag behind their proper thermodynamic values. Hence the enthalpy will differ from its equilibrium value, and the continuum flow velocity must adjust itself to maintain the energy balance. The streamwise velocity profile in a high speed flow thus can be influenced by the relative rates of adjustment of the various α's and β's of the gas. Unfortunately, many of these are incompletely understood at present, and only a few generalizations of their typical behavior can be made.

β_r, β_r^+: Rotation adjusts almost as rapidly as translation, and is fully excited at all temperatures above a few degrees Kelvin.

ELECTROTHERMAL ACCELERATION

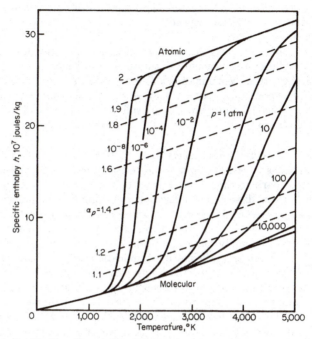

Fig. 6-2 The effect of dissociation on the enthalpy of hydrogen.

β_v, β_v^+: The rate of vibrational adjustment depends strongly on the particular molecule and mode involved. Some vibrational modes are several orders of magnitude slower to adjust than translation or rotation. Even at equilibrium, vibration may be only partially excited for the temperature ranges of propulsion interest.

β_j, β_k, β_l, β_m: Population and depopulation of excited electronic states occur by a variety of radiative and collisional processes, like those described in Chap. 4. The rates of these processes are strongly dependent on gas density and on the surrounding environment; thus no general statement can be made. Unless the gas is optically "thick," thereby trapping the bulk of the radiation, statistical theory for the equilibrium state will require modification for radiation energy loss. Fortunately, in many cases the electronic levels are relatively high compared with the gas temperatures of interest, and the total energy contained in the finite number of these excited states is small enough to be neglected.

α_1, α_2: Dissociation and recombination require very special types of collisions (low-energy, three-body, radiative, etc.), and hence are usually quite slow to adjust and the rates are strongly density-dependent. Equilibrium values depend heavily on the ratio of dissociation energy to gas temperature.

α_1^+, α_2^+: Like dissociation, ionization involves large energy transfer on collision to surmount the threshold, and hence may be slow to adjust. Ion-electron recombination also is an inefficient atomic-scale process, and the overall rate may be very slow. Equilibrium values provided by the Saha relation [Eq. (3-26)] will require modification if radiation is significant.

Based on these general tendencies, it is to be expected that in those portions of the flow where the temperature is changing rapidly on the local particle time scale, certain of the slower internal modes, such as vibration, dissociation, ionization, or recombination, may lag significantly behind their equilibrium levels, and consequently the enthalpy, and thence the flow velocity, will depart from the equilibrium values. Such a situation is commonly termed *frozen flow*.

To illustrate the possible implications of this tendency, return again to the problem of the hydrogen flow through our simple accelerator. Assume that the gas has reached complete equilibrium in the heating chamber at a temperature $T_c = 3000°K$, and pressure $P_c = 0.01$ atm. At these conditions hydrogen has much of its vibration excited but very little electronic excitation. It is, however, about 60 percent dissociated (Fig. 6-2). That is, $\alpha_2 \approx 0.4$, $\alpha_1 \approx 1.2$, hence $\alpha_p \approx 1.6$; all other α's ≈ 0. Hence we may roughly approximate its enthalpy by the expression

$$h_c = N_0[\alpha_2(\tfrac{9}{2}kT) + \alpha_1(\tfrac{5}{2}kT + \tfrac{1}{2}\varepsilon_d)] \tag{6-16}$$

Next assume that when the gas leaves the nozzle, all degrees of freedom have reached equilibrium at the exit temperature T_e except dissociation, which retains some fraction ξ of the value it had at T_c. To the approximation of negligible exhaust temperature, the only enthalpy retained by the gas at the exit is that tied up in this frozen dissociation:

$$h_e = \tfrac{1}{2}\xi\alpha_1 N_0 \varepsilon_d \tag{6-17}$$

If we again neglect u_c^2 in comparison with u_e^2, the exit velocity is given by

$$\tfrac{1}{2}u_e^2 = h_c - h_e = N_0 kT\left[\tfrac{9}{2}\alpha_2 + \tfrac{5}{2}\alpha_1 + \alpha_1(1-\xi)\frac{\varepsilon_d}{2kT}\right] \tag{6-18}$$

Inserting for hydrogen $\varepsilon_d = 4.5$ ev = $52,000°K \cdot k$, $N_0 k = 4.16 \times 10^3$ joules/(kg)(°K), and α_2 and α_1 as given above, we compute the two extreme

ELECTROTHERMAL ACCELERATION

values for the exhaust speed, depending on the frozen flow fraction ξ:

$$u_e = \begin{cases} 1.95 \times 10^4 \text{ m/sec} & \xi = 0 \\ 1.10 \times 10^4 \text{ m/sec} & \xi = 1 \end{cases} \quad (6\text{-}19)$$

Two important points are demonstrated by this calculation: First, the dissociation of the propellant achieved in the heating duct greatly increases its heat capacity, i.e., the enthalpy which can be imparted to it at the limiting wall temperature. Second, a predominant part of this advantage is lost if the gas does not substantially recombine before it leaves the nozzle. With respect to the former, note that dissociation, like ionization, is favored by lower pressures. (Indeed, dissociation can be described by a Saha type of statistical equation [6].) By reducing the chamber pressure from 1 atm in our original example to 0.01 atm here, we increased c_p at 3000°K by a factor of 3.

Clearly, it is of considerable importance to reduce ξ to a minimum, not merely to raise u_e, but to reduce the energy loss associated with the frozen exhaust flow. Three possibilities suggest themselves: (1) protract the nozzle length to provide more time for molecular recombination; (2) operate at higher pressure levels to increase the recombination rate (and lower the chamber dissociation level); and (3) use other propellants with less tendency to frozen flow losses.

It is well known from practical experience that protraction of the nozzle normally reduces frozen flow losses less than it increases viscothermal losses. High-pressure operation is a more attractive solution, and may improve performance in other respects, as we shall discuss later. Figure 6-3 illustrates the effect of pressure on *frozen flow efficiency*, defined as

$$\eta_f = \frac{h_c - h_e(\xi = 1)}{h_c} \quad (6\text{-}20)$$

i.e., the ratio of enthalpy converted to kinetic energy of the jet to that imparted in the chamber, assuming no recombination throughout the nozzle.

Doubtless, the most important factor in controlling frozen flow losses, however, is the selection of the proper propellant gas for a given range of operation. In practice, this choice actually involves several somewhat conflicting factors. In addition to the demonstrated desirability of low molecular mass and fast internal modes to maximize specific heat capacity, and a low tendency toward frozen flow losses, it is essential that the gas be readily storable in space and that it not be excessively corrosive or tend to enhance erosion of the chamber or nozzle. Also, if the heating involves

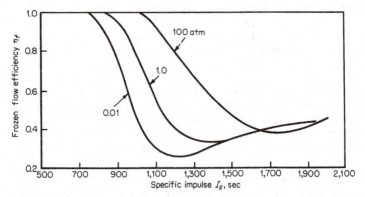

Fig. 6-3 Theoretical variation of frozen flow efficiency with specific impulse for hydrogen at various pressures. *(From J. R. Jack, Theoretical Performance of Propellants Suitable for Electrothermal Jet Engines, ARS J., vol. 31, p. 1685, 1961.)*

an electrical discharge, the gas must have satisfactory ionization and electrical conduction characteristics. Briefly, one might consider the following possibilities (Table 6-1 and Fig. 6-4).

Table 6-1 Physical properties of possible electrothermal propellants

Propellant	Molecular weight, amu	Specific heat at constant pressure, 10^3 joules/(kg)(°K)		Boiling point, °K at 1 atm	Melting point, °K at 1 atm	Critical pressure, atm	Critical temperature, °K
		1000°K	3000°K				
Hydrogen (H_2)	2.016	15.0	18.4	20	14	12.8	33
Helium (He)	4.003	5.20	5.20	4		2.3	5
Lithium (Li)	6.94	3.00	3.14	1500	460		
Beryllium (Be)	9.01	2.31	2.33	1800	1600		
Boron (B)	10.82	1.92	1.92	2800	2600		
Carbon (C)	12.01	1.73	1.80	4500	3800		
Ammonia (NH_3)	17.03	3.20	4.51	240	196	111.3	406
Nitrogen (N_2)	28.02	1.17	1.32	77	63	33.5	126
Hydrazine (N_2H_4)	32.05	2.76		387	275	145	653
Pentaborane (B_5H_9)	63.13	4.03	5.08	332	226		

SOURCES: JANAF Thermochemical Tables, The Dow Chemical Company, Midland, Mich., 1965, and International Critical Tables of Numerical Data: Physics, Chemistry, and Technology, McGraw-Hill Publications for the National Research Council, McGraw-Hill Book Company, New York, 1926–1930.

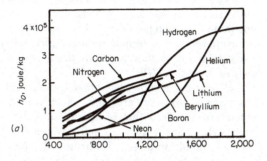
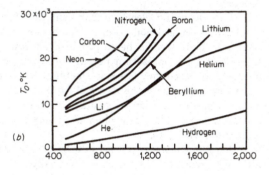
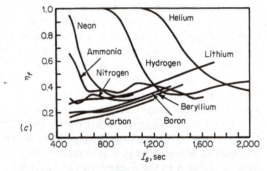

Fig. 6-4 Stagnation enthalpy (a), stagnation temperature (b), and frozen flow efficiency (c) of various propellants ($p_c = 1$ atm). (From J. R. Jack, Theoretical Performance of Propellants Suitable for Electrothermal Jet Engines, NASA Tech. Note D-682, Mar. 19, 1961, and L. E. Wallner and J. Czika, Jr., Arc-jet Thrustor for Space Propulsion, NASA Tech. Note D-2868, June 7, 1965.)

Hydrogen has a very high specific heat and thermal conductivity, can be stored cryogenically with some difficulty ($\approx 20°K$), and does not contribute to erosion problems. It is well behaved in electrical discharges. Unfortunately, its molecular recombination rate is so slow that it suffers nearly total frozen flow losses in the nozzle expansion for the desired range of operation [7–9].

Helium has a somewhat lower specific heat than hydrogen, but its heat transfer properties are good. Since it is monatomic, its first important internal mode is that of ionization, and this potential is very high (24.46 ev). Thus it does not suffer frozen flow losses until much higher levels of specific impulse. Unfortunately, its liquefaction temperature is so low ($\approx 4°K$) that its storage in space seems an insurmountable problem.

Lithium would clearly circumvent the storage problem since it is a solid at standard conditions. For a monatomic substance, it has a relatively high specific heat, provided by low-lying excited electronic states and a low ionization potential. These unfortunately also introduce severe frozen flow losses in the temperature range of interest. It is necessary to pregasify the material for flow through a passive heater; in an arc it gasifies and ionizes readily. Condensation shocks may arise as it cools in the nozzle. It is chemically very active, and must be handled and stored with care.

Beryllium, *boron*, and *carbon* have characteristics similar to lithium, although somewhat less active chemically. They are of successively larger atomic mass, hence lower specific heat, to the point that they are of marginal interest for electrothermal propulsion.

Ammonia (NH_3) is also attractive from the standpoint of storage, since the liquid phase requires no refrigeration. Although it is a rather heavy molecule, when heated it dissociates into low-molecular-mass constituents, which improves its specific heat capacity but introduces obvious frozen flow losses. Attainable thrusts, specific impulses, and efficiencies are not badly inferior to pure-hydrogen operation, but its chemical activity tends to enhance nozzle and heater erosion.

Hydrazine (N_2H_4) and other compounds displaying exothermic dissociation reactions have been proposed on the basis that higher performance may be achieved by judicious release of chemical energy in the heater duct, in addition to the electric input. This will be advantageous only if programmed in such a way that the heat transfer problems at the nozzle and chamber walls are not intensified and if the chemical erosion remains tolerable.

Pentaborane (B_5H_9) has been proposed as an example of a combination of a light metallic element with a hydrogen carrier gas, which, theoretically, is superior to pure hydrogen under certain conditions [10].

This concept regards the heat of vaporization of the solid as an available thermal degree of freedom, and thus appears to be restricted to low-pressure operation.

The final selection of an electrothermal propellant will normally be based primarily on specific impulse and frozen flow considerations in the assigned range of operation, but this choice may be modified somewhat by broader considerations of the overall system performance, e.g., propellant storage, feed, deterioration, etc., on the particular mission involved.

This section has dealt in some generality with the frozen flow loss because of its fundamental import on all electrothermal thrusters. More specific details of this loss and of the heater and nozzle losses will be covered in the discussions of particular types of thrusters to follow.

6-3 RESISTOJETS

The simplest of all electric propulsion devices is the resistojet, wherein the propellant gas is heated by passing it over an electrically heated solid surface. Many configurations of resistojets have been conceived and developed, and some versions have evolved to the status of practical space thrusters. Heater elements have been constructed of coils of wire aligned parallel to the flow (Fig. 6-5a), of a succession of similar coils deployed transversely to the flow (Fig. 6-5b), of a bed of tungsten spheres heated by passing current through their contact resistance (Fig. 6-5c), and of contiguous knife-edges or sharp points carrying the heating current through similar contact resistance (Fig. 6-5d), or the heating-chamber walls themselves have been resistively heated (Fig. 6-5e). These devices have been run radiatively or regeneratively cooled, on ac or dc supplies, at power levels from a fraction of 1 watt to 60 kw, over a broad range of terminal voltage. Many propellant gases have been tried, both in steady and in pulsed flow.

In addition to the frozen flow losses discussed above, the basic problems in the development of a resistojet concern the heat transfer from the resistance element to gas stream, the radiation losses from the complete assembly, and the high-temperature materials technology. Analysis of the first two may be approached by classical heat transfer techniques, but tend to become cumbersome in the geometries and temperature ranges involved in these devices. Since the flow in the chamber is usually laminar, the heat transfer to the fluid stream is primarily by conduction, and closed-form solutions for simple geometries would be possible if the gas flow were calorically ideal. Unfortunately, the specific heat, thermal conductivity, and gas density, all vary substantially with temperature in the range considered; hence iterative or similar procedures are required to achieve self-consistent solutions [11]. Actually, detailed

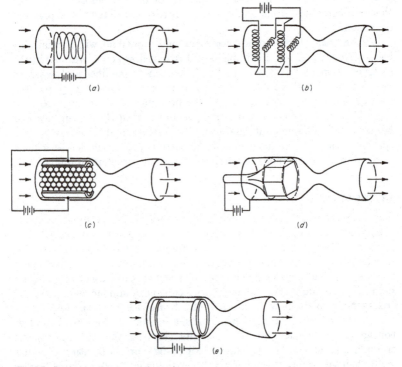

Fig. 6-5 Resistojet heater configurations.

solutions of the gasdynamic heat transfer problem are seldom critical to implementation of a particular resistojet concept. Generally, a few experimental surveys using the desired propellants with various heater and chamber dimensions will lead rather directly to an adequate optimization of the geometry and bulk flow parameters for a given device, without the necessity for detailed understanding of the heat transfer pattern.

A similar situation prevails with respect to the thermal radiation losses from resistojets. In principle, these detract from the performance of the device to an extent which can be calculated from basic elements of radiant heat transfer. In practice, empirical common sense normally will suffice to design and construct a configuration wherein these losses are reduced to comparative unimportance to the overall system. For example, the active heat transfer duct may be surrounded with insulation or reentrant gas flow passages, or both, to a sufficient extent so that negligible heat is radiated from the body of the composite thruster (Fig. 6-6a)

ELECTROTHERMAL ACCELERATION

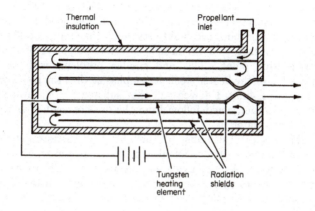

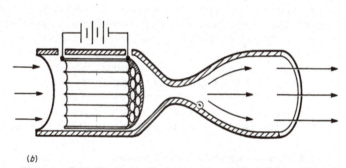

Fig. 6-6 Thermal insulation of resistojets (a) by reentrant flow passages; (b) by parallel flow passages.

[12]. Alternatively, many heater ducts may be honeycombed in parallel, and the array surrounded with insulation (Fig. 6-6b). Such cavalier solutions to the radiation problem are permissible simply because the total weight of this type of thruster is by nature a trivial fraction of that of the power supply needed to drive it. Even after such insulation procedures, there remains some unavoidable radiation out of the nozzle mouth from the hot propellant gas, heater cavity, and nozzle throat. This loss will depend on the limiting optical aperture, i.e., the nozzle throat, and this is invariably a small fraction of the chamber dimension.

A major portion of the resistojet development effort is concerned with practical problems of preparing, fabricating, and maintaining the high-temperature conductor and insulator materials which must retain vacuum seals and electrical integrity in the 2500-to-3000°K environment desired. It is found, for example, that a tungsten conductor and a boron nitride insulator, both acceptably stable by themselves at 3000°K, when placed in contact tend to form a eutectic compound of substantially lower melting point. This and similar problems with the thermal degradation of the heater elements emphasize that the resistojet is truly a temperature-limited device and that substantial improvements in its level of operation could follow from the development of superior high-temperature materials.

The choice of chamber pressure for a given resistojet may be an important factor in its overall efficiency, and is determined by balancing several factors. As we have seen, operation at high pressures reduces frozen flow losses by lowering the dissociation level in the chamber and increasing recombination rates in the nozzle. In addition, it improves heat transfer to the flow from the heater surfaces, reduces radiation losses by increasing the optical depth of the hot gas, and permits a smaller chamber and nozzle for a given mass flow. Counteracting these advantages are the increased stress on the hot chamber walls and the increased nozzle throat erosion. The latter process has frequently been found to be the limiting factor on the lifetime of a thruster of this type. With the present heater concepts, the best operational compromise seems to lie in the range of 1 to 5 atm of chamber pressure.

Figures 6-7 to 6-11 display photographs of various resistojet thrusters, and Table 6-2 summarizes some typical performance characteristics. It is seen that these admirably simple devices have already achieved excellent

Table 6-2 Performance of typical resistojet thrusters

Input power, kw	0.01	1.0	3.0	3.0	12.3	30.0
Propellant	NH_3	H_2	H_2	H_2	NH_3	H_2
Heater configuration	Single tube	Concentric contact	Concentric tubes	Transverse coils	Concentric contact	Concentric contact
Thrust, newtons	5×10^{-4}	0.176	0.652	0.534	2.93	6.04
Specific impulse, sec	250	729	840	838	423	846
Thrust efficiency†		0.63	0.88	0.74	0.50	0.85
Chamber pressure, atm	0.3	3.9	8.8	2.4	2.4	4.1
Laboratory	AVCO	Giannini	Marquardt	AVCO	Giannini	Giannini

† $\eta = TI_s g_0/2P$; neglects cold-flow power.

SOURCES: R. J. Page et al., 3-kw Concentric Tubular Resistojet Performance, *J. Spacecraft Rockets*, vol. 3, p. 1669, 1966; A. C. Ducati, E. Muehlberger, and J. P. Todd, Resistance-heated Thrustor Research, *Giannini Scientific Corp. Tech. Rept.* AFAPL-TR-65-71, July, 1965; and private communications.

Fig. 6-7 Three-kilowatt regeneratively cooled resistojet; hydrogen propellant; five transverse tungsten-rhenium heater coils. (*AVCO Corporation, Space Systems Division, Wilmington, Mass.*)

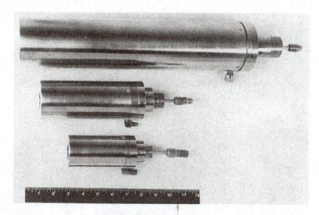

Fig. 6-8 Contact-heated resistojet thrusters: 30, 6, and 1 kw. (*Giannini Scientific Corporation, Santa Ana, Calif.*)

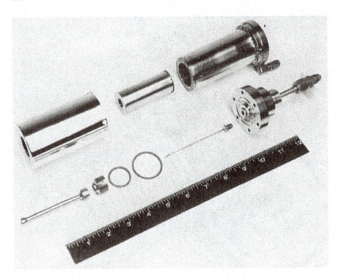

Fig. 6-9 Exploded view of 1-kw contact-heated resistojet. (*Giannini Scientific Corporation, Santa Ana, Calif.*)

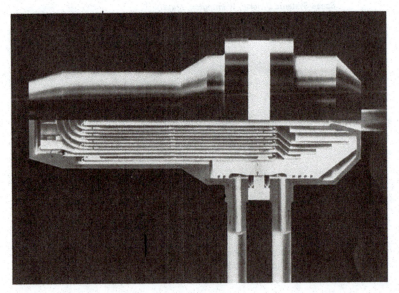

Fig. 6-10 Three-kilowatt concentric tubular resistojet. (*Marquardt Corporation, Van Nuys, Calif.*)

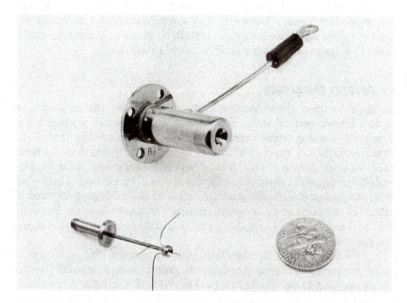

Fig. 6-11 Fifty-watt pulsed resistojet; ammonia propellant; molybdenum duct heater. (*AVCO Corporation, Space Systems Division, Wilmington, Mass.*)

overall efficiencies and that our original crude estimate of 1,000-sec specific impulse for a surface-heated flow is within range. These thrusters have been developed with particular space applications in mind, particularly as elements of satellite control systems, and doubtless will see increasing use in this capacity.

As a point of historical interest, the first space operation of an electrothermal propulsion unit of any kind took place on Sept. 19, 1965, when a tiny resistojet was fired successfully for 30 min to adjust slightly the position of a Vela nuclear-detection satellite. This device, constructed by TRW Systems, was reported to be 6 in. long by 1½ in. in diameter, to weigh 0.6 lb, and to consume 90 watts of power. Its propellant was nitrogen, and its heater a helical resistance rod which reached a temperature of 1000°F. It developed a total thrust of 0.042 lb at a specific impulse of 123 sec [13]—surely a modest beginning for the space application of electrothermal propulsion! (See Prob. 6-4.)

In conclusion, the summary of practical experience with resistojet operation to this time confirms the expectations of the elementary concepts of these devices. Namely, they are mechanically simple, compact, and easy to start and stop; they have a wide latitude of control and high overall efficiency; and they are adaptable to a variety of power supplies and propellant gases. However, they are fundamentally temperature-limited by the available structural materials, which at present constrain them to a range of exhaust velocities below 10^4 m/sec.

6-4 GASEOUS DISCHARGES

To improve upon the exhaust velocity attainable by the resistojet, or surface-heater, technique, an electrothermal device must produce a gas flow in its chamber whose average temperature is higher than that of the chamber walls; i.e., the core of the gas flow must be hotter than the layer near the duct surface. This in turn implies that the heat must be generated and deposited within the central gas flow. One means of achieving this is to pass an electric arc through the gas in a suitable geometry and to allow the radiative and convective transport of energy from the intensely hot discharge column to establish the desired temperature profile across the gas stream.

Before discussing the details of such arc heaters, it is important to recognize certain physical properties of gaseous arcs in general [14–16]. Using the electric circuit sketched in the insert of Fig. 6-12, it is possible to trace out a voltage-current characteristic for any particular gaseous discharge. The results obtained depend strongly on many properties of the specific discharge gap, notably the type, pressure, and flow velocity of the gas in the gap; the electrode material, shape, and spacing; any constrictions on the discharge pattern; and any external radiation sources; but in most cases the behavior qualitatively resembles that illustrated in Fig. 6-12. With reference to the notation of that figure, we may identify various characteristic regimes of behavior of the discharge.

Regime O–A. The electric field between the electrodes collects any stray charges created by ionization of the gas molecules by external radiation and any charges emitted by the electrodes due to that same radiation. The current saturates at a value determined by the strength of the external source, which may be only a very weak cosmic-ray background or may be purposely intensified, as in vacuum and gaseous photocells, x-ray detectors, radioactive counters, etc.

Regime A–B (first Townsend region). The stray electrons formed as above acquire enough energy from the electric field between collisions to ionize other atoms by collision. The secondary electrons in turn produce others, etc. The current is thus linear in the external source intensity, and exponential in the ionization cross section, neutral density, and gap spacing, for one-dimensional electrode geometries.

Regime B–C (second Townsend region). The positive ions acquire enough energy from the field between collisions to emit some electrons

ELECTROTHERMAL ACCELERATION

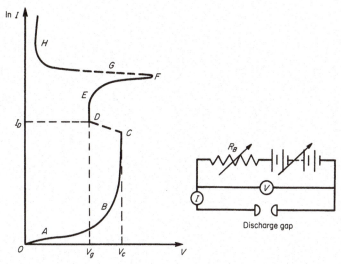

Fig. 6-12 Typical voltage-current characteristic for a gaseous discharge.

from the cathode by bombardment. Internal radiation may also contribute to photoemission from the cathode.

Regime C–D. Beyond the point C, denoted by the *sparking potential* V_c, the ion bombardment of the cathode and/or the radiation of the cathode by the discharge gas become sufficiently intense that the discharge becomes momentarily unstable and "runs away" to a new, lower-voltage mode. V_c is strongly dependent on all the properties of the discharge gap listed above. For given electrodes in a given gas, V_c is found to be a function of the product of gas pressure p and gap spacing d (Paschen's law). Typical breakdown profiles, V_c versus pd, are shown in Fig. 6-13. The significance of the product pd may be rationalized by noting that it is roughly proportional to the number of collisions an electron experiences in traversing the electrode gap.

Beyond C, the discharge behavior depends on the voltage source, the shape of the electrodes, and the gas pressure. If the electrodes have sharp points and if the gas pressure is high, a corona discharge will develop. If they are smooth and the gas pressure is low, a "normal" glow discharge sets in, of current density determined by the capacity of the source. (In the special case in which the source cannot supply the minimum glow current I_D, the gap will spark, return to C, spark again, etc.) In particular, since the gap voltage is nearly constant in this range at some value V_g, the glow current may be adjusted to any value between D and E on the figure by setting the ballast resistor R_B.

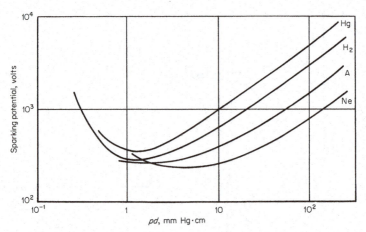

Fig. 6-13 Sparking potential for various gases (Paschen's law). (*From A. von Engel, "Ionized Gases," chap. 7, p. 172, Oxford University Press, Fair Lawn, N.J., 1955.*)

The glow discharge, sustained by ion-bombardment emission from the cathode, is one of the most fascinating phenomena in all physics, and has been studied extensively. Because its current densities (10^{-3} to 1 amp/cm^2) and gas temperatures ($<100°C$) are quite low, it is of little primary significance to propulsion; but many of its details are relevant to the higher current discharges. Figure 6-14 displays axial profiles of various properties of a simple glow discharge. The spatial extent and relative importance of the different regions again depend on the specific details of the environment. Note that the bulk of the discharge space is taken by the "positive column," in which the gas atoms are found to have a maxwellian distribution of energies at a very low temperature, the ions at slightly higher temperature, and the electrons at very high temperatures ($\approx 10^4$ to $10^{5}°K$). The field is small and uniform along the column, and the bulk of the voltage drop occurs at the cathode (cathode fall). In short, unlike the Townsend discharges, the glow discharge is not axially uniform, is not in thermal equilibrium, and does not maintain charge neutrality.

Returning to the general discharge characteristic (Fig. 6-12):

Regime E–F ("abnormal" glow). Most of the increase of voltage here appears across the cathode-fall region, where the ion current density and bombardment energies now increase enough to heat the cathode substantially.

Regime F–G (glow-arc transition). At the critical point F, a new, far more prolific set of emission mechanisms take over at the cathode.

ELECTROTHERMAL ACCELERATION

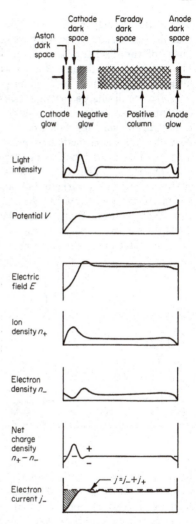

Fig. 6-14 Axial profiles of various glow discharge properties. (*From J. D. Cobine, "Gaseous Conductors," p. 213, Dover Publications, Inc., New York, 1958.*)

The cathode-fall region becomes sufficiently hot that the cathode emits electrons thermionically, aided to some extent by the strong fields here, and perhaps photoelectrically by the intense luminosity of the adjacent gas. The cathode-fall voltage then drops, and a high current arc sets in.

Regime G–H. The arc resistance now drops more rapidly than the current rises, and unless protected by a ballast resistor, the arc will "run away" to currents of 1,000 amp or more, vaporizing its electrodes and becoming extemely hot (>10^{4}°K). The desired operating level again must be established by the adjustment of R_B. It is this last, high current phase of the discharge that interests us for propulsion. Much less is known about the details of such arcs than about the glow discharges, although the essential participating processes have been identified [17].

An arc discharge through a gas is distinguished by relatively large currents (several amperes or greater) and relatively low electrode voltages (less than 100 volts for short arcs). The cathode fall is typically less than 20 volts, but the cathode surface emits electrons prolifically (10^3 to 10^7 amp/cm^2) by a combination of thermionic, photoelectric, and field emission processes. The anode fall is roughly the same magnitude and spatial extent as the cathode fall, and in both these regions the principal ionization processes involve the field-accelerated electrons. There is a net negative space charge near the anode, positive near the cathode. Thus, in several respects, these are nonequilibrium regions. A typical voltage profile along the arc is sketched in Fig. 6-15.

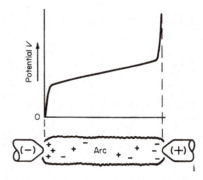

Fig. 6-15 Axial potential profile for a high current arc (schematic).

The positive column, which occupies all the available interval between the two electrode-fall regions, is, in contrast to them, a good approximation to a thermal plasma. It consists of a strongly radiating mixture of electrons, ions, and neutral atoms at nearly the same temperatures, say, from 5000 to 50,000°K, having corresponding degrees of ionization from a few percent to essentially 100 percent. The random thermal velocities of the electrons in this region far exceed their mean drift velocities in the weak electric field along the column, and thus the predominant ionization mechanisms are thermal electron collisions and photoionization. In this column and in the anode sheath, the current conduction is mainly via a diffusion-dominated drift of the electrons. Only near the cathode do the ions contribute significantly to the current during their acceleration through the cathode fall.

ELECTROTHERMAL ACCELERATION

In the absence of significant gas flow, the arc temperature is set by an energy balance between the electric power input, primarily via ohmic heating, and the radial radiative and conductive losses from the column, plus the heat transferred axially to the electrode regions. The latter serves both to heat the electrode material and to create the current carriers in these sheaths and at the electrode surfaces.

The high gas temperatures thus attained in the positive column may be further amplified by a magnetic self-constriction of the conduction channel to a relatively small cross section, the so-called *pinch* effect. To illustrate this process, we shall idealize the arc as a uniform cylinder of current density j over a radius r_1. From Maxwell's relation for the magnetic field from a steady current [Eq. (2-32)],

$$\nabla \times \mathbf{H} = \mathbf{j} \qquad (6\text{-}21)$$

it follows that the associated magnetic field is azimuthal and proportional to the radial dimension inside the column, inversely proportional to it outside:

$$H = \begin{cases} \dfrac{jr}{2} & r \leq r_1 \\ \dfrac{jr_1^2}{2r} & r \geq r_1 \end{cases} \qquad (6\text{-}22)$$

This "self-field" reacts on the current that produces it with a force density

$$\mathbf{F} = \mathbf{j} \times \mathbf{B} = \frac{\mu j^2 \mathbf{r}}{2} \qquad r \leq r_1 \qquad (6\text{-}23)$$

In equilibrium, this magnetic force can balance a radial gradient in the gas-kinetic pressure in the current column:

$$\left. \begin{aligned} \frac{dp}{dr} &= -\frac{\mu j^2 r}{2} \\ p &= \frac{\mu j^2}{4} (r_1^2 - r^2) + p_1 \end{aligned} \right\} \; r \leq r_1 \qquad (6\text{-}24)$$

where p_1 is the ambient gas pressure outside of the arc edge, r_1 (Fig. 6-16a).

The gas temperature within the arc is related to the pressure profile by an appropriate equation of state. For large current densities, the temperature attainable at the center of the column is substantially higher because of this magnetic constriction (Prob. 6-5). Clearly, any real arc will have a smoother radial variation of current density, with corresponding changes in the magnetic field and magnetic force variations (Fig. 6-16b), but the essential pinch mechanism still operates.

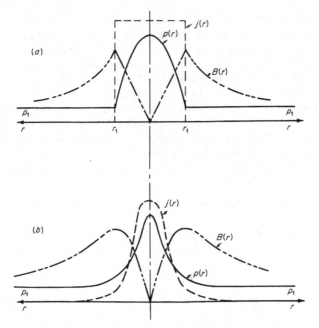

Fig. 6-16 Radial variation of pressure through a pinched arc column. (a) Uniform current density; (b) nonuniform current density.

Unfortunately, regardless of the current density distribution, this otherwise convenient self-constriction mechanism embodies intrinsic dynamic instabilities. For example, should the column develop any "sausage" constrictions (Fig. 6-17a), the magnetic pinching force at the small-radius portions would exceed that elsewhere, thus enhancing the constriction until the column actually became severed. In the same way, any "kinks" in the plasma column (Fig. 6-17b) tend to amplify because of the excessive magnetic forces on the concave sides.

If very high current densities are to be maintained against such intrinsic instabilities, it is necessary to invoke some external stabilizing mechanism. One possibility is to apply external magnetic fields preferentially to constrain the plasma against such distortions. This approach has been commonly employed in ground-based plasma generators, such as those which have been used in the fusion program. The magnetic fields needed are large, however, and the associated magnet weight is unattractive for a space engine.

For propulsion purposes, a more promising method for stabilization of an arc follows from a technique reported in 1909 by Schonherr [18]

ELECTROTHERMAL ACCELERATION

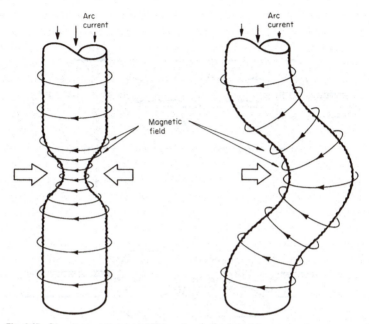

Fig. 6-17 Sausage and kink instabilities of a pinched electric arc.

and later by Gerdien and Lotz [19]. This involves surrounding an arc column with a swirling vortex of gas, or even liquid, injected tangentially at the periphery of the arc chamber (Fig. 6-18a). The rotational motion of the injected fluid has the multiple benefits of centrifugally constraining the hot gas discharge column to the axis of the vortex, of cooling the electrodes and chamber walls, and—far more important to present propulsion applications than to those original researches—of bringing the injected gas into better and longer contact with the arc column, thereby heating it to a more significant fraction of the arc temperature. Conversely, the injected gas cools the arc, thereby lowering its conductivity and permitting much higher arc voltages and corresponding power inputs to the arc.

A second gasdynamic stabilization technique was developed empirically during earlier experimental research on arc thrusters and gas heaters. It involves the passage of the arc and the gas flow to be heated through a relatively long, narrow tube between the electrodes (Fig. 6-18b). Such *constricted arcs* are found to be stable against radial kinking without the necessity for swirling the gas flow, and also embody sufficiently steep temperature gradients to protect the constrictor walls from softening. In heating the gas, this arrangement seems to be at least as effective as

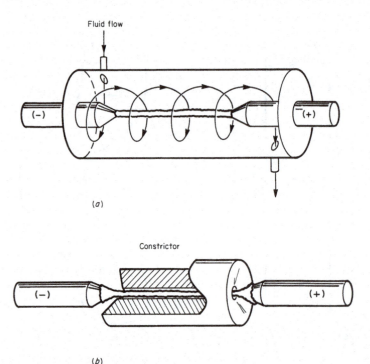

Fig. 6-18 Stabilization of the arc column (a) by fluid vortex; (b) by material constrictor.

the vortex-stabilized arcs, and its comparative gasdynamic simplicity leaves some hope of relevant theoretical analysis.

From this brief background of electric-arc phenomenology, we return to the central topic of electrothermal propulsion via arc heating of a propellant stream. As we shall see, the coupling of the physics of high current arcs to three-dimensional compressible gas flows presents formidable analytical difficulties, and much of the progress in this field must be ascribed to the patient empirical experiments and physical intuition of the investigators.

6-5 ARCJET OPERATION AND ANALYSIS

The gas vortex-stabilized arcs mentioned above were the immediate ancestors of the first arcjet plasma sources, which were developed primarily for materials testing and reentry simulation. These typically used a central conical cathode and a coaxial annular anode which formed an arc chamber

ELECTROTHERMAL ACCELERATION

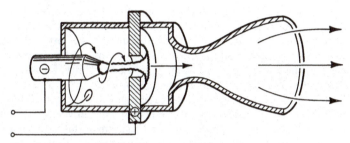

Fig. 6-19 Vortex-stabilized arcjet plasma source.

of cross section considerably larger than the arc column itself (Fig. 6-19). This chamber was followed by a plenum, or mixing volume, which in turn supplied a supersonic nozzle. The presumption was that the mixing volume would provide sufficient residence time for the swirling, highly excited and ionized gas which had passed through the arc chamber to come to thermal equilibrium, whence it could then drive the exit nozzle to a maximum exhaust velocity. Subsequent experience revealed that a large fraction of the gas reaching the mixing chamber was essentially unheated and that residence times in the chamber long enough to equilibrate the gas could not be attained without excessive heat transfer losses at that stage.

As interest in arcjets for space propulsion arose, these inefficiencies became intolerable, and a great variety of other arc-chamber configurations were constructed and tested in empirical attempts to improve on the original design. By this process, mixing chambers were largely eliminated, nozzles shortened, and electrode configurations substantially modified, yielding distinctly new breeds of these devices, such as the constricted arcs mentioned above. During this evolution, much was learned about high-temperature materials behavior and techniques for assembly and cooling of the arc devices, in addition to the physical behavior of the arcs themselves. Whereas, in the construction of a resistojet, a poor choice of material or a faulty assembly would result in reduced lifetime or a simple heater failure, a similar error in construction of an arcjet heater carrying hundreds of amperes of current to and through the gas flow could precipitate abrupt destruction of an electrode, nozzle, or other vital element of the device. It was not at all unusual in such cases to see large pieces of molten electrode or insulator material disgorged from the jet orifice as the arc devoured the vulnerable element.

Gradually, and almost entirely by trial and error, serviceable techniques for cooling, vacuum sealing, gas handling, and materials selection and conditioning were developed, to the point that stable arc-heater operation could be routinely sustained in the power ranges of interest. Attention could then be turned to the efficiency of particular heater configurations for propellant acceleration. Rather than attempting to review the history of this development further, or to survey the many arcjet thrusters that have emerged from this effort, let us examine one typical model that has had the benefit of a certain amount of basic study, for the purpose of discussing the details of the gas-heating process.

The model selected is a 30-kw radiation-cooled constricted-arc device which has demonstrated attractive performance in the 1,500-sec specific impulse range [20]. A photograph and diagram of this thruster are shown in Figs. 6-20 and 6-21, and typical operating conditions are listed in Table 6-3.

Fig. 6-20 Thirty-kilowatt radiation-cooled arcjet operating in vacuum tank. (*AVCO Corporation, Space Systems Division, Wilmington, Mass.*)

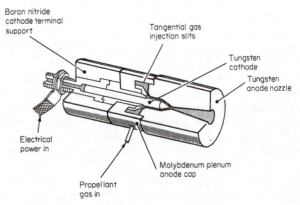

Fig. 6-21 Diagram of 30-kw arcjet shown in Fig. 6-20.

Table 6-3 Operating conditions for 30-kw example thruster

Propellant	Hydrogen
Mass flow	0.10 g/sec
Chamber pressure	1 atm
Power input	30 kw
Electrode current	150 amp
Electrode voltage	200 volts
Thrust	147 newtons
Specific impulse	1,500 sec
Thrust power	11 kw
Radiation loss	3 kw
Efficiency	37 percent

SOURCE: R. R. John et al., Arc Jet Engine Performance: Experiment and Theory, *AIAA J.*, vol. 1, no. 11, p. 2517, 1963.

By various optical observations it is established that the arc column in this device emerges from a relatively small portion of the cathode tip (≈ 1 mm^2), extends along the axis through the entire constrictor section in a laminar column considerably smaller in diameter than the constrictor channel, and attaches in a diffuse, axially symmetric distribution on the nozzle mouth, which serves as the anode (Fig. 6-22). A strong radial temperature profile exists in the constrictor, involving an intensely hot central core and a substantially cooler layer of gas flow separating it from the constrictor wall. This radial temperature profile is found to extend far down into the exhaust nozzle, thereby amply fulfilling the criterion

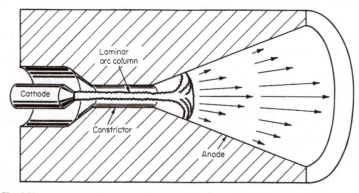

Fig. 6-22 Core-flow pattern in a constricted arcjet.

mentioned earlier, that the mean gas temperature should substantially exceed the wall temperature if the resistojet performance is to be surpassed. Indeed, the attainment of a 1,500-sec specific impulse from a tungsten engine guarantees that such a "core-flow" process must be occurring (Prob. 6-6).

Theoretical analysis of the operation of such a device can be approached only after bold approximation of the many complex thermal and electrical processes which participate. In principle, one requires simultaneous solution of statements of mass flux, momentum balance, and energy conservation for the gas flow, Maxwell's equations and the constitutive relations for the electrical processes, and a heat conduction statement for the solid portions of the device, where the electrical, thermal, and viscous transport parameters involved may be sensitive functions of local temperature and density. Fortunately, all self-induced electric and magnetic effects are easily dismissed for steady operation at this current level, leaving a scalar Ohm's law as the only essential electrical statement. Beyond the algebraic simplifications provided by the evident cylindrical symmetry, further reduction of the complexity of the analysis formulation depends on identification of the dominant mechanisms for heat transfer from the arc column to the surrounding cooler gas flow. The observations support assumption of a laminar flow, wherein convection processes are primarily axial. The importance of radiation from the column is less simply assessed. At the prevailing pressures and temperatures, radiation, although a significant energy loss from the arc column, probably cannot heat the outer gas flow effectively by direct absorption. This cool gas is largely transparent to the radiation, passing it on to the surrounding constrictor walls. Here, however, it is absorbed, and these walls, thus heated, return some of the energy flux to the gas flow by conduction through the boundary layer. Radiative heat transfer to the gas is thus an indirect process, depending for its effectiveness on the constrictor operating temperature, material, and geometry.

A particularly simple analytical model may be constructed on the following assumptions:

1. All electrical energy input occurs in the tight arc column, which fills only a small fraction of the constrictor cross section, and hence involves little mass flow. This power input per unit volume is computed from a scalar Ohm's law with a constant electrical conductivity σ:

$$P_{in} = \mathbf{j} \cdot \mathbf{E} = \sigma E^2 \qquad (6\text{-}25)$$

2. The cathode is heated by the cathode-fall region of the arc; the cathode in turn preheats the incoming gas flow.

ELECTROTHERMAL ACCELERATION

3. The constrictor wall is heated by radiation from the arc column; this wall in turn heats the outer gas flow region.
4. The upstream part of the nozzle is heated by the anode-fall region of the arc; the downstream part is heated by the exhaust flow.
5. The hot core flow mixes with the outer flow only in the nozzle.

To the approximation that the constricted arc column is a transparent, cylindrically symmetric plasma in local thermal equilibrium, with negligible axial gradients, its energy balance may be written

$$\sigma E^2 = P_r - \frac{1}{r}\frac{d}{dr}\left(r\kappa \frac{dT}{dr}\right) \qquad (6\text{-}26)$$

where the electrical and thermal conductivities σ and κ and the power radiated per unit volume P_r are each functions of the gas type, temperature, and pressure and are tabulated for some common gases, including hydrogen [8]. Enforcing boundary conditions $T = T_0$, $dT/dr = 0$ on the centerline, the central temperature T_0 may be estimated for various voltage gradients along the column E. For example, at a pressure of 1 atm and an arc current of 150 amp, the central temperature of a hydrogen arc reaches values of 20,000, 40,000, and 60,000°K for voltage gradients of 25, 100, and 250 volts/cm, respectively. The same calculations predict an effective arc diameter of about 1 or 2 mm, in accordance with visual observation [20].

The concomitant heat transfer processes within the thruster block and from the block back into the gas flow can be estimated under similarly rough approximations:

1. The cathode and anode surfaces receive a given fraction of the electric power deposited in the cathode- and anode-fall regions of the arc, computed as the product of arc current with cathode and anode voltage drops, respectively.
2. The constrictor surface receives radiation from the arc column and transmits heat to the cool outer flow by conduction through the laminar boundary layer.
3. The mixing gases in the nozzle transmit heat to it convectively, via some empirical model.
4. Outer surfaces of the thruster block radiate as a "gray body," $P_s \propto T^4$.

The results of such a calculation for the example thruster are shown in Fig. 6-23. In this particular case it happens that some 6 kw of the total 30 kw is transmitted to the thruster block by radiation and by the

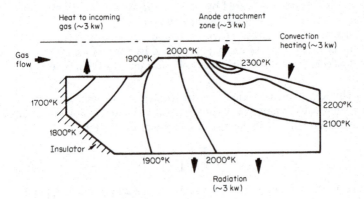

Fig. 6-23 Temperature profiles in 30-kw arcjet thruster block. (*AVCO Corporation, Space Systems Division, Wilmington, Mass.*)

arc terminations. Of this about 3 kw returns to the cool gas flow and 3 kw is lost by radiation from the outer surfaces.

The importance of the regenerative heat transfer from the thruster block to the cooler portions of the gas has been nicely illustrated by a series of experiments wherein engine performance is monitored shortly after the electric arc is interrupted (Fig. 6-24). Persistence of a large

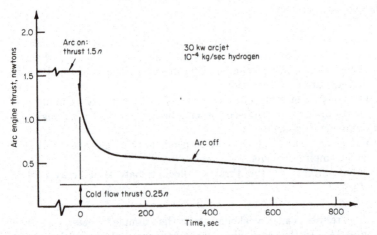

Fig. 6-24 Arcjet thrust decay after arc extinction. (*From R. R. John et al., Arc Jet Engine Performance: Experiment and Theory, AIAA J., ser. 11, vol. 1, p. 2517, 1963.*)

fraction of the thrust indicates that substantial energy input to the propellant occurs via the block walls; i.e., in a sense the device functions partially like a resistojet.

The core-flow model of the arcjet outlined above can be further manipulated to yield transverse profiles of the flow properties of interest. As might be anticipated, the bulk of the mass flow ρu is in the cool outer flow, while the enthalpy per unit mass is much larger in the arc column. The thrust density profile ρu^2 thus tends to be more uniform across the jet than either of these.

To the extent that detailed diagnostic measurements can be made in the hostile environment of the arc, and to the extent that overall performance measurements of thrust, specific impulse, and efficiency can be related to this analysis, the core-flow model displays sufficient relevance to the actual operation of a constricted arc thruster to make it useful for extrapolation of test results to other geometries, propellants, and ranges of operation. It fails, however, to provide much insight about the axial enthalpy flux in the arc column itself, or about the detailed participation of the electrode-fall regions in the heating process.

An alternative simple model can be cast in terms of a more uniform volumetric heat addition process wherein all incoming gas is presumed to participate in the arc, radiation is neglected, and an axial enthalpy gradient is allowed [21]:

$$\sigma E^2 = \rho u \frac{\partial h}{\partial z} - \frac{1}{r}\frac{\partial}{\partial r}\left(r\kappa \frac{\partial T}{\partial r}\right) \qquad (6\text{-}27)$$

By assuming linear relations among h, σ, and $\int \kappa\, dT$, it is possible to construct a map of the enthalpy $h(r,z)$ over the constrictor volume whereby a variety of other properties of interest, such as heat losses, voltage gradients, mean enthalpy vs. axial position, and electrical conversion efficiency may be calculated. These calculations also can be shown to fit reasonably well with certain experimental data of arcjet operation, but also fail to describe the details of the electrode processes accurately [22].

To attempt more general analysis of the arc heating process one must also allow for axial heat conduction and work done by pressure gradients:

$$\sigma E^2 = P_r - \frac{1}{r}\frac{\partial}{\partial r}\left(r\kappa \frac{\partial T}{\partial r}\right) - \frac{\partial}{\partial z}\left(\kappa \frac{\partial T}{\partial z}\right) + \rho u \frac{\partial h}{\partial z} + \rho v \frac{\partial h}{\partial r} - u \frac{\partial p}{\partial z} \qquad (6\text{-}28)$$

To this energy equation must be added a momentum statement,

$$\rho u \frac{\partial u}{\partial z} + \rho v \frac{\partial u}{\partial r} = -\frac{\partial p}{\partial z} + \frac{1}{r}\frac{\partial}{\partial r}\left(r\mu \frac{\partial u}{\partial r}\right) \qquad (6\text{-}29)$$

where the viscous force $\mu(\partial u/\partial r)$ will in general be important in the constrictor because of the small duct diameter; a mass flux statement,

$$2\pi \int_0^{r_e} \rho u r \, dr = \text{const} \tag{6-30}$$

an appropriate equation of state; and suitable boundary conditions on the arc axis and at the channel walls. It is doubtful that a complete set of such relations will ever be solved to yield a nozzle outflow. They do serve, however, as a basis for other simplifications of the gas-heating model, appropriate to other domains of arcjet operation.

6-6 ARCJET THRUSTERS

Most of the arcjet thrusters in current operation strongly resemble the example discussed in the preceding section. Figures 6-25 to 6-28 show a few diagrams and photographs of other typical models, and Table 6-4 summarizes sample performance figures which illustrate the range of operation of this class of device. Detailed descriptions of the construction and operation of these thrusters are available in the References; here we shall remark only on certain common features of their design and on

Fig. 6-25 Thirty-kilowatt radiation-cooled hydrogen arcjet operating in vacuum tank. (*Giannini Scientific Corporation, Santa Ana, Calif.*)

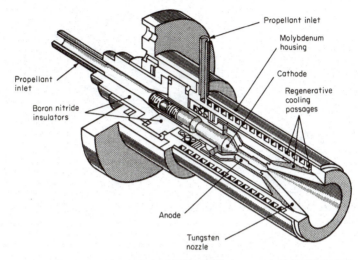

Fig. 6-26 Cutaway views of 30-kw regeneratively cooled arcjet thruster. (*Giannini Scientific Corporation, Santa Ana, Calif.*)

128   ELECTRICAL ACCELERATION OF GASES

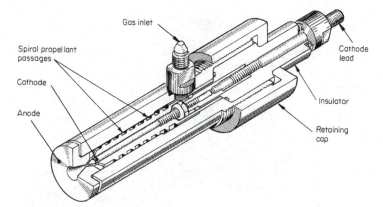

Fig. 6-27 Two-kilowatt regeneratively cooled arcjet. (*Giannini Scientific Corporation, Santa Ana, Calif.*)

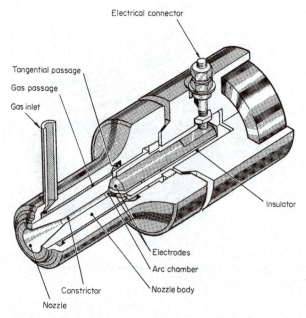

Fig. 6-28 Thirty-kilowatt three-phase ac arcjet thruster. (*From R. Richter*, Development of a 30 KW Three-phase AC Arc Jet Propulsion System, *General Electric Co. Rept., NASA* CR-54112, Aug. 4, 1964.)

Table 6-4 Performance of typical arcjet thrusters

Input power, kw	1	30	30	30	30 AC	200
Propellant	H_2	H_2	H_2	NH_3	H_2	H_2
Thrust, newtons	0.044	3.35	1.77	2.37	2.26	6.80
Specific impulse, sec	1,100	1,010	1,520	1,012	1,020	2,120
Thrust efficiency	0.35	0.54	0.44	0.39	0.38	0.35
Laboratory	Giannini	Giannini	AVCO	AVCO	G.E.	AVCO

SOURCE: L. E. Wallner and J. Czika, Jr., Arc-jet Thrustor for Space Propulsion, *NASA Tech. Note D-2868*, June, 1965.

the motivation for certain differences. It will first be noted that in each of the dc models the cathode is a cylindrical tungsten rod with a conical tip, aligned on the axis. The conical tip serves to intensify the electric field at the surface and to provide a localized hot spot to amplify the thermionic emission of electrons needed to carry the large arc currents. With the exception of this hot spot, the bulk of the cathode is cooled by the incoming gas flow and by conduction to the large insulators which mount it in the thruster assembly. The anodes, in contrast, tend to be rather massive blocks of tungsten, out of which are carved the desired nozzle and constrictor profiles. In arcs of this type, the heat liberated at the anode fall and attachment region exceeds that at the cathode, and the large anode mass and exterior surface are necessary for adequate distribution and radiation of the heat deposited on the inner nozzle surface. The insulators which separate cathode and anode are typically boron nitride, one of the best high-temperature dielectrics, and also are as massive as the electrode configuration and gas flow passages will permit. Remaining elements of the housing which bear the gas inlet connections, mounting flanges, etc., may also be fabricated from tungsten, or perhaps from molybdenum, which is more easily machined and has adequate thermal properties for the cooler portions of the assembly. The gas flow is led into the arc chamber through passages cored into the housing in accordance with the desired preheating profile (see below). Injection is typically made with a large tangential component to establish a swirling flow through the constrictor. The necessity of a strong vortex flow to stabilize a long, highly constricted arc is somewhat dubious, but this vestige of early arcjet technology seems to persist, perhaps justified by slightly better heat transfer characteristics in the chamber.

Beyond these common elements of design, certain variations in operation of arcjet devices have been tried, in the hope of ameliorating specific limitations of their performance. These have included regenerative gas flow preheating to improve thruster cooling, use of ac rather than dc power to simplify power-conditioning equipment, use of reactive or hybrid propellants to improve energy release, and control of the arc geometry by applied magnetic fields. As a general rule, modifications which involve substantial complication of the thruster design or of the auxiliary equipment have not justified themselves from an overall performance standpoint, but the experience of the various attempts has proved instructive, and is worth a brief discussion.

THRUSTER COOLING

From the earlier discussion of the characteristics of high current arcs, it is clear that more is involved in the cooling of an arcjet engine than simply retaining the structural integrity of the components and restraining electrode and nozzle erosion to tolerable levels. The arc itself is sensitive to the temperature of its electrode surfaces, and injudicious cooling can thus produce poor performance of the essential element of the device. For example, if the cathode is allowed to become too hot, it will erode too rapidly and change its shape and spacing from the anode. If it is excessively cooled, however, it will be unable to sustain the level of electron emission required by the arc.

One may identify three modes of cooling: radiative, regenerative, and external. The latter, consisting of a separate cooling fluid and circulating system, is immediately unattractive for space engines from weight and bulk considerations, although it is the conventional method for cooling almost all high power laboratory arc-heater devices, such as high performance wind tunnels and reentry simulators. Early in the development of the field, there was considerable discussion about the relative merits of the regenerative vs. the radiative modes of cooling a thruster, but in the light of later work, such as the engine-shutdown tests described above, much of this discussion seems now to be academic. It is clear that a constricted arcjet inherently performs substantial regenerative heating inside the arc chamber, if not in exterior coring, and hence the concept of a pure radiation-cooled engine is an inappropriate abstraction. The significant distinctions thus are among the details of the flow and chamber geometries employed to optimize the efficiency of heat transfer to the propellant, rather than between two basic modes of operation.

AC ARCJETS

Motivation for ac arcjets stems primarily from the (questionable) anticipation that large space nuclear power supplies will lend themselves more readily to the generation of ac power. Conversion to direct current will then require the added weight of a rectifying unit. In addition, the ballast resistor normally needed to control a dc arc could be replaced by an inductor which ideally would dissipate considerably less heat. Various disadvantages of ac operation stem from the inherent directionality of an arc electrode system and the problem of reignition on phase reversal. As mentioned above, the requirements on cathode and anode in terms of emission, heat load, etc., are substantially different, as manifested in the design of the electrodes in the basic dc arcjet. It is difficult to conceive of an electrode design which would provide optimum cathode and anode operation on both halves of an ac cycle. In addition, low-frequency ac arcs may encounter reignition difficulties. In a dc arc, the initial ignition of the discharge is a troublesome point in the operation, requiring large voltages or external excitation, followed by a period of stabilization. In an ac arc, one is asking the discharge to reignite twice each cycle. It seems reasonable that for sufficiently high frequencies the gas would remain sufficiently ionized, and the electrode tips hot enough so that this problem is considerably relieved, but in view of the atomic-scale relaxation processes involved, this probably involves kilocycle alternation.

Only a few ac arcjet thrusters have been constructed or tested. In isolated cases, claims have been made of lower electrode erosion, better heat transfer to the gas and less to the electrodes, and easier starting than the comparable dc engines. A few interesting studies have been made of devices which operate on three-phase power, using four coaxial ring electrodes or other symmetrical electrode arrangements [24]. Despite these occasional successes, the ac arcjet program has so far generated little organized enthusiasm, presumably because it superimposes further physical complications on the already complex arc-heating process.

CHOICE OF PROPELLANT

In addition to those properties of the propellant that were shown to be important for electrothermal thrusters in general—high specific heat and thermal conductivity, low frozen flow losses, storability, and lack of corrosiveness to thruster parts—it is important to consider the electrical and thermochemical behavior of the substance within the arc itself. From the electrical standpoint, the ionization potential and electrical conductivity should permit the establishment of a stable protracted arc column of the desired voltage gradient, at the power level of interest. On the thermochemical side, one must deal with a new spectrum of physical and chemical processes which can proceed in the intense thermal environment of the arc. If these processes involve "fast" degrees of freedom, they can increase the effective specific heat of the propellant, and thereby the attainable exhaust speed; if "slow" modes are excited, additional frozen flow losses will result. For example, ammonia (NH_3), which is storable in space as a liquid with no refrigeration, readily dissociates in the arc into low molecular mass species, and thus provides a high specific heat capacity, albeit with substantial frozen flow losses. The attainable thrust, specific impulse, and efficiency of ammonia arcjets are not badly inferior to those of hydrogen devices, and may be preferable for systems where cryogenic hydrogen storage would be objectionable [25].

Other high-temperature propellant possibilities include injection of light alkali metals into a carrier stream of hydrogen or other common gas. Whereas the use of materials like lithium in a resistojet is of questionable benefit because of the problems of gasifying these substances, such species are readily vaporized in the arc when entrained in particulate form in the carrier gas or when introduced via consumable electrodes impregnated with the desired substance. Involvement of such materials in the arc will change its physical characteristics noticeably. Because of the lower ionization potentials, the arc will run satisfactorily at lower temperatures and display lower electrode falls. Other bipropellant combinations of this sort may provide similar benefits [26].

As a final possibility, one may consider the use of exothermically reacting propellants to increase the enthalpy release in the heating chamber above the purely electrical input. A variety of monopropellants and bipropellant combinations have been considered for this process. For example, hydrazine (N_2H_4), like ammonia, is readily storable and dissociates into a tolerably low average molecular mass in the arc, and in so doing liberates a significant reaction energy which in principle is available to the flow. Clearly, this chemical energy release will be advantageous only if programmed in such a way that the heat transfer and erosion problems at the chamber and nozzle walls are not intensified [27].

MAGNETICALLY DIFFUSED AND FOCUSED ARCJETS

It has been mentioned that high current arcs will constrict themselves somewhat by their own magnetic field, but that further constraint on the arc column by an external magnetic field is unattractive for propulsion because of the excessive magnet weight associated with the required high field strengths. However, certain other benefits in arc behavior can be obtained by application of much more modest fields that can reasonably be generated by coils in series with the arc itself (Fig. 6-29) or by small permanent magnets. The two most notable improvements that can be achieved this way are in the anode attachment of the arc and in a focusing of the plasma exhaust stream.

In the example thruster discussed in Sec. 6-5, the arc was shown terminating on the anode nozzle in a diffuse, axially symmetric pattern, which is desirable to minimize erosion and to maximize heat transfer to the gas stream in this region. Many arcjet configurations tend to display an instability in this mode of attachment, wherein the arc collapses into a radial "spoke" in this region and makes contact with the anode surface at only one narrow angular position. This point of attachment then

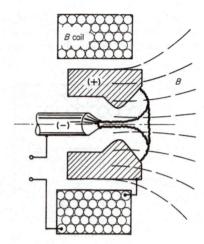

Fig. 6-29 Stabilization of arcjet by magnetic field of series coil.

rapidly erodes the anode surface and becomes the limiting factor in the lifetime of the thruster. Indeed, in very high current arcjets, a collapse of the anode attachment of this sort can destroy the thruster instantaneously.

Fortunately, this tendency to spoke instability can usually be inhibited by application of a modest axial B field (≈ 0.1 weber/m^2) from a coaxial coil [28]. The details of this stabilization mechanism are not completely clear, but appear to be related to an azimuthal swirling of the radial current filaments by their interaction with the axial field.

If the same coil that supplies this stabilizing field is positioned axially so that its fringing field is divergent outward in the region of plasma exhaust, it can act as a magnetic nozzle, reducing both the divergence angle of the exhaust stream and its thermal and viscous interaction with the material nozzle. This focusing process is somewhat akin to the confinement of plasmas in "mirror machines," wherein the magnetic field preferentially deters migration of the charged particles across the field lines (Prob. 6-7).

In a sense, devices which invoke magnetic body forces for arc diffusion or containment are no longer purely electrothermal devices, but rather are hybrid thrusters involving elements of both electrothermal and electromagnetic interactions. Indeed, very similar devices have evolved from totally different origins, notably, the Hall-current electromagnetic accelerators (Sec. 8-6), which use arcjets as the ionized gas source. Likewise, some of the highest performance crossed-field accelerators, the *magnetoplasmadynamic arcs* (Sec. 8-9), actually originated as arcjets which were driven to very high current levels at low chamber pressures.

This cooperative interplay of more than one electrical interaction to produce the desired acceleration is nicely illustrated by a particular low thrust device called the *magnetic expansion thruster* (Fig. 6-30).

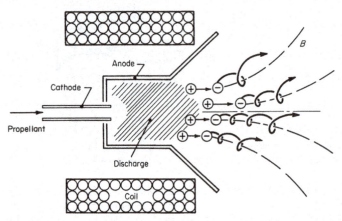

Fig. 6-30 Magnetic expansion thruster.

A glow discharge of roughly conical shape is drawn between a filament or hollow cathode and an annular anode in a region impressed with a diverging magnetic field like that described above. As in most glow discharges, the electrons acquire energies of a few electron volts from the electric field (an electrothermal process), but the ions remain relatively cold. Because of the magnetic field geometry, these energetic electrons are constrained to expand axially outward (an electromagnetic process), and in so doing they establish an electric field between themselves and the cold ions left behind, which in turn accelerates these ions to follow the electrons (an electrostatic interaction). The net effect is to convert electrical input energy from the glow discharge circuit into an axial streaming of the cold plasma, but the detailed process clearly does not submit to any single electrical classification. As we proceed, we shall find many other examples of cooperative interaction of various electrodynamic processes. Indeed, it seems highly likely that the ultimate electric space thrusters will be hybrid devices, rather than accelerators relying exclusively on one type of electrical interaction.

6-7 ELECTRODELESS-DISCHARGE ACCELERATION

The elements of the arcjet engine most susceptible to erosion are, of course, the electrodes, subjected as they are to vigorous electronic and

ELECTROTHERMAL ACCELERATION

ionic bombardment and to radiative heating by the adjacent anode- and cathode-fall regions of the arc. It is reasonable, therefore, to speculate that electrodeless discharges of one sort or another might have some advantages as gas heaters for propulsion if they could be made to impart significant enthalpy to a flowing gas stream. Electrodeless discharges are well known in various other applications, such as lighting fixtures and spectroscopic Geissler tubes, but these are usually low pressure, highly nonequilibrium discharges that impart little thermal energy to the heavy gas particles.

All electrodeless discharges operate on ac power input, but the nature of the electromagnetic coupling to the gas discharge, and the technique for establishing it, vary considerably with the frequency of the source. In the kilocycle range, the discharge is typically established by electrodes or coils physically insulated from the gas, but in close proximity to it to maximize the coupling. One normally distinguishes here between E-type discharges, wherein an alternating electric field is capacitively coupled to the gas by two axially separated "electrodes" (Fig. 6-31a), and H-type discharges, wherein an alternating magnetic field is established by an external coil, and this in turn induces an interior electric field in accord with Maxwell's $\nabla \times \mathbf{E}$ relation (Fig. 6-31b). In devices of this latter class, some elements of both interaction may be present, because of unavoidable potential drops along the coils.

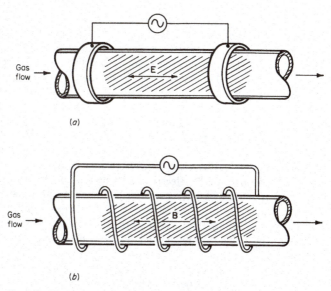

Fig. 6-31 Electrodeless discharges. (a) E type; (b) H type.

As one goes to higher frequencies, into the hundred-kilocycle and megacycle rf bands, the coupling becomes more localized, self-propagating radiation fields become significant, and it is possible to withdraw the electrodes or antennae farther from the body of gas. In the microwave range, 10^9 to 10^{11} cps, a discharge may be excited by placing the gas sample within a waveguide or at the focus of a horn antenna (Fig. 6-32). Essentially the same interaction can be obtained at optical frequencies, using intense laser sources.

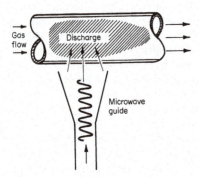

Fig. 6-32 Microwave heating of a gas flow.

Regardless of the particular frequency and field geometry, the bulk of the electrical energy input is initially absorbed in kinetic energy of the free electrons, and must subsequently be transmitted to the heavy particles. For a pure electrothermal device, this involves a collisional heating of the ions and neutrals. As we saw in Chap. 4, energy transfer by elastic collision between electrons and heavy particles is inefficient to the order of the mass ratio, whereas inelastic collisions in ranges of energy resonance may be highly effective in exciting electronic states of the atoms and ions. Since this excitation energy is rapidly radiated away by spontaneous photon emission, it is not unusual for low pressure discharges of this sort to dispense most of the input electric energy as optical radiation, rather than in heating the gas.

Improvement of the electrical heating may be attempted in various ways. Operation at higher gas pressures increases the electron collision frequency, thereby lowering the mean thermal speed and improving the ratio of elastic energy transfer to excitation losses. If this is overdone, however, the ionization level, and thereby the conductivity of the gas, will suffer.

Judicious use of magnetic fields can also enhance the heating process. Energetic electrons can be magnetically constrained to orbit within the

ELECTROTHERMAL ACCELERATION

hot gas zone longer before they are lost to the chamber walls, thereby increasing the plasma density. Also, the discharge exhaust pattern can be magnetically confined to a narrower exit cone, thereby protracting the zone of effective collisional energy transfer farther along the flow pattern. In a bit more subtle application, magnetic fields can provide various cyclotron or gyro resonances in the plasma. The electron cyclotron resonance (Sec. 5-1),

$$\omega_B^- = \frac{eB}{m} \quad (6\text{-}31)$$

is of no great help in a pure heating process since the energetic electrons thus obtained would have the same tendency to inelastic collisions mentioned above, but the ion gyro resonance,

$$\omega_B^+ = \frac{eB}{M_+} \quad (6\text{-}32)$$

offers a possibility for coupling the electric energy directly into the heavy particles. Basically, the concept is to constrain the positive ions to spiral through many oscillations of the applied ac electric field, increasing their energy incrementally, until finally striking neutral particles and efficiently yielding their energy to them. The choice of magnetic field strength must be a compromise among the associated magnet mass, the necessity to keep ω_B^+ much larger than the collision frequency, and the need to keep the ion orbit radii small compared with reasonable chamber dimensions. In addition, the ionization density in the discharge must be kept low enough that the electromagnetic fields are not excluded from the interior of the discharge. Simultaneous satisfaction of these requirements within reasonable limits on total system mass is a challenging engineering problem (Prob. 6-9). As an alternative approach, certain rather sophisticated techniques for ion heating of dense plasmas, involving the damping of imposed electromagnetic waves, have been demonstrated in the laboratory [31], but have not yet been directed toward propulsion application.

Electrodeless discharges seem to offer better promise for propulsion via certain hybrid interactions, such as those mentioned in the preceding section. Returning to electron-cyclotron heating again, since the basic difficulty of a pure electrothermal mode is the collisional transfer of energy from the electrons to the heavy particles, it may be better to accomplish this coupling electrostatically. For example, if the magnetic field constraining the electron orbits is axially nonuniform, the swarm of gyrating electrons will tend to drift toward regions of lower field intensity, thereby establishing a space-charge electric field to accelerate the ions in the desired axial direction. Devices of this type, called *cyclotron-resonance thrusters*, typically operate in the microwave frequency bands, with magnetic fields of several tenths of a weber per square meter (Fig. 6-33; [29,30]). A circularly polarized microwave beam of sufficient intensity

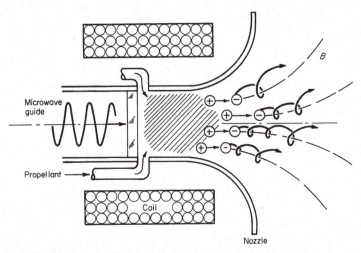

Fig. 6-33 Cyclotron resonance accelerator.

to ionize the incoming gas flow is launched from a conventional waveguide into an axially symmetric, divergent magnetic field, via a suitable dielectric window. The free electrons are driven into transverse cyclotron orbits which drift axially away from the source toward the weaker field regions, dragging ions after them via the space-charge separation field. Optimum conversion of microwave power to plasma energy occurs near the cyclotron resonance frequency, provided the gas density is sufficiently low that the electron neutral collision frequency is much smaller than this value. Unlike pure electrothermal thrusters, there is no necessity to minimize the molecular weight of the propellant; rather, it is preferred to optimize the ratio of ionization potential to kinetic energy of the exhaust ions, i.e., to use heavy, easily ionized, yet noncorrosive substances. A typical device of this class might employ a few kilowatts of X-band (8.35×10^9 cps) microwave power, producing rf field intensities of 10^4 to 10^5 volts/m in xenon at 10^{-4} torr (10^{-7} atm). Efficiency of transmission of rf power into plasma enthalpy can be quite high (60 to 70 percent), but overall conversion to directed kinetic energy of the exhaust beam is lowered by frozen flow and beam divergence losses.

In general, research on all types of electrodeless thrusters has been

ELECTROTHERMAL ACCELERATION

depressed by the problems attending the preparation of high-frequency power in space, which are even more severe than for dc or low-frequency units. The generation of megacycle or gigacycle fields at high power levels is normally accomplished by some type of electronic oscillator, such as a magnetron or klystron, which is inefficient in its power conversion from the primary source. Consequently, the specific weight of radio-frequency power supplies is an order of magnitude higher than the corresponding dc supplies. It is probably safe to presume that only on spacecraft committed to generation of large amounts of rf power for other purposes, e.g., communication, will electrodeless electric thrusters find any reasonable application.

PROBLEMS

6-1. Estimate the error introduced in Eq. (6-1) by neglecting the gas flow velocity in the chamber, u_c.

6-2. Compute the specific enthalpy of hydrogen at 2000 and 4000°K at 0.01 and 1.0 atm pressure.

6-3. Compute the chamber temperatures that would be needed in the ideal electrothermal accelerator to achieve 1,000 sec specific impulse with helium, lithium, and carbon, assuming no frozen flow losses and a chamber pressure of 1 atm. Repeat, assuming completely frozen flow through the nozzle.

6-4. Compute the specific impulse of the first resistojet used on the Vela satellite from the given heater temperature and from the power-to-thrust ratio. Compare with the value quoted in the text. What are the advantages and disadvantages in using nitrogen as the propellant?

6-5. For a cylindrical arc column of uniform current density and radius 2 mm, what total current is needed to produce a pressure on the arc axis double that of the ambient gas outside the arc? Estimate the corresponding temperature on the arc axis.

6-6. What arc-chamber temperature is needed to produce a 1,500-sec specific impulse with frozen flow hydrogen if transverse temperature profiles are uniform throughout the thruster and the chamber pressure is 1 atm? If profiles are parabolic and the limiting wall temperature is 2500°K, what temperature is needed on the chamber axis to achieve the same exhaust speed?

6-7. Describe in terms of charged-particle trajectories how an axisymmetric diverging magnetic field channels an ionized gas stream.

6-8. Compute the rate at which a charged particle in cyclotron resonance with an rf electric field of amplitude E_0 gains energy prior to a collision. If collisions occur at a frequency $\nu_c \ll \omega_B$, what is the mean energy of the charged particle over many collision periods?

6-9. Find a reasonable combination of magnetic field strength, gas type and pressure, and electric field amplitude and frequency wherein ion-cyclotron heating may proceed to useful electrothermal chamber temperatures. Comment on the feasibility of this heating mode for electrothermal propulsion.

6-10. Discuss the possibility of a hybrid electrothermal-electrostatic acceleration process, using the electron-plasma frequency resonance, rather than the electron-cyclotron resonance, as the rf input mode.

REFERENCES

An extensive list of additional references is provided in Ref. 22, below.

1. Boucher, R. A.: Electrical Propulsion for Control of Stationary Satellites, *J. Spacecraft Rockets*, vol. 1, no. 2, pp. 164–169, March–April, 1964.
2. Liepmann, H. W., and A. Roshko: "Elements of Gasdynamics," John Wiley & Sons, Inc., New York, 1957.
3. Fowler, R., and E. A. Guggenheim: "Statistical Thermodynamics," chap. 5, Cambridge University Press, New York, 1952.
4. Tables of Thermodynamic Properties of Gases, *Natl. Bur. Stand. (U.S.) Circ.* 564, Nov. 1, 1955.
5. Atomic Energy Levels, *Nat. Bur. Stand. (U.S.) Circ.* 467, June 15, 1949.
6. Lighthill, M. J.: Dynamics of a Dissociating Gas, Part I, Equilibrium Flow, *J. Fluid Mech.*, vol. 2, no. 1, 1957.
7. Jack, J. R.: Theoretical Performance of Propellants Suitable for Electrothermal Jet Engines, *ARS J.*, vol. 31, p. 1685, 1961.
8. King, C. R.: Compilation of Thermodynamic Properties, Transport Properties, and Theoretical Rocket Performance of Gaseous Hydrogen, *NASA Tech. Note* D-275, April, 1960.
9. Hall, J. G., A. Q. Eschenroeder, and J. J. Klein: Chemical Non-equilibrium Effects on Hydrogen Rocket Impulse at Low Pressure, *ARS J.*, vol. 30, no. 2, pp. 188–189, February, 1960.
10. Glassman, I., R. F. Sawyer, and A. M. Mellor: Propellant Potential of Vaporized Metals in Temperature-limited Rocket Systems, *AIAA J.*, vol. 2, no. 11, p. 2049, November, 1964.
11. Ducati, A. C., E. Muehlberger, and J. P. Todd: Resistance-heated Thrustor Research, *Giannini Scientific Corp. Tech. Rept.* AFAPL-TR-65-71, July, 1965.
12. Howard, J. M.: The Resistojet, *ARS J.*, vol. 32, no. 6, p. 961, June, 1962.
13. Jackson, F. A., et al.: An Operational Electrothermal Propulsion System for Spacecraft Reaction Control, 5th Electric Propulsion Conference, San Diego, Calif., March 7-9, 1966, *AIAA Paper* 66-213.
14. Cobine, J. D.: "Gaseous Conductors," chaps. 7-9, Dover Publications, Inc., New York, 1958.
15. Loeb, L. B.: "Basic Processes of Gaseous Electronics," University of California Press, Berkeley, Calif., 1960.
16. Von Engel, A.: "Ionized Gases," Oxford University Press, Fair Lawn, N.J., 1955.
17. Finkelnburg, W., and H. Maecker: Electric Arcs and Thermal Plasma, *Handbuch der Physik*, vol. 22, 1956.
18. Schonherr, O.: Process of the Badische Anilin- und Sodafabrik for the Fixation of Atmospheric Nitrogen, *Elektrotech. Z.*, vol. 30, pp. 365–369, 1909.
19. Gerdien, H., and A. Lotz: *Wiss. Veroeffentl. Siemens-Konz.*, vol. 2, p. 489, 1922, and *Z. Tech. Phys.*, vol. 4, p. 157, 1923.
20. John, R. R., et al.: Arc Jet Engine Performance: Experiment and Theory, *AIAA J.*, vol. 1, no. 11, p. 2517, 1963.
21. Stine, H. A., and V. R. Watson: The Theoretical Enthalpy Distribution of Air in Steady Flow along the Axis of a Direct-current Electric Arc, *NASA Tech. Note* D-1331, 1962.
22. Wallner, L. E., and J. Czika, Jr.: Arc-jet Thrustor for Space Propulsion, *NASA Tech. Note* D-2868, June, 1965.
23. Todd, J. P.: Thirty KW Arc-jet Thrustor Research, *Giannini Scientific Corp. Rept.* FR024-10338 (APL-TDR-64-58), March, 1964.

24. Richter, R.: Development of a 30 KW Three-phase AC Arc Jet Propulsion System (*NASA* CR-54112), General Electric Co., Aug. 4, 1964.
25. John, R. R., et al.: Arc Jet Engine Performance—Experiment and Theory, II, IAS-ARS Joint National Meeting, Los Angeles, Calif., June 13–16, 1961, *ARS Preprint* 61-101-1795.
26. Noeske, H. O., and R. R. Kassner: Analytical Investigation of a Bipropellant Arc Jet, *ARS J.*, vol. 32, no. 11, p. 1701, November, 1962.
27. Bender, R. W.: A Chemical Arc-jet Rocket Feasibility Study, in E. Stuhlinger (ed.), "Electric Propulsion Development," vol. 9 of "Progress in Astronautics and Aeronautics," pp. 95–120, Academic Press Inc., New York, 1963.
28. Mayo, R. F., and D. D. Davis, Jr.: Magnetically Diffused Radial Electric-arc Air Heater Employing Water-cooled Copper Electrodes, in E. Stuhlinger (ed.), "Electric Propulsion Development," vol. 9 of "Progress in Astronautics and Aeronautics," pp. 147–162, Academic Press Inc., New York, 1963.
29. Miller, D. B., and G. W. Bethke: Cyclotron Resonance Thrustor Design Techniques, *AIAA J.*, vol. 4, no. 5, pp. 835–840, May, 1966.
30. Hendel, H., T. Faith, and E. C. Hutter: Plasma Acceleration by Electron Cyclotron Resonance, *RCA Rev.*, vol. 26, no. 2, June, 1965.
31. Stix, T. H.: "The Theory of Plasma Waves," McGraw-Hill Book Company, New York, 1962.

7
Electrostatic Acceleration

The basic thermal limitations on attainable exhaust speeds and thrusts associated with the heating and subsequent expansion of a propellant gas through a nozzle can be circumvented if the gas is directly accelerated by electric body forces. The simplest concept for the application of such electric body forces to a propellant stream is the ion thruster, wherein a collisionless beam of positive atomic ions is accelerated by a suitable electrostatic field. The essential elements of such a device are sketched in Fig. 7-1. A stream of ions, liberated from some source surface, is accelerated by an electric field established between the source and a negative grid electrode. Subsequently, a stream of electrons joins the ion stream, producing a beam of zero net charge, which leaves the accelerator with a velocity determined by the total potential drop between the source and the exit electrode and by the charge-to-mass ratio of the ions employed. The thrust attainable in this manner depends only on this exhaust speed, on the mass of the ion, and on the total ion flux that can be accommodated by the source-accelerator-neutralizer system. Strictly, some new thermal limitations must now be associated with the

ELECTROSTATIC ACCELERATION

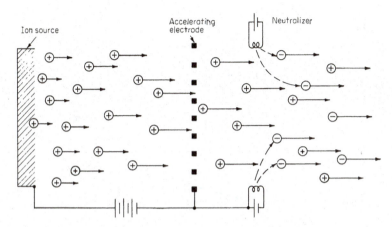

Fig. 7-1 Schematic diagram of ion thruster.

preparation of the ions at the source, but these pose far less severe constraints on attainable exhaust speed and efficiency than do those inherent in electrothermal accelerators.

The ion thruster is the most fully developed concept in electric propulsion, and a substantial body of technical literature exists on the subject ([1–10] and the references listed therein). In particular, Ref. 1 examines in considerable detail the interrelated aspects of ion thruster design, propulsion system dynamics, space power sources, and low thrust space flight mechanics which bear on the ultimate feasibility of ion-propelled spacecraft. Here we shall be concerned primarily with the basic processes of ion production, acceleration, and neutralization which must be accomplished in ion thrusters, rather than with technological details of their construction or implementation in a space vehicle.

7-1 ONE-DIMENSIONAL SPACE-CHARGE FLOWS

The interrelation of the electrical and dynamical parameters in an ion thruster can be illustrated by a simple one-dimensional model (Fig. 7-2). Let x be the streamwise coordinate; $x = 0$ the position of the source, which is at potential $V = V_0$; and $x = x_a$ the position of the accelerating grid, which is at potential $V = 0$. The potential V, the electric field $E = -(dV/dx)$, the ion density N, and the ion velocity v are all functions of x. In the steady state, the current density, $j = Nqv$, is a constant over x. The ion velocity at any position follows from conservation of energy, or equivalently, from integration of Newton's law of motion.

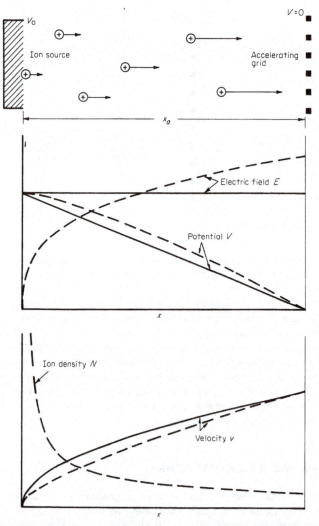

Fig. 7-2 One-dimensional ion acceleration: profiles of potential V, electric field E, velocity v, and charge density N for negligible charge density (———) and for space-charge limit (— — —).

ELECTROSTATIC ACCELERATION

For ions of charge q and mass M emitted with negligible velocity at the source,

$$v(x) = \left[\frac{2q(V_0 - V)}{M}\right]^{1/2} \tag{7-1}$$

The profile of the potential function, $V(x)$, is related to the profile of the ion density, $N(x)$, by Poisson's relation (2-16):

$$\frac{d^2V}{dx^2} = -\frac{Nq}{\epsilon_0} = -\frac{j}{\epsilon_0 v} = -\frac{j}{\epsilon_0}\left[\frac{M}{2q(V_0 - V)}\right]^{1/2} \tag{7-2}$$

The equation may be integrated simply when multiplied by $2(dV/dx)$, yielding

$$\left(\frac{dV}{dx}\right)^2 - \left(\frac{dV}{dx}\right)_0^2 = \frac{4j}{\epsilon_0}\left[\frac{M(V_0 - V)}{2q}\right]^{1/2} \tag{7-3}$$

The term $\left(\dfrac{dV}{dx}\right)_0^2$ is the square of the electric field at the ion source, E_0. If the ions are emitted with negligible velocity, E_0 cannot be negative if any current is to be drawn. Rather, it may assume any value in the range

$$0 < E_0 < \frac{V_0}{x_a} \tag{7-4}$$

The upper limit is the pure electrostatic field which would prevail in the absence of any space charge and which would be constant over the one-dimensional gap. The lower limit is approached as the current is increased to some maximum value consistent with a monotonic voltage profile between the two boundary values. This case is said to be *space-charge limited* in the sense that here the accelerating field has just been neutralized at the source plane by all the distributed intervening charge (Fig. 7-2). In this space-charge limited case,

$$\frac{dV}{dx} = 2\left(\frac{j}{\epsilon_0}\right)^{1/2}\left[\frac{M(V_0 - V)}{2q}\right]^{1/4} \tag{7-5}$$

which can be integrated again by separating the variables:

$$V = V_0 - \left[\frac{3}{2}\left(\frac{j}{\epsilon_0}\right)^{1/2}\left(\frac{M}{2q}\right)^{1/4} x\right]^{4/3} \tag{7-6}$$

Insertion of $V = 0$ at $x = x_a$ yields the space-charge limited current density for this particular gap:

$$j = \frac{4\epsilon_0}{9}\left(\frac{2q}{M}\right)^{1/2}\frac{V_0^{3/2}}{x_a^2} \tag{7-7}$$

This relation is called Child's law, originally determined for electron current in a vacuum diode, and represents one fundamental limit on the current which can be drawn across a given plane gap by a given potential difference ([11–15]; Prob. 7-1).

The restriction on ion current density expressed by (7-7) implies a corresponding limit on the thrust density attainable by a given one-dimensional ion accelerator. From (7-1), (7-7), and the definitions of thrust and current,

$$\frac{T}{A} = \dot{m}v_a = \frac{jMv_a}{q} = \frac{8\epsilon_0}{9}\left(\frac{V_0}{x_a}\right)^2 \qquad (7\text{-}8)$$

where $\dot{m} = N_a M v_a$ is the mass flow rate per unit area. Note that for fixed electrode spacing and voltage, the thrust density is independent of the charge-to-mass ratio of the ions. The exhaust velocity does depend on this parameter, however:

$$v_a = \left(\frac{2qV_0}{M}\right)^{1/2} \qquad (7\text{-}9)$$

as does the power required per unit area:

$$\frac{P}{A} = \frac{Tv_a}{2A} = \frac{4\epsilon_0}{9}\left(\frac{2q}{M}\right)^{1/2}\frac{V_0^{5/2}}{x_a^2} \qquad (7\text{-}10)$$

Relations (7-8) to (7-10) provide some definition of the range of utility of the ion engine concept. For example, (7-8) emphasizes the sensitive dependence of engine performance on the electric field strength which can be sustained across the gap. In a hard vacuum, fields as high as 10^8 or 10^9 volts/m can be established between smooth isolated conductors. In an ion thruster, however, where sharp electrode structures must be mounted on insulators and are subjected to ion bombardment and ultraviolet radiation and surrounded by neutral propellant atoms and stray ions, the practical limit must be considerably lower, say, 10^7 volts/m at best. Observing from (7-5) that the maximum field in the gap occurs at the accelerating electrode:

$$E_a = -\left(\frac{dV}{dx}\right)_a = \frac{4}{3}\frac{V_0}{x_a} \qquad (7\text{-}11)$$

it follows that there will be the corresponding limit in thrust density (Prob. 7-2):

$$\frac{T}{A} = \tfrac{1}{2}\epsilon_0 E_a^2 \approx 440 \text{ newtons/m}^2 \approx 0.064 \text{ lb/in.}^2 \qquad (7\text{-}12)$$

In practice, this limit may need to be lowered because of the high voltages it demands for realistic gap spacings, x_a. Experience has shown that for a sufficiently sturdy electrode structure to survive long-term space operation without significant distortion, $x_a = 0.002$ m is probably the minimum value, and $x_a = 0.005$ m a far more comfortable spacing. From (7-11) the corresponding accelerating voltages for breakdown-limited performance are 15,000 and 37,500 volts, respectively. Aside from the possible power generation and transmission problems these voltages might entail, they are undesirable from the standpoint of the excessively high ion exhaust speeds they predicate for the available atomic ions.

For example, because of its combination of low ionization potential and high atomic mass, cesium is an attractive propellant for ion thrusters. With an atomic weight of 133, the charge-to-mass ratio of a cesium single ion is 7.24×10^5 coul/kg. Thus the exhaust speeds corresponding to 15 and 37.5 kv acceleration are about 1.5×10^5 m/sec and 2.3×10^5 m/sec, respectively. These values are considerably higher than optimum for most lunar and near-planetary missions, and hence would encumber a spacecraft with a needlessly large power plant (Sec. 1-4 and Prob. 7-3).

Three possibilities for reducing the exhaust speed suggest themselves:

1. Operate at lower voltages than the breakdown limit. For a given gap, this reduces the thrust density quadratically, while v_a falls off only as $V_0^{1/2}$.
2. Use ions of larger mass-to-charge ratio. A few other easily ionized atomic species, such as mercury, have slightly higher M/q, but to effect substantial improvement by this route one must employ charged heavy molecules or colloidal particles. Their use would provide lower exhaust speeds at a given voltage without penalty in thrust density, or alternatively, yield higher thrusts for given exhaust speed. This possibility is discussed in Sec. 7-8.
3. Install a third electrode to decelerate the ions somewhat before ejection, thereby lowering the exhaust speed without reducing the ion current density. This scheme has the additional very important benefit of restraining the neutralizing electrons from accelerating upstream to bombard the ion source. It is discussed in Sec. 7-5.

Regardless of the particulars of implementation, the foregoing one-dimensional considerations serve to identify the electrostatic thruster as a low thrust density, high exhaust speed device, requiring a corresponding low current high voltage power supply. Compared with an electrothermal thruster of the same beam power, for example, it will produce nearly an order of magnitude higher exhaust speed at correspondingly lower thrust and will draw two orders of magnitude less current at correspondingly higher voltage from the power plant. Its efficiency of conversion of electric power into thrust power can be very high, and is primarily limited by the processes of preparation of the ions, some examples of which are discussed in the following section.

7-2 PRODUCTION OF POSITIVE IONS

From the simple one-dimensional thruster model discussed in the previous section, three desirable specifications for practical ion sources can be assigned immediately:

1. The source should be capable of sustained emission of ions at current density levels corresponding to space-charge limited flow at the desired voltage over the given gap. For the extreme case cited above—15 kv across 0.002 m—the required level of emission, from (7-7), would be about 2,000 amp/m². In long-term operation, actual space thrusters will probably operate in a more conservative regime requiring between 200 and 500 amp/m² of beam current.
2. Whatever the ionization mechanism, a certain average amount of energy must be expended to create each ion at the source. This is by no means simply the ionization potential of the atom, but includes all thermal and radiative losses associated with maintaining the environment in which the ions are created. If the thruster is to be efficient, this average energy expenditure per ion created must be much less than the kinetic energy it derives from the accelerating field.

$$\frac{\mathcal{E}_i}{V_0} \ll 1 \qquad (7\text{-}13)$$

For multikilovolt accelerating voltages, several hundred electron volts per ion could thus be tolerated for $\mathcal{E}_i$.

3. The source must be highly selective in restricting its emission to ions rather than neutral particles. Small fractions of the latter, unaffected by the accelerating field, migrate relatively slowly through the gap and tend to disturb the ion trajectories. Specifically, ions from the source may suffer charge-exchange collisions with these ambient neutrals, thereby producing ions which are "out of focus" in the accelerating system and fast neutrals which are uncontrolled by the fields. Either of these may then strike the accelerating electrode, causing sputtering erosion and contamination of its surface. Even at the modest prevailing temperatures, an alkali-contaminated accelerator surface may emit electrons, which then return to the ion source, causing a current drain on the power supply,

with no corresponding useful thrust. From detailed consideration of such processes it becomes clear that the ratio of ion emission to neutral emission from the source should be better than 100:1.
4. The foregoing characteristics of adequate current density, low energy expenditure per ion, and low neutral emission must be maintained during continuous or intermittent operation over very long periods of time. Most missions for which electric propulsion may be considered involve flight times of the order of a year (10^4 hr) or more, over which time the thruster must operate, presumably unattended, without serious degradation in its performance. On such a time scale, even very slow processes of materials decay and erosion which would be unimportant to any laboratory ion source can become intolerable for a thruster.

By these standards, two distinctly different types of ion source have shown some promise of utility for ion engines. One, the *contact source*, invokes the tendency of alkali atoms to transfer their valence electrons to metal surfaces of high work function. The other, the *electron bombardment source*, utilizes the ions formed in low density electric discharges. These sources are sufficiently different in their operational requirements and in the nature of the ion beam they produce to influence strongly the design of other elements in the accelerator and details of the electric power supply needed to drive it. For this reason electrostatic engines are normally classified in terms of their source, e.g., cesium contact engines, mercury bombardment engines, etc.

SURFACE-CONTACT IONIZATION SOURCES

It was originally demonstrated by Langmuir and others in 1913 that a heated surface of a metal of high electronic work function ϕ immersed in a vapor of an alkali metal of ionization potential ε_i lower than ϕ would convert some of the alkali atoms to positive ions [16]. Many such combinations of metallic work functions and alkali ionization potentials exist, but the most serviceable for ion source purposes has proved to be a surface of tungsten ($\phi = 4.3$ to 4.6 ev, depending on the crystal surface) immersed in cesium vapor ($\varepsilon_i = 3.9$ ev) [17].

On an atomic scale, the process seems to proceed through the following sequence: A neutral alkali atom from the vapor strikes the metal surface and adheres to it by polarization in the surface field established by the conduction electrons within the metal. For positive values of $\phi - \varepsilon_i$, the alkali atom now has a strong probability of yielding its valence electron to the metal. Later, after some time dependent on the surface temperature, the residual ion leaves the surface and reenters the vapor. In the overall steady-state situation, then, there is a rate of arrival of neutral alkali atoms to the surface determined by the local vapor pressure, an equal rate of departure of alkali ions plus neutrals, in a ratio dependent on $\phi - \varepsilon_i$, and a density of transient alkali particles on the surface dependent on the surface temperature and rate of alkali influx.

The degree of steady-state surface coverage is highly important to the entire process, since very small levels of surface contamination seriously lower the work function, thereby lowering the electron-transfer probability. For example, a thoroughly cesiated tungsten surface may have a work function as low as 1.5 ev, far below the cesium ionization potential. For steady-state alkali coverage of only about 7 percent, the work function will be reduced below ε_i, and most of the atoms will leave the surface un-ionized. (Actually, neutral coverage in excess of 0.5 percent can provide a surface unstable to sudden transient increase in neutral coverage beyond the critical 7 percent condition.) This then predicates the need for getting the alkali particles off of the surface quickly after their ionization, i.e., a high surface temperature. For a given surface temperature T_m, there will be a maximum ion current density beyond which the emission will become seriously contaminated with neutrals. A typical dependence of ion current, or ionization efficiency, on surface temperature is shown in Fig. 7-3. For the cesium-

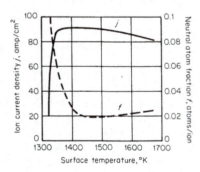

Fig. 7-3 Typical dependence of ion current density j and neutral fraction f on surface temperature. [From G. Kuskevics and B. L. Thompson, Comparison of Commercial, Sphercial Powder and Wire Bundle Tungsten Ionizers, AIAA Paper 63016, 1963, or W. R. Mickelsen and H. R. Kaufmann, Status of Electrostatic Thrustors for Space Propulsion, pp. 28 and 101 (fig. 36). NASA Tech. Note D-2172, May, 1964.]

tungsten combination, T_m occurs at about 1400°K for a current density of 500 amp/m², 1300°K for 100 amp/m², etc. The slight decrease in ion current at temperatures above T_m is a consequence of thermionic emission of electrons from the hot metal surface, which neutralizes some fraction of the ion emission. This same effect causes the ratio of emitted neutrals to ions to be higher at higher current density operation, since the surface must then be kept at higher temperature.

Operation at high current and high temperature also suffers larger thermal radiation losses from the incandescent metal surface, which in a typical contact thruster will have an unobstructed view out the exhaust port. This radiation loss is the largest contribution to the effective ionization energy defined earlier, and thus may be one of the most serious sources of inefficiency in thrusters using this type of source. Clearly, the designer must optimize the conflicting thermal requirements to minimize neutral emission and radiation losses at the desired ion current densities.

Implementation of this surface ionization process into a serviceable ion source for a particular thruster presents several practical difficulties. In addition to the obvious requirements for a system to store, vaporize, and transmit the corrosive alkali material to the ionizer surface, the geometry of the alkali vapor flow to and over the surface is critical in maintaining an adequately low neutral emission fraction. On the one hand, every entering vapor atom must be guaranteed at least one collision with the surface during its passage through the ionizer chamber. On the other hand, incoming vapor atoms cannot be allowed to obstruct the extraction of the ionized particles by the accelerating field. Many arrangements of ionizer surfaces and vapor flows have been studied, none of which performs flawlessly. One of the most popular involves the forced diffusion of cesium vapor through porous wafers or plugs of tungsten (Fig. 7-4; [18–31]).

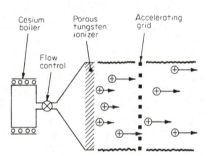

Fig. 7-4 Porous-tungsten, cesium-contact ionization source.

Porous tungsten plugs are normally prepared by forcibly bonding small tungsten particles at an elevated temperature (sintering), although other techniques, such as the juxtaposition of many fine wires, have been used. Typically, the pores may range from a fraction of a micron to several microns in aperture, with a distribution of as many as 10^6 cm^{-2}, an arrangement which represents a compromise between the desire to maximize the contact of the cesium flow with the pore walls and end surface and the need to retain a stable structure over long operation times. The physical properties of such fine-pored specimens differ somewhat from those of solid tungsten, as do the observed relations among the critical temperature, current density, and neutral fraction.

The flow of cesium through porous tungsten and the process of its ionization therein have received extensive experimental and theoretical study. Elaborate photomicrographic techniques have been developed for detailed observation of the emitted ion flow patterns from such sources. The results have provided valuable design criteria and some insight into the prevailing flow processes. In the theoretical studies it is usually presumed that the flow through the pores proceeds in two ways: a surface creep or diffusion layer and a "free-stream" flow, which in view of the low prevailing densities is essentially free-molecular (Fig. 7-5).

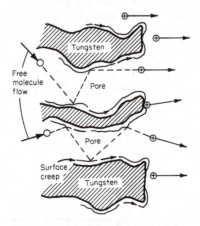

Fig. 7-5 Flow of cesium through porous-tungsten capillaries.

Although the equilibrium coverage of the pore walls with cesium may be quite high, there is evidence that the flow is at least partially ionized within these channels. The remainder of the ionization is presumed to occur on the front surface of the plug, near the pore mouth, where the creep flow must be of primary importance.

Such studies have established the need for closely spaced micron or submicron pores for low neutral emission at the high flow rates desired to optimize the ion flux relative to the radiation loss. In this way, neutral particle efflux has been reduced to less than 1 percent, and in some cases to less than 0.1 percent. Unfortunately, specimens of very small pore size tend to degrade in performance over long-term operation. At the elevated temperatures needed to reduce surface contamination, a further sintering of the tungsten granules occurs which closes some pores, widens others, and possibly opens relatively large fissures, which then emit high neutral fractions. The restraint of this process to tolerable levels for long-term operation seems to be primarily a materials problem, and has been the subject of considerable metallurgical research.

A porous-plug contact ionizer of the type described above clearly requires input power at a level sufficient to vaporize the cesium at the

ELECTROSTATIC ACCELERATION

desired rate of flow and to maintain the necessary surface temperature against the radiation loss and the energy carried away by the ion beam. Note that these needs could be met by a direct thermal input, without necessarily drawing electric power. By strict accounting, however, the radiation loss from the ionizer must be charged to the effective ionization energy, and, particularly for the lower exhaust speed range of operation, can substantially lower the overall efficiency of an ion thruster. Typically, a porous tungsten ionizer will expend some 500 to 700 ev for each ion created, which amounts to some 7 to 10 percent of the beam energy at 10,000 sec operation and 30 to 40 percent at 5,000 sec.

ELECTRON BOMBARDMENT SOURCES

Gaseous discharges of one sort or another have long been used as sources of ions for physical experiments, such as mass spectrometry and atomic beams, but the requirements on beam density, neutral fraction emission, and source lifetime have been far less stringent than for electrostatic thruster applications. Upgrading of such sources to propulsion utility is not a trivial task, and has been notably successful for only one type of discharge, the electron bombardment source [32,33]. This source, sketched in Fig. 7-6, essentially derives from the classical magnetron discharge. Electrons, emitted by a thermal filament at the center of the ionization chamber, are attracted toward a concentric cylindrical anode, but are prevented from reaching it directly by an externally applied weak axial magnetic field, which causes them to spiral axially back and forth in the chamber until they collide with a propellant atom. Depending on the relative elastic and inelastic cross sections of the propellant atoms, and the electron energy distribution, some fraction of these collisions produce electron-ion pairs, which themselves may participate in subsequent collisions. In the steady state, then, the chamber contains a mixture of ions, neutrals, and primary and secondary electrons. A strong electric field is established between a pair of perforated electrodes at one end of the chamber, which serves to extract preferentially any ions which wander close to the inner grid. This same field may provide the primary acceleration of the ion beam. The inner grid and opposite end wall of the chamber are kept at cathode potential to prevent axial loss of electrons from the discharge region.

The propellant for such a device is selected for its combination of low ionization potential, high atomic weight, and handling and storage properties. Mercury vapor (atomic weight 200) is the most common choice, although cesium vapor [34, 35] and several noble gases have also been used. Anode potential in the ionization chamber is established by the need to accelerate the electrons to the energy range where the ionization cross section is a maximum, say, 10 to 50 ev. The axial magnetic

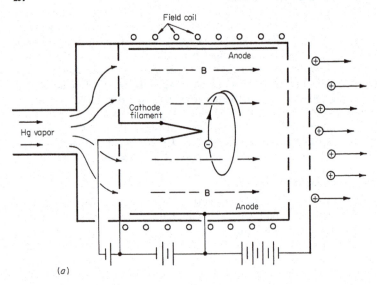

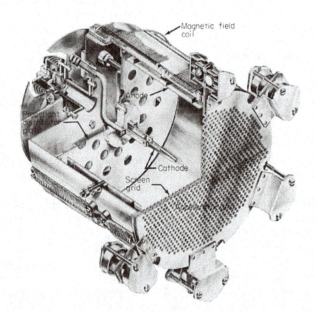

Fig. 7-6 Electron bombardment ion source. *(a)* Schematic; *(b)* cutaway drawing. *(NASA-Lewis Research Center, Cleveland, Ohio.)*

field is set to match the electron gyro radius to the chamber size, and is of the order of 10^{-3} weber/m² for a 0.10-m chamber and increases inversely with chamber diameter. A field of this magnitude can easily be supplied by a modest permanent magnet, as well as by an external solenoid; the former has the advantages of requiring no power and dissipating no heat, and it need not necessarily involve more total mass [36]. The potential across the extracting-accelerating electrodes is determined by the same space-charge flow considerations developed in Sec. 7-1, and is typically several thousand volts.

Electron bombardment sources expend about 450 to 600 ev per ion in ionizing about 80 percent of the injected mercury propellant vapor. They can be made to ionize 90 percent of the propellant flow at an expense of about 800 ev per ion by varying the discharge parameters somewhat. In addition to this high neutral-fraction efflux, they suffer somewhat from an unavoidable production of a small percentage of double ions, which tends to lower both the efficiency and thrust density of the accelerated ion beam.

The primary difficulty of the electron bombardment source, however, is erosion of the electron-emitting cathode. At the thermionic temperatures necessary to emit sufficient electron current to sustain the desired discharge, most pure metals have a significant sublimation rate. More important, the same field which accelerates the electrons outward drives the ions inward, and the latter, essentially unaffected by the magnetic field, focus on the cathode surface. At the potentials involved, serious sputtering damage can ensue, which gradually erodes the emitter surface. As an added detriment, contamination of the emitter surface by occluded propellant material may impair its electron emission and cause higher cathode-fall potentials, which in turn increase the sputtering damage, etc. [37].

The cathode sputtering and contamination problems can be avoided by various *autocathode* techniques. For example, if the cathode for a mercury-vapor discharge were simply a pool of liquid mercury, contamination and sputtering damage would clearly be irrelevant. Liquid mercury cathodes are commonly used in various electronic switches, such as ignitrons, but their application to bombardment ion thrusters clearly is complicated by the zero gravitational field inertial environment in which they must function. Various techniques for controlled bleeding of mercury onto a backing surface have been studied for this purpose [38]. In some versions, the liquid surface cathode can be made to supply not only the discharge electrons, but the propellant vapor as well.

The autocathode concept may be more simply implemented if cesium vapor is used instead of mercury for the propellant. In this case the cathode may be a cesiated metal, which again is insensitive to contamination by the propellant. Cesiated cathodes have the additional advantage of very low electronic work function, which reduces the cathode fall to less than 10 volts, thereby essentially eliminating all sputtering damage. Indeed, the work function is so low that once the discharge has been initiated, adequate electron emission can be sustained simply by the ion bombardment heating, without the need for a separate cathode-heating circuit [34]. A source of this sort also expends considerably less energy per ion created at comparable mass utilization fractions than the mercury discharge—perhaps 350 ev at 80 percent, 600 ev at 90 percent—but this advantage is roughly offset by the lower atomic weight of cesium, which reflects itself in correspondingly lower beam power density. That is, the energy to ionize a given mass of propellant is about the same for cesium and mercury.

A somewhat different approach to the cathode erosion problem utilizes the relatively low work functions of the alkaline earth carbonates, such as $BaCO_3$, $SrCO_3$, and $CaCO_3$, in an oxide-coated structure. Although such compounds are not adequately conducting in the pure state, satisfactory cathodes can be constructed by imbedding generous amounts of these materials in an inert metal matrix [39]. These hybrid cathodes then emit the required electron current densities at substantially lower temperatures than the pure metals, and thereby reduce the radiative heat loss. It is desirable to make such oxide-coated cathodes quite massive to withstand the long-term ion bombardment erosion, yet the composite structure must be sufficiently sturdy to withstand the dynamic and thermal stresses of the space operation.

Depending on the details of its design, then, the electron bombardment source is seen to have terminal power requirements that are substantially different from those of the surface-contact source. There is again a need for thermal input to vaporize the propellant, but there is less radiation loss to be compensated. Low voltage high current power is required for the cathode filament (except for cesiated cathodes, which need only a startup heater), a similar supply for the magnetic field windings (unless a permanent magnet is used), and another to sustain the actual discharge. It is interesting that, despite the dissimilarities in physical process and power requirements, these two sources are roughly competitive in the ion beams they produce, from an overall standpoint of current densities, mass utilization, energy per ion, and adaptability to a thruster. Table 7-1 compares typical operating performance of contact and bombardment ion sources.

OTHER DISCHARGE ION SOURCES

A modification of another classical discharge, the *Penning tube*, has also been employed as an ion source for an electrostatic thruster [40–43].

ELECTROSTATIC ACCELERATION

Table 7-1 Typical performance characteristics of contact and bombardment ion sources

	Cesium-tungsten contact source	Electron bombardment source	
		Mercury	Cesium
Maximum current density, amp/cm^2	0.010–0.020	0.005–0.010	0.005–0.010
Ion-creation energy, ev/ion	500–700	450–600	300–400
Neutral fraction	0.001–0.01	0.01–0.02	0.01
Lifetime, years	0.5–3	1–3	1–3
Handling and storage	Highly corrosive	Less corrosive, less volume	Highly corrosive

SOURCES: G. R. Brewer, Current Status of Ion Engine Development, AIAA 2d Propulsion Joint Specialist Conference, Colorado Springs, Colo., *AIAA Paper 66-564*, June 13–17, 1966; and W. R. Mickelsen and H. R. Kaufman, Status of Electrostatic Thrustors for Space Propulsion, *NASA Tech. Note* D-2172, May, 1964.

This device involves a tubular anode and two cathodes, one an electron-emitting filament, the other a tube, placed on opposite sides of the anode along a common axis (Fig. 7-7). Electrons, emitted by the filament

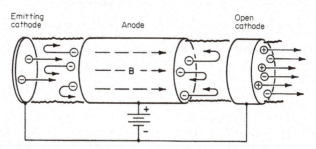

Fig. 7-7 Oscillating electron discharge ion source (schematic).

cathode, oscillate back and forth in the potential well established by the electrode configuration, and are constrained from radial migration by an imposed axial magnetic field. Propellant ions are generated by electron collision, and are extracted through the tubular cathode, which can also serve as the primary accelerating electrode. Electrode erosion is found to be quite low, and under some circumstances of operation it appears that enough electrons leak through the tubular cathode to neutralize the ion beam automatically, without the necessity for a special neutralizer electrode. This property, like the overall source effectiveness, however, seems to be sensitively coupled to the testing environment, and the overall efficiency of conversion of electrical energy input to beam energy may not be as high as the electron bombardment thruster in actual space operation.

Higher ion current densities can be achieved using a magnetically confined low pressure arc of the von Ardenne type ([44–48]; Fig. 7-8).

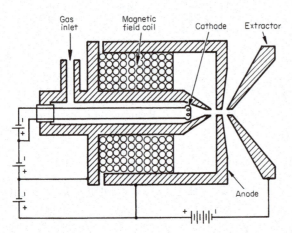

Fig. 7-8 Duoplasmatron arc ion source. *(From M. von Ardenne, Über ein Ionenquellen-System mit Messenmonochromator für Neutronengeneratoren, Phys. Z., ser. 2, vol. 42, p. 91, 1942.)*

The tubular anode and axial magnetic field create an intense column of plasma from which ions can be extracted through a second cathode, much like the Penning discharge. Ion fluxes in excess of 2,000 amp/m² have been obtained in this way, with extremely low neutral emission, but the energy expenditure per ion is substantially higher than that of the electron bombardment source, and sputtering erosion is substantial. Actually, the level of current density inherently required for optimum operation of any arc source may be too high for the other components of an electrostatic thruster to handle over long periods of time.

Finally, one might consider a high-frequency inductive discharge for preparation of the needed ions. Such discharges are well known for their efficiency and cleanliness in ionizing a gas, since they impart high energies to the electrons without the electrode heating and erosion problems of the dc discharges. Unfortunately, the same drawbacks of rf space power-supply weight and complexity and inefficient coupling to the discharge load mentioned earlier (Sec. 6-7) have suppressed research in this area. What work there has been indicates that rf discharges, confined in external magnetic fields, are at least comparable with the other schemes in ionization efficiency and energy per ion considerations, once the power generation and application difficulties have been allowed. It is possible that the coupling of rf fields to a Penning discharge may combine the advantages of a low electrode erosion with a high current density capability [49].

7-3 DESIGN OF THE ACCELERATING FIELD

In the matter of the design of its electrodes to provide the desired accelerating field, the ion engine presents a well-defined boundary value problem in classical electrodynamics. Namely, what should be the geometry of the source surface and accelerating electrodes in order that an ion beam of given current density and cross section be accelerated to a given exit velocity with optimum uniformity and with a minimum of impingement on the electrode structure? The latter condition is clearly critical to the lifetime of the engine, for at the ion energies obtained in the accelerating fields, sputtering erosion upon ion impact on the electrodes is high. Beam impingement in excess of one part in 10^5 probably invalidates the device as a space thruster.

Considerable theoretical and experimental experience in the techniques of electrode design was acquired in connection with the acceleration and focusing of electron beams for vacuum tubes and cathode-ray oscilloscopes [50]. Many of these concepts and results may be directly transcribed to the ion beam problem. For example, all the analytic techniques for electrode design are based upon appropriate simultaneous solution of statements of conservation of ion flux, Newton's law, and Poisson's relation over the relevant domains:

$$\nabla \cdot (n\mathbf{v}) = 0 \tag{7-14}$$

$$(\mathbf{v} \cdot \nabla)\mathbf{v} = -\frac{q}{M}\nabla V \tag{7-15}$$

$$\nabla^2 V = -\frac{nq}{\epsilon_0} \tag{7-16}$$

For given beam dimension at the source and given total potential drop along the beam (i.e., desired exhaust velocity), one searches for equipotential surfaces in the region exterior to the beam, of suitable magnitude and position so that physical electrodes may there be placed of a shape corresponding as closely as possible to the theoretical contours and charged to the indicated potentials. It is then presumed that this real electrode structure will extract from the source an ion beam closely resembling that derived in the ideal analytical model.

As one example of the technique, consider an ion beam of uniform density j, emitted from a flat-strip source, very much broader in one dimension, z, than in the other, y. Let us ask what electrode structure is needed to accelerate this beam to a desired velocity in the x direction while retaining it parallel and of uniform density at any cross section (Fig. 7-9). Inside the beam we may immediately require $\partial V/\partial z = 0$

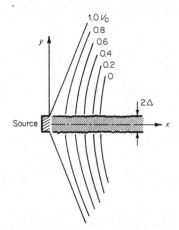

Fig. 7-9 Equipotential contours for one-dimensional ion beam acceleration. *(From J. R. Pierce, "Theory and Design of Electron Beams," D. Van Nostrand Company, Inc., Princeton, N.J., 1954.)*

because of the large z dimension, and $\partial V/\partial y = 0$ because the ion trajectories are to be rectilinear. Therefore Poisson's equation reduces to the readily integrable one-dimensional form of (7-2), which yields the space-charge limited profile of (7-6),

$$V = V_0 - \left(\frac{9j}{4\epsilon_0}\right)^{2/3}\left(\frac{M}{2q}\right)^{1/3} x^{4/3} \qquad (7\text{-}17)$$

Outside the beam we can no longer require $\partial V/\partial y = 0$, but here there is no free charge, and (7-16) reduces to Laplace's equation. In addition, (7-17) provides boundary conditions along the planes $y = \pm \Delta$, where Δ is the beam half-width, which in conjunction with the requirement of continuity of $\partial V/\partial y$ there adequately determines the problem.

In this case an analytic solution for the exterior region can be obtained by utilizing the properties of complex analytic functions in Laplace's equation. Namely, given (7-17) as the solution of $\nabla^2 V = 0$ along the plane $y = \Delta$, the solution over the entire positive exterior domain is generated by the replacement of x in (7-17) by $x + i(y - \Delta)$ and extraction of its real part:

$$x^4 - 6x^2(y-\Delta)^2 + (y-\Delta)^4 = \left(\frac{4\epsilon_0}{9j}\right)^2 \frac{2q}{M}(V_0 - V)^3 \qquad (7\text{-}18)$$

ELECTROSTATIC ACCELERATION

The contours of equipotential described by (7-18) and their images in the lower domain are sketched in Fig. 7-9. Note that each contour is normal to the beam edge, with the exception of the source contours V_0, which have a singularity there. Even many beam widths from the centerline, the equipotentials are predominantly normal to the beam edge.

The orientation of these equipotentials makes this particular beam profile difficult to utilize in an ion thruster. To simulate adequately the potential conditions at the beam edge with material electrodes would require several closely spaced conducting surfaces, or a lesser number extending many beam widths on each side of the beam. The former situation predicates an excessively complicated electrode structure; the latter would demand a thruster cross section many times larger than the beam area. To improve the ratio of usable source area to total thruster cross section, it is advantageous to focus the beam somewhat as it passes through the accelerating electrode (Fig. 7-10). A small y dependence

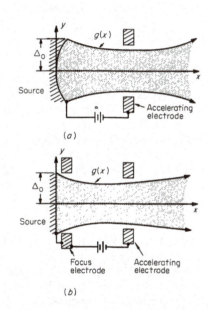

Fig. 7-10 Mildly focused ion beams using (a) contoured emitter, (b) auxiliary focusing electrode (schematic).

of the ion trajectories of this sort can be adequately analyzed by expansion and similarity methods. As a second example, consider such a mildly focused beam, inside of which the potential may be represented [51]:

$$V(x,y) = V(x,0) + y\left(\frac{\partial V}{\partial y}\right)_0 + \frac{y^2}{2}\left(\frac{\partial^2 V}{\partial y^2}\right)_0 + \cdots \qquad (7\text{-}19)$$

where the first-order term is usually zero from symmetry of the beam about its axis. The problem may then be cast into similarity form by nondimensionalizing in terms of the beam half-width at the source, Δ_0,

$$X = \frac{x}{\Delta_0} \qquad Y = \frac{y}{\Delta_0} \tag{7-20}$$

and requiring

$$Y = Y_0 g(X) \tag{7-21}$$

where Y_0 is the transverse coordinate at the source $X = 0$, and $g(X)$ are the boundary trajectories of the beam. If the x dependence of the potential is assumed gentle enough that $(dg/dX)^2 \ll 1$, then

$$V(X,Y) = V^0 + \frac{Y^2}{2g}\left(\frac{dg}{dX}\frac{dV^0}{dX} + 2\frac{d^2g}{dX^2}V^0\right) \tag{7-22}$$

where $V^0 \equiv V(X,0)$ satisfies the ordinary differential equation

$$g\frac{d^2V^0}{dX^2} + \frac{dg}{dX}\frac{dV^0}{dX} + 2\frac{d^2g}{dX^2}V^0 = \left(\frac{8\epsilon_0^2 q}{j^2 M}\right)^{1/6}\left(\frac{1}{V^0}\right)^{1/2} \tag{7-23}$$

In this form, one may assume a reasonable potential profile along the beam axis $V^0(X)$ and solve (7-23) for the outer ion trajectory $g(X)$. If this beam shape is acceptable, Eq. (7-22) and its derivative will then provide the potential variation and its normal derivative along $g(X)$, which are the required boundary conditions for solution of Laplace's equation for the potential outside the beam. Solutions of this sort can be found which permit simpler implementation with realistic electrode structures. In particular, several parallel beams of this type can be produced by interspersed electrode elements of relatively small dimensions (Fig. 7-11).

Application of these analytic techniques to beams of two-dimensional cross section, such as square or circular beams, involves considerably greater mathematical complexity, although certain special cases, such as the cylindrical beam, can be solved by asymptotic methods and judicious interpolation. The extent to which these solutions, like those discussed above, actually represent the beam produced by electrodes placed along the indicated equipotentials depends on many factors. First, actual electrode structures cannot have zero thickness or extend to infinity as the theoretical equipotentials do. Also, the beam aperture in the electrodes allows a distortion of the axial potential gradient and a divergence of the beam not included in the theoretical model. Small departures of the actual ion source from the assumptions of a uniform emitter of zero-velocity ions can have significant effects on the downstream behavior of the beam. To check on these uncertainties, as well as to indicate solutions to more difficult beam geometries and parallel beam arrays, detailed

ELECTROSTATIC ACCELERATION

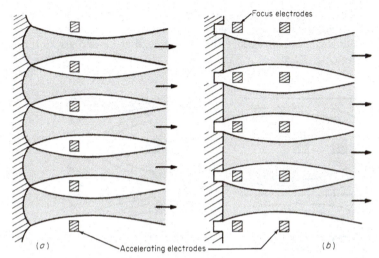

Fig. 7-11 Multibeam ion accelerator using (a) contoured emitter, (b) auxiliary focusing electrode (schematic).

computer programs have been developed which iterate through the space-charge densities, space-charge fields, ion trajectories, etc., for realistic electrode configurations [52]. In addition, a variety of analog methods, such as resistance network simulators, electrolytic plotting tanks, and rubber-sheet analogs, have also been borrowed from earlier work on electron-gun design [53,54].

In practice, it has been found necessary to compromise the optimized electrode designs suggested by the academic procedures outlined above, with certain concessions to the ease of fabrication and maintenance of the structure. Many of the ideal analytic solutions predicate electrode arrays too complex in contour or too delicate in form to be implemented in an operational thruster. A more practical approach is to search, analytically or experimentally, through families of easily fabricated electrode geometries for those which yield satisfactory, albeit imperfect, beam geometries. Many of the contact ion thrusters, for example, employ simple pierced plate, parallel strip, or hexagonal honeycomb-type accelerating electrodes (Figs. 7-20 to 7-27). Some also provide a concave *Sastrugi* contouring of the emitter surface to aid the focusing [55].

The electron bombardment engine also can use very simple electrodes, such as plates pierced with cylindrical holes or arrays of parallel wires. In these devices, the plasma in the ionization chamber is found fortuitously to assume a concave sheath profile near the screen-accelerator extracting electrode which aids in focusing the ion beams through the electrode apertures (Fig. 7-12a; [56]). This happy property prevails only for certain levels of ion current density. At higher currents, the sheath profile becomes convex and the ion beams diverge and strike the electrodes (Fig. 7-12b).

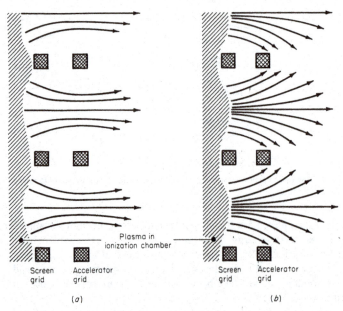

Fig. 7-12 Ion beam focusing in electron bombardment accelerator. (a) Concave plasma sheath, proper focusing; (b) convex plasma sheath, misfocusing.

Two empirical ratios of some utility in comparing the characteristics of various ion accelerators are the *aspect ratio* and *perveance*. The former may be defined by rewriting the expression for the thrust of a space-charge limited accelerator in terms of the diameter D of a circle of area equal to that of the ion beam cross section A, whatever its shape [Eq. (7-8)].

$$T = \frac{8\epsilon_0}{9}\left(\frac{V_0}{x_a}\right)^2 A = \frac{2\pi\epsilon_0}{9} V_0^2 \left(\frac{D}{x_a}\right)^2 \qquad (7\text{-}24)$$

The ratio of this effective beam diameter to the interelectrode spacing is called the aspect ratio,

$$\mathcal{R} = \frac{D}{x_a} \qquad (7\text{-}25)$$

ELECTROSTATIC ACCELERATION

which uniquely relates the thrust to the square of the applied voltage.

This aspect ratio, in turn, is related simply to the perveance, another concept borrowed from vacuum tube technology, which expresses total current in terms of applied voltage:

$$\mathcal{P} = \frac{J}{V_0^{3/2}} \qquad (7\text{-}26)$$

In an ion accelerator operating at space-charge limited current,

$$\mathcal{P} = \frac{jA}{(V_0)^{3/2}} = \frac{4\epsilon_0}{9} A \left(\frac{2q}{M}\right)^{1/2} \left(\frac{1}{x_a}\right)^2 = \frac{\pi\epsilon_0}{9}\left(\frac{2q}{M}\right)^{1/2} \mathcal{R}^2 \qquad (7\text{-}27)$$

Note that the perveance is a purely electrical ratio whose value depends on the charge-to-mass ratio of the current carriers, as well as on the field configuration, and is definable for any accelerator system, whereas the aspect ratio concept is restricted to a plane diode. It remains useful, however, for more complex systems in defining an "equivalent plane diode," for a known perveance. For example, it follows directly from the above relations that $\mathcal{R} \gg 1$ is a requirement for useful thrust levels at the desired exhaust speeds. From more detailed two- and three-dimensional space-charge flow analyses, however, it becomes apparent that $\mathcal{R} \lesssim 1$ is a practical limit for any one ion beam accelerated by an electrode completely external to it. Beyond this value, radial gradients in the potential associated with the finite electrode apertures badly distort the flow. To produce useful thrust levels, therefore, an ion thruster must comprise many parallel beams, each of width of the order of the interelectrode spacing.

7-4 NEUTRALIZATION OF THE ION BEAM

To produce useful levels of thrust, an ion engine must emit many amperes of positive ion current. Yet the total electrical capacitance of a typical ion-propelled spacecraft will probably not exceed 10^{-9} farad (Sec. 2-1); hence, if no provision for neutralization of the ion beam were made, the craft would acquire a negative potential at a rate

$$\frac{dV}{dt} = \frac{1}{C}\frac{dQ_s}{dt} = \frac{J}{C} \approx 10^9 \text{ volts/(sec)(amp of ion current)} \qquad (7\text{-}28)$$

This gross charging could be inhibited by the emission of an identical electron current from any convenient location on the craft. Somewhat more subtle, however, is the need for detailed neutralization of the ion beam itself before it gets very far from the exit electrode, lest the positive space-charge potentials within the beam cause it to stall or reflect upon itself. The essence of this problem can be illustrated by returning to a one-dimensional model, which again involves several obvious idealizations.

Consider a system whose source plane is at potential V_0 and whose accelerator is at zero potential and at a separation x_a from the source. Assume that by some means the beam has been neutralized everywhere beyond a third plane, Δx from the accelerator, so that the potential is also zero there, and at all larger x (Fig. 7-13). If the ion current is space-

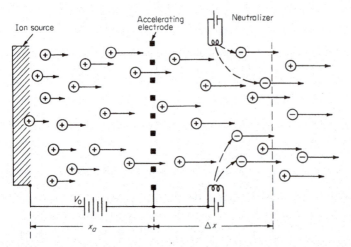

Fig. 7-13 Neutralization of one-dimensional ion beam.

charge limited over the accelerating gap x_a, its density is fixed by Child's relation (7-7). The potential profile in the region between accelerator and neutralizer is then determined in accordance with Poisson's equation by this current density, and by the gap size, or for a particular V_0, by the ratio of gap sizes, $\Delta x/x_a$. The problem is complicated by the mathematical admissibility of negative-flowing as well as positive-flowing ion currents, which introduce a duality, or hysteresis, in the potential profile solutions. Physically, this can arise only if the potential in the second gap rises to a value of V_0 at some intermediate point, whereby some fraction of the incident beam may be reflected. Indeed, this is the basic beam-stalling condition. Because of this hysteresis, the results of the analysis must be presented in a self-consistent sequence [1,57].

If $\Delta x/x_a$ is imagined to be increased gradually from zero, $V(x)$ is found to be symmetrical about $x_a + \Delta x/2$, where it attains a maximum value $\hat{V}$, whose magnitude increases monotonically with Δx (Fig. 7-14a).

ELECTROSTATIC ACCELERATION

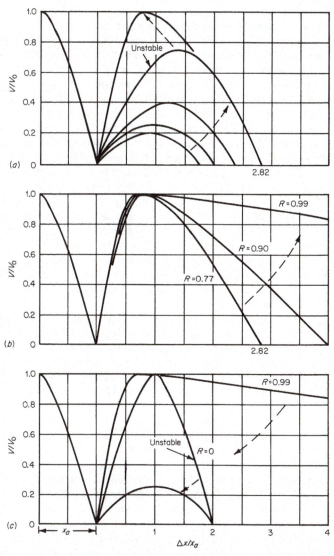

Fig. 7-14 Potential profiles for one-dimensional ion beam neutralized a distance Δx from accelerator grid. (a) Δx increased from 0 to $2.82x_a$, no beam reflection; (b) Δx increased from $2.82x_a$ to ∞, increasing beam reflection; (c) Δx reduced from ∞ to $2.00x_a$, decreasing beam reflection.

At $\Delta x/x_a = 2.82$, $\hat{V}$ reaches the value $\frac{3}{4}V_0$, beyond which point no solution involving only positive-flowing ion current exists. If negative-flowing current is admitted, a new, asymmetric branch of solutions is available, each of which displays a maximum of $\hat{V} = V_0$, at which some fraction R of the beam is reflected and returns to the source (where it is presumed to be absorbed), thereby contributing twice to the space charge in the intervening region. For the position $\Delta x/x_a = 2.82$, where the simple symmetric solutions cease, the maximum $\hat{V}$ appears at $x = 1.75x_a$ and reflects 0.77 of the positive-flowing beam. Physically, this jump in solutions must correspond to the beam-stalling condition we wish to avoid. Experimentally, the beam is found to develop oscillations at this condition [58]; these oscillations have been derived theoretically by more elaborate dynamical calculational treatments [59,60].

As $\Delta x/x_a$ is now increased from 2.82 to ∞, the asymmetrical profile is retained, but the fraction of the beam reflected increases from 0.77 to 1.00, and the maximum regresses slightly to position $1.707x_a$ (Fig. 7.14b). If $\Delta x/x_a'$ is now decreased, the discontinuous jump in profile is not simply reversed. Rather, the maximum $\hat{V}$ retains the value V_0, but shifts its relative position to the right until, at $\Delta x/x_a = 2$, it resides at $x = x_a + \Delta x/2$ again, and the fraction of the beam reflected is zero (Fig. 7.14c). From here the profile jumps discontinuously to a symmetric one of maximum $\hat{V} = \frac{1}{4}V_0$. For all smaller $\Delta x/x_a$, the solutions retrace those of the first symmetric branch.

The physical conclusion to this highly idealized problem is thus that the beam must be neutralized at least within a distance of $2.82x_a$ after it leaves the accelerator, and even then it must be protected against starting fluctuations in current or voltage that might drive it irreversibly beyond the critical value, into the asymmetric reflecting situations. To be perfectly safe, the neutralization would have to be accomplished within $2.00x_a$ of the exit.

More elaborate three-dimensional calculations are needed to derive the corresponding neutralization requirements for ion beams of finite cross sections, such as pencil or strip beams. These are found to be considerably less severe, since finite beams have the freedom to expand their cross section in response to the space-charge forces. However, if such beams are to be ganged in parallel for larger thrusts, their mutual interaction inhibits much expansion, and the one-dimensional criteria regain some relevance.

The requirement to neutralize the ion beam within a few accelerator gap widths does not necessarily demand placement of the neutralizing electrode within this distance. From a more general analysis than that attempted above, it can be shown that although the electron emitter is placed farther downstream, some electrons tend to migrate upstream, and thus achieve neutralization somewhat ahead of the actual electron source [15]. This process is particularly significant in self-decelerating ion beams, discussed in Sec. 7-5.

Conceptually, the simplest method of beam neutralization would be to inject a stream of electrons into the ion stream closely behind the exit electrode, such that the electrons had the same density and streaming velocity distribution as the ions. In this way the ion and electron currents would be equal, and preservation of both the spacecraft potential and the charge neutrality of the beam would be guaranteed. Unfortunately, the typical thermionic emitter liberates electrons with mean velocities much higher than that of the ion beams of interest. To reduce the mean velocity of the emitted electrons to the desired level would require a very cool filament whose emission current density would be intolerably low. In addition, whereas the ion engine emits a collimated beam of ions with relatively little spread in velocity in comparison with their directed motion, the usual thermionic electron emitters supply an essentially isotropic, broad distribution of electron velocities. One thus faces the choice of constructing much more elaborate sources to prepare monoenergetic low velocity electron beams, or of injecting the electrons with a broad velocity distribution and hoping for a rapid automatic adjustment within the mixed beam. Estimation of the effectiveness of such internal adjustment is a challenging theoretical problem.

The mixing of the injected electrons with the ion stream may proceed through a variety of physical mechanisms, and theoretical analysis of the mixing process must proceed at a level of sophistication appropriate to the number and complexity of these mechanisms that are admitted. First, there are obvious electrostatic space-charge forces tending to enhance mixing on a macroscopic scale. These include not only the axial potential gradients discussed in the one-dimensional models above but, for ion beams of finite cross section, transverse electric fields commensurate with the transverse space-charge profiles which tend to reflect and scatter electrons within the beam. On an atomic scale, two-body elastic collisions between the electrons and ions within the beam would also tend to mix them, but at the beam densities permitted by space-charge limitations in the accelerating gap, these are quite rare (Prob. 7-5). Finally, on an intermediate scale, certain collective effects of the electron-ion plasma, such as two-stream instabilities, may arise to hasten the mixing process.

In the simplest one-dimensional formulation, one can impose a monoenergetically injected electron stream on a uniform background stream of ions and require a simultaneous satisfaction of Poisson's equation and conservation of current flux and total energy, in the absence of any small-scale interactions. The typical result is the appearance of time-independent striations of alternate positive and negative space charge of amplitude and wavelength dependent on the speeds of the two beams at injection (Fig. 7-15; [61]). A similar exercise can be performed

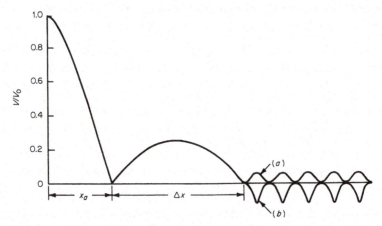

Fig. 7-15 Potential oscillations in one-dimensional ion beam neutralized by monoenergetic longitudinal electron injection (schematic not to scale). *(a)* Electron-injection speed less than ion speed; *(b)* electron-injection speed larger than ion speed.

in a two-dimensional framework, allowing the injected electron beam to have some transverse component of initial velocity [62]. The results are more complex, but again display patterns of interior space charge sufficiently severe to disturb the beam.

It might be thought that these space-charge striations are peculiar to monoenergetic electron injection and would tend to wash out when more realistic electron distributions were allowed. To repeat the calculation allowing a distribution of electron speeds at injection representative of a real thermal emitter, the simple energy statement must be replaced by an appropriate Boltzmann equation. If the problem is still regarded as collisionless, this reduces to the time-independent Vlasov relation for the electrons:

$$v \frac{\partial f}{\partial x} - \frac{q}{m} \frac{dV}{dx} \frac{\partial f}{\partial v} = 0 \qquad (7\text{-}29)$$

where $f(x,v)$ is the electron distribution function. Numerical solution of this relation, along with the appropriate Poisson equation, again reveals the possibility of various modal structures of space-charge striations, which, if physically realized, would be intolerable from the standpoint of successful neutralization of a real exhaust beam [63].

Considerable encouragement can be gained from the admission of collective plasma interactions into the problem, however. A conventional linear stability analysis of the time-independent solutions quoted above reveals that the space-charge striations found there are violently unstable on a time scale short compared with the relevant times of the macroscopic problem [64]. The implication is that the collective plasma processes will act rapidly to mix the electron and ion beams.

Further amelioration of the problem is suggested by a more rigorous treatment of the internal reflection of the electrons in the beam at its transverse boundaries. In consequence of the finite thermal energy of these electrons, a sheath is formed at the beam edge of thickness of the order of the Debye length. Internal reflections at this sheath can be shown to be quite effective in randomizing the electron motion. The composite image is that of a "plasma bottle," whose walls are the sheath, within which the injected electrons are trapped [65].

Depending on the dominant physical mixing processes which actually pertain, some choice of the practical means for supplying the neutralizing electrons must be made. Four methods have been used, each of which has certain disadvantages. The electron emitter can be placed exterior to the beam, and the electrons drawn into the beam by a potential difference established between the beam and emitter (Fig. 7-16a). This method succeeds only if the effective perveance of the arrangement matches that of the ion beam, i.e., if sufficient electron current is drawn across the emitter-beam gap at the applied potential. This potential ideally should be kept below 20 volts in order that any ions attracted to the emitter not acquire sufficient energy to cause significant sputtering damage. For ion beams of low peripheral densities, this neutralizer perveance condition may not be readily satisfied below 50 or 100 volts negative bias, however, and neutralizer filament lifetime may be a problem.

The electron emission at a given neutralizer bias can be substantially improved by immersing the source in a denser portion of the ion beam (Fig. 7-16b). Now, however, the emitter is subject to direct bombardment by the main ion beam, and either must itself be extremely rugged in construction to withstand the sputtering erosion over the desired lifetime of the engine, or must be protected by a passive shield which is.

As an alternative, the source can be in the form of an electron gun, which independently accelerates the electrons to a desirable speed at the desired density prior to injection (Fig. 7-16c). The additional complexities on engine structure and power supply requirements which this technique predicates tend to overbalance any slight physical advantages it has.

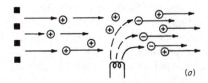

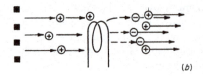

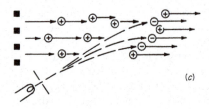

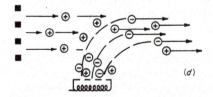

Fig. 7-16 Ion beam neutralization techniques. *(a)* Thermionic emitter adjacent to beam; *(b)* thermionic emitter immersed in beam; *(c)* electron-gun injector; *(d)* plasma-bridge method.

Finally, a "plasma-bridge" technique may be employed, wherein the electron source, like the autocathodes discussed earlier, simultaneously vaporizes atoms of the propellant used, e.g., a cesiated tungsten or mercury-film emitter. These atoms are ionized by the emitted electrons, and a plasma bridge is formed between the emitter and ion beam (Fig. 7-16d). Like the autocathodes in bombardment ionization chambers, little sheath voltage is required (≈ 10 to 20 volts), and emitter erosion is negligible.

ELECTROSTATIC ACCELERATION

In each of these techniques, subtle variations in geometry, physical location, operating potential, etc., are possible, and may or may not be significant, depending on the vigor of the internal mixing processes within the beam.

The approximations necessary in the various analyses, the ambiguities in their solutions, and the variety of possible emitter characteristics strongly urge experimental observation of the actual neutralization behavior. Unfortunately, ground-based studies in even the largest available vacuum tank facilities are also somewhat inconclusive because of the unavoidable interaction of the exhaust beam with the tank walls. For example, ion beams for which no neutralization provision has been made at the engine are found to neutralize themselves immediately after their initial impingement on the far wall of the vacuum tank, presumably by upstream migration of the electrons emitted there under the ion bombardment. Hence little information about the effectiveness of a particular thruster neutralizer in space operation can be obtained in such a tank environment. One possible exception is a notable sequence of transient beam experiments in which the neutralization process was studied within the short interval before any part of the ion beam contacted the tank walls [66]. These seemed to indicate that neutralization could be achieved with relatively simple electron-emitter configurations, but the extrapolation of this behavior to steady-state operation in space is not entirely convincing.

The myriad of conflicting indirect indications outlined above regarding this fundamentally important process placed a high premium on an early space flight test which would allow unambiguous study of the neutralization function. In the latter half of 1964, such a test was accomplished, which demonstrated successful steady-state neutralization with relatively simple electron-emitter arrangements (Sec. 7-7). Thus this specter of beam stalling which had haunted the ion engine concept since its origin was apparently laid to rest, and much of the extensive earlier work outlined above abruptly became academic. Its tutorial value is substantial, however, and not without side applications.

7-5 THE ACCELERATION-DECELERATION CONCEPT

Throughout the foregoing discussion of beam neutralization, the possibility of the injected electrons migrating upstream past the accelerating electrode was ignored. Such migration cannot be tolerated, for once beyond this electrode, the electrons would be vigorously accelerated toward the ion source plane by the same electric field which drives the ions downstream. This electron flux would constitute a current drain on the power supply, with no corresponding thrust power; it would distort the potential profile in the acceleration gap from the simple ion space-charge configuration; and it could damage the ion source and disturb the emission process. To preclude upstream electron migration into the acceleration gap, a region of increasing potential aft of the accelerator seems indicated (Fig. 7-17).

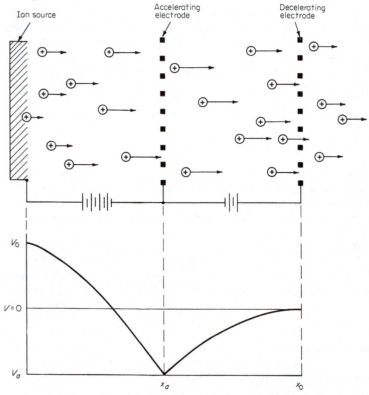

Fig. 7-17 One-dimensional ion accelerator with decelerating stage (schematic).

It was mentioned in Sec. 7-1 that the addition of a decelerating stage to the axial potential profile could also provide a means of reducing ion exhaust speeds without a corresponding loss in space-charge current, and thereby preserve higher thrust densities at lower specific impulse levels. Actually, these two functions—restraint of upstream electron migration and control of ion exhaust speed—can be implemented by the same cusped potential profile, commonly called an *accel-decel* system.

ELECTROSTATIC ACCELERATION

Returning once again to a simple one-dimensional configuration, imagine a source plane at potential V_0, an accelerating grid of potential $-V_a$ at position x_a, and a decelerating grid of potential zero at position x_0. Space-charge limited ion current is drawn from the source at a density determined by the full accelerating field,

$$j = \frac{4\epsilon_0}{9}\left(\frac{2q}{M}\right)^{1/2}\frac{(V_0 + V_a)^{3/2}}{x_a^2} \tag{7-30}$$

but is then decelerated to an exit velocity determined by the net potential drop V_0:

$$v(x_0) = \left(\frac{2qV_0}{M}\right)^{1/2} \tag{7-31}$$

The thrust thus produced is the hybrid product of the mass flow rate extracted by the accelerating field and the exhaust speed set by the net field:

$$\frac{T}{A} = \frac{jMv(x_0)}{q} = \frac{8\epsilon_0}{9}\frac{(V_0 + V_a)^{3/2}(V_0)^{1/2}}{x_a^2} \tag{7-32}$$

As such, it is less than that produced by monotonic space-charge acceleration through the full potential, $V_0 + V_a$, but greater than that corresponding to monotonic acceleration through V_0. The beam power required at a given level of thrust is clearly less for the accel-decel system (Prob. 7-6).

The position and potential of the decelerating electrode are not completely arbitrary. Presumably, a minimum separation from the accelerating electrode is established by the same fabrication tolerances which govern the minimum separation of the latter from the source plane. In addition, the potential difference between accelerating and decelerating electrode cannot exceed that corresponding to the space-charge limit for the given ion current flow. From one-dimensional analysis much like that used in neutralization discussion, it can be shown that if the three electrodes are equally spaced ($x_0 = 2x_a$), V_0 may have any positive value; i.e., the beam may be reduced to arbitrarily low exit velocity. For larger deceleration gaps, however, V_0 has a minimum positive value for full beam transmission, which increases rapidly as x_0 approaches $3.283x_a$. Beyond this separation, no beam deceleration is possible without stalling. As in the neutralization problem, these larger gap profiles are unstable to small perturbations in beam current, and may revert to profiles containing intermediate maxima of V_0, with attendant beam reflection. In typical operation of ion thrusters of this kind, the ion flow is not space-charge limited in the deceleration gap, and hence is less sensitive to the precise position of the decel electrode. In this way the decel electrode is free to function as an electron baffle, and as a control grid on the exhaust speed if desired, without disturbing the ion flux.

It is possible to incorporate accel-decel potential profiles into the electrode design calculations outlined in Sec. 7-3 without fundamental change in the approach. For example, Fig. 7-18 shows the equipotential

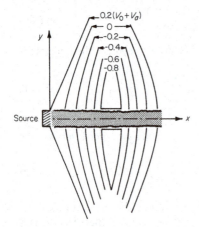

Fig. 7-18 Equipotential contours for parallel ion beam in an accel-decel field.

contours for a two-dimensional parallel beam passing through an accelerating potential $V_0 + V_a$, followed by a decelerating region $-V_a = 0.8(V_0 + V_a)$ [1]. Focused beams can also be designed, as indicated in Sec. 7-3, by searching for axial potential profiles $V^0(x)$ which display the desired accel-decel ratio.

In some cases, physical separation of the decelerating electrode and neutralizing filament may be unnecessary. That is, the neutralized ion beam plasma created near the electron-emitting filament may serve as a virtual deceleration electrode. The potential of this plasma is established essentially by that of the filament, modified by its small sheath drop to the plasma, where the filament itself may not necessarily be immersed in the beam, nor be as close to the accelerating electrode as the effective neutralization plane. In this scheme, the beam may become space-charge-neutralized before it is current-neutralized, i.e., before the plane of the filament which completes the main circuit of the device (Fig. 7-19). Virtual deceleration schemes of this sort show some tendency to defocus the beam, but in parallel beam arrays this disadvantage is probably less significant than the convenience of no decelerating electrode structure.

To summarize the acceleration-deceleration concept, then, one first extracts ions from the source by a relatively high potential difference, at a correspondingly high power drain from the source circuit to maximize

ELECTROSTATIC ACCELERATION

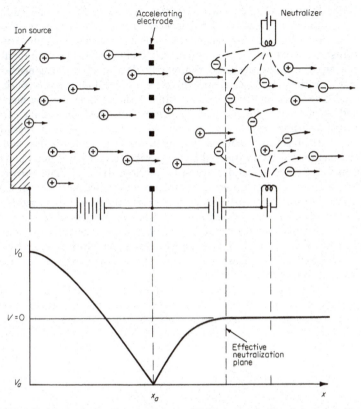

Fig. 7-19 Virtual deceleration electrode provided by upstream migration of neutralizing electrons (schematic).

the current density. One then decelerates this high density beam by a third electrode, real or virtual thereby returning some power to the external circuit without reducing the current density of the ion flux. The net effect is a higher thrust device than the single-stage accelerator of the same exhaust speed, and one which excludes neutralizer electrons from the accelerating gap.

7-6 ION THRUSTER DESIGN AND PERFORMANCE

Careful combination of the three functional elements discussed above—ion source, accelerating electrode structure, and neutralizer—yields an ion thruster, a device which converts electric power into kinetic energy of an essentially collisionless neutral beam of propellant ions and electrons.

Properly designed and operated, ion thrusters can be highly efficient long-lived propulsion units, qualified in every respect for a variety of space missions. Probably because of their relative conceptual simplicity, ion thrusters have been the most intensively studied of all electric propulsion concepts, and are the most thoroughly developed and engineered for incorporation into operational spacecraft. The bulk of this work has tended to focus on electron bombardment thrusters employing mercury, and to a lesser extent cesium, as propellant, and on cesium contact thrusters using porous tungsten ionizers. The concepts of accel-decel electrode structures, oxide-coated filaments and autocathodes, and the rudimentary neutralization schemes described above have found practical implementation in various versions of these devices. The theoretical limits on attainable thrust density and conversion efficiency discussed earlier have been approached in short-term operation, but in most cases have had to be compromised somewhat to achieve desired operating lifetimes without excessive electrode erosion or other structural degradations. At the present time, a variety of serviceable ion thrusters literally wait on the shelf for coupling with suitable space power supplies and assignment to appropriate space missions.

A few typical ion thrusters are shown schematically and in photograph in Figs. 7-20 to 7-27. A summary of example performance data is displayed in Table 7-2 and in the graphs of Figs. 7-28 and 7-29. Doubt-

Fig. 7-20 First low density electron bombardment thruster, circa 1960. (*NASA-Lewis Research Center, Cleveland, Ohio.*)

Fig. 7-21 Fifteen-centimeter mercury electron bombardment ion engine; thrust ≈0.035 newton at 5,000 sec. Power input ≈1 kw, permanent magnet focusing. *(Hughes Aircraft Company, Malibu, Calif.)*

Fig. 7-22 Mercury electron bombardment thruster used in SERT I space test. *(NASA-Lewis Research Center, Cleveland, Ohio.)*

Fig. 7-23 Ten-centimeter electron bombardment thruster with permanent magnet; thrust ≈0.03 newton at 4,000 sec. (*NASA-Lewis Research Center, Cleveland, Ohio.*)

Fig. 7-24 Cesium electron bombardment engines of various sizes up to 1 kw. (*Electro-Optical Systems, Inc., Pasadena, Calif., a subsidiary of Xerox Corporation.*)

ELECTROSTATIC ACCELERATION

Fig. 7-25 Cesium contact ion microthruster; thrust $\approx 5 \times 10^{-4}$ newton. The circular beam from a button ionizer can be deflected in any transverse direction by means of the four segments of accel electrode shown. (*Hughes Aircraft Company, Malibu, Calif.*)

less, the continued refinement of ion thruster technology will make these illustrations and data transitory, but a few general trends in the comparison of performance of the three types of thruster will probably persist. Namely, although the overall efficiency of ion thrusters is quite good at high levels of specific impulse, it degrades significantly at lower exhaust speeds, mainly because of the increase in relative importance of the

Table 7-2 Characteristics of typical long-life ion thrusters

Ion source	Cs-W contact	Hg bomb	Hg bomb	Cs bomb
Laboratory	Hughes	NASA-Lewis	NASA-Lewis	E.O.S.
Size	8.5 × 13.6 cm	15 cm diam	30 cm diam	12.7 cm diam
Weight, kg	8.9	3.0	19.7	2.2
Specific impulse, sec	9,000	5,000	9,000	5,000
Thrust, newtons	0.023	0.033	0.50	0.026
Jet power, kw	1.0	0.8	22.0	0.7
Ion current, amp	0.2	0.27	2.2	0.4

SOURCE: W. R. Mickelsen, Future Trends in Electric Propulsion, *AIAA Paper* 66-595, 1966.

effective energy expenditure for ionization. This degradation of efficiency is more pronounced in the cesium contact thrusters than in the bombardment devices. Compensating this is a superiority of contact sources in current density capability and restraint of neutral particle emission, but these in turn are compromised somewhat by the higher charge-exchange cross section of cesium, which aggravates electrode sputtering problems. Also inherent in cesium devices is a more complex storage and feed system for the highly corrosive propellant, and greater tendency to surface contamination and high voltage arc breakdowns in the accelerating chamber.

Fig. 7-26 Cesium contact ion engine; thrust ≈0.05 newton with a zero-g propellant feed system. *(Electro-Optical Systems, Inc., a subsidiary of Xerox Corporation.)*

Fig. 7-27 A nine-engine 30-kw array of electron bombardment thrusters. *(NASA-Lewis Research Center, Cleveland, Ohio.)*

Fig. 7-28 Comparison of efficiency vs. specific impulse for various typical ion thrusters. *(From G. R. Brewer, Current Status of Ion Engine Development, AIAA Paper 66-564, 1966.)*

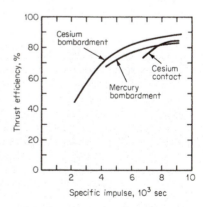

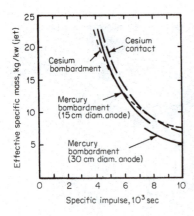

Fig. 7-29 Effective specific mass of ion thruster systems for primary propulsion, including power plant, power conditioning, controls, and propellant tankage. (From W. R. Mickelsen, Future Trends in Electric Propulsion, AIAA Paper 66-595, 1966.)

The ion thruster concept has been implemented over several orders of magnitude of total thrust, from microthrusters yielding 10^{-5} newton to large modular arrays of thrusters capable of almost 1 newton. This versatility is achieved primarily by parallel ganging of smaller thruster modules and does not, of course, obviate the fundamental bound on thrust density imposed by the space-charge limited ion flux. Indeed, it is this inherent limit which may become the most serious problem in the application of ion thrusters to high-power primary propulsion in the megawatt range (Prob. 7-7).[1]

7-7 FIRST FLIGHT TEST

On July 20, 1964, a four-stage Scout rocket was launched from the NASA facility at Wallops Island, Va., bearing aloft the SERT I (Space Electric Rocket Test) spacecraft. For 47 min the spacecraft followed a ballistic trajectory over the Atlantic Ocean, providing two ion thrusters mounted on it an opportunity to function in the space environment. One of these, a cesium contact ion thruster, was unable to operate because of a high voltage short circuit. The other, a mercury electron bombardment ion thruster, operated for over 31 min, thereby qualifying as the first successful electric space thruster of any kind.

The announced primary purpose of this flight test was to verify that the ion exhaust beam could indeed be neutralized in space. Secondary objectives included a search for any unforeseen difficulties in space operation of the thruster, including difficulties in radio communication with the ion-propelled spacecraft and study of any differences in space per-

[1] However, preliminary studies with a very large single-module electron bombardment thruster have recently been reported by the NASA-Lewis Research Center. (S. Nakanishi and E. V. Pawlik, Experimental Investigation of a 1.5 Meter Diameter Kaufman Thruster, *AIAA Electric Propulsion and Plasmadynamics Conference*, Colorado Springs, Colo., Sept. 11–13, 1967, paper 67-725.)

formance. The primary purpose—the check on effective beam neutralization—was to be determined by monitoring the thrust produced by the ion beams.

The flight test had been preceded by a lengthy period of design, development, and ground testing of the spacecraft, including the actual thrusters, power supplies, and telemetry components. This study culminated in a series of flight simulation tests in which the entire SERT capsule was mounted in a large vacuum tank, to provide a basis for comparison of space results.

Figures 7-30 and 7-31 show a photograph and schematic electrical diagram of the successful thruster, developed at the NASA Lewis Research

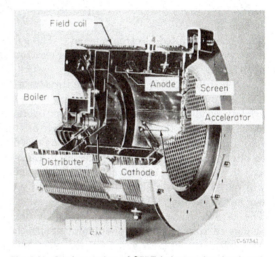

Fig. 7-30 Sectioned view of SERT I electron bombardment thruster. (*NASA-Lewis Research Center, Cleveland, Ohio.*)

Center [32]. Details of its operation on the SERT flight are presented in several NASA reports [67–69]. Briefly, the mercury propellant vapor was prepared in an electrically heated boiler maintained at about 450°K by a 10-watt ac heater. The magnetic field for the ionization chamber, supplied by an external helical coil, was about 0.003 watt/m^2, and the discharge potential was about 45 volts. An accelerating potential of 4,500 volts was applied between the screen and accelerator grids, followed by a virtual deceleration stage of 2,000 volts established by the neutralizer, which was maintained at spacecraft ground. The neutralizer was a simple tantalum filament projecting partially into the beam and electrically heated to a temperature sufficient to supply the required

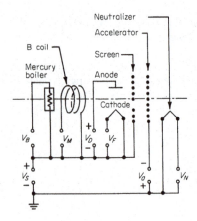

Fig. 7-31 Electrical schematic of SERT I thruster. (*From R. J. Cybulski et al., Results from SERT I Ion Rocket Flight Test, NASA Tech. Note D-2718, March, 1965.*)

space-charge limited electron current. The entire body of the thruster was surrounded by a shield grid to deter electron backstreaming from the neutralizer.

The prime power source for the thruster was a 56-volt battery pack which drove transistorized high voltage supplies. These were appropriately regulated, and contained current overload sensors which turned off all outputs in the event of high voltage arcing in the thruster.

In flight, the thrust of the ion engine was determined by allowing it to alter the spin rate of the spacecraft. Each engine was mounted off axis in the capsule and aligned to thrust tangentially. The spacecraft was initially spin-stabilized at 85 rpm, and by the thrust of the one ion engine over about a 30-min period, this rotation was increased to 94 rpm. The capsule contained three independent devices for monitoring the change in spin rate: two photovoltaic solar-cell devices, each with its own telemetry link to ground-based electronic counters, followed the rotation of the capsule with respect to the sun; in addition, a sensitive accelerometer was mounted in a radial direction in the capsule to sense angular motion dynamically.

In addition to the thrust-measuring instrumentation, the spacecraft contained other devices to study particular aspects of the engine operation. All pertinent voltages and currents and the vaporizer temperature were monitored continuously throughout the flight. A hot-wire calorimeter probe [70] was suspended so that it could be swung through the ion beam 18 cm aft of the accelerator electrode to provide contours of beam power density. A rotating-vane electric field meter was mounted in the capsule to search for electrostatic fields exterior to the thruster. Figure 7-32 illustrates some details of the capsule assembly.

From these measurements it was found that the engine did develop thrust, at a level corresponding to complete beam neutralization, with the

ELECTROSTATIC ACCELERATION

Fig. 7-32 SERT I flight-model spacecraft. *(NASA-Lewis Research Center, Cleveland, Ohio.)*

simple neutralizing filament biased over a wide range of potentials relative to the spacecraft shield. Spin-sensor data indicated that the engine produced thrusts of about 0.02 newton (0.0045 lb) with the neutralizer on. With the neutralizer switched off, the ion current was collected by the accelerating electrode, thereby overloading and interrupting the power supply. The measured neutralizer current agreed well with the measured ion current. When the ion beam was turned off, the neutralizer current also dropped to zero. These observations, along with those of the beam-spreading contours and potential profiles, agreed with those found in ground-based vacuum tank tests of the same type, to within the modest sensitivity of the space flight instrumentation, thereby implying that significant neutralization experiments are possible in ground facilities. By way of additional information, the test revealed that there was no interference with the radio communication with the spacecraft and that there was little tendency for the spacecraft to precess about any axis other than its major spin axis.

Other flight tests of electric thrusters have since followed SERT I and will continue to follow this event in steady succession. Different types of thrusters will be exercised; more elaborate instrumentation will be provided; eventually, the thrusters will perform useful functions. Although these later results will supersede those just described, the historical, political, and psychological values of the SERT I success establish it as a milepost in the development of electric propulsion.

7-8 HEAVY-PARTICLE ACCELERATORS

Certain of the fundamental limitations on the performance of electrostatic thrusters discussed in Sec. 7-1 can be relieved by the use of ions of large mass-to-charge ratio. The advantages of heavy-ion acceleration may be expressed in various ways, depending on the most stringent prevailing limitation for the particular application. For example, if it is the relative energy expenditure per ion created [Eq. (7-13)] which limits the utility of a given accelerator, this ratio can be improved by accelerating a much heavier ion, of roughly the same effective ionization energy. If accelerated to the same final velocity, this heavy ion will possess exhaust kinetic energy larger by the mass ratio. Alternatively, if it is an excessive exhaust speed which limits the applicability of the accelerator, this clearly can be reduced by using the heavier ion in the same accelerating field [Eq. (7-9)]. In a broader sense, if a given accelerator could generate and accelerate a range of various mass-to-charge ions, it would qualify as a variable specific impulse thruster, and this degree of freedom would considerably enhance its overall mission performance [7].

Further review of the relations and arguments of Sec. 7-1 provide a general image of a heavy-ion accelerator which involves lower current densities, larger electrode gaps, lower exhaust speeds, less energy expenditure per ion, and larger accelerating voltages than its atomic-ion counterpart operated at the same input power. All these features, save the last, are in its favor. The tendency to higher accelerating potentials (but not necessarily higher electric fields) will pose a new limit on attainable performance, but probably not below the range of 10^5 volts. With this empirical limit, the interesting range of heavy ions is seen to lie between 10^3 and 10^5 atomic mass units per electronic charge (amu/e) (Prob. 7-8).

Despite their obvious electrodynamical advantages, heavy-particle accelerators have been very slow to develop. The essential problem is the preparation of appropriate mass ions in sufficient quantity at adequate propellant utilization ratios and having sufficiently uniform mass-to-charge ratio. The two classes of particles most evidently available are the macromolecules and colloids. No satisfactory means of preparation of the former have yet been displayed. Several methods for preparing charged colloids have been established, all of which tend to produce a rather broad spectrum of mass-to-charge ions, of magnitude somewhat higher than optimum for electrostatic thrusters.

Electric spraying. Under certain circumstances, fluid particles emitted from a simple "atomizer" nozzle will carry a net electrostatic charge, probably acquired by a triboelectric process at the nozzle tip. The electrification process is greatly enhanced if the fluid is extracted from a hollow needle tip by application of a strong axial electric field. Here the charging process is more likely a combination of dielectrophoresis, or polarization of the incipient droplet, followed by rupture of the droplet into two portions of net charge, one of which remains on the tip. Both conducting and dielectric fluids can be charged in this manner, although the characteristics of the resulting ion streams differ considerably. The metallic ion particles tend to be excessively large; the dielectric droplets can be formed in usefully small sizes, but the rate of droplet emission from a given needle tip is very low, thereby limiting the effective ion beam current [71,72]. Slightly conducting mixtures may provide the best compromise. Reference 73 quotes a value of about 7 μamp per needle for glycerol doped with sulfuric acid to a resistivity of 30 ohm-m, yielding an M/q spectrum centered around 5×10^4 amu/e.

Vapor condensation. A stream of supersaturated vapor may be caused to condense in an appropriately contoured supersonic nozzle or by passing through a condensation shock wave. The condensation follows the familiar particle-nucleation process, and the rate and range of final particle sizes may be controlled by the flow conditions in the stream. The colloidal particles thus prepared may then be ionized by electron bombardment from a thermionic gun or by electron attachment in a corona discharge. Colloids of mercurous chloride, mercuric chloride, aluminum chloride, and other halides have been prepared in this manner. A broad spectrum of mass-to-charge ratios is again observed, but values as low as 10^5 amu/e have been obtained. Unfortunately, the mass utilization efficiency seems inherently poor [74].

Ion nucleation. If the supersaturated vapor stream is first partially ionized, either photoelectrically or in a discharge or electron beam, the ions will act as agglomeration nuclei, and some condensation will proceed about them even in the absence of cooling or condensation shocks. Application of a retarding electric field at some point in the stream slows the ion motion and allows more time for particle growth. The heaviest mass-to-charge particles thus formed will progress farthest against this field, and thereby may be selectively extracted.

Surface condensation. Under proper conditions, a stream of supersaturated vapor will agglomerate on a solid surface. Application of a strong electric field normal to that surface can then polarize the condensed particles and extract them bearing a net charge. The ion current density attainable by this technique again seems to be fundamentally restricted to very low levels by the polarization rates of the individual droplets.

Preformed particles. Solid particles of the desired size have been prepared by various agglomeration processes, and subsequently ionized by electron bombardment [75]. The agglomeration may follow a vaporization of a bulk solid or result from chemical reaction of two gases or liquids. Preparation of satisfactory solid colloids by the grinding or fracturing of larger particles has not yet been achieved, although the ultimate desirability of such a technique is evident. Interest in solid colloids is enhanced by the prospect of spacecraft refueling with planetary or lunar surface material, booster upper-stage tankage, or biowastes from the crew, which would then be rendered in an on-board "colloid mill" [6].

It seems highly probable that other, superior means of obtaining ionized colloid streams will ultimately be found [76–79]. Regardless of the specific means of preparation, certain general electrical and mechanical limitations restrict the attainable mass-to-charge ratio of particles of given material and size. In particular, the particle must possess adequate structural strength to withstand the forces of electrostatic repulsion of its own charges tending to rupture it, and its self–electrostatic field must not be high enough to cause field emission of electrons or ions from its surface. The structural strength of a liquid droplet lies in its surface tension, and the minimum mass-to-charge ratio thus scales as particle radius to the $3/2$ power (Prob. 7-9; [1]):

$$\frac{M}{q} \geq A_\sigma \rho \sigma_t^{-1/2} r^{3/2} \qquad (7\text{-}33)$$

where ρ = liquid density
σ_t = surface tension
A_σ = a dimensional coefficient

For solids, the strength is reflected by the bulk modulus β, and the minimum M/q is linear in particle radius:

$$\frac{M}{q} \geq A_\beta \rho \beta^{-1/2} r \qquad (7\text{-}34)$$

The restriction on field emission of ions or electrons from the particle surface also yields a linear relation between M/q and radius:

$$\frac{M}{q} \geq \frac{A_E \rho r}{\hat{E}} \qquad (7\text{-}35)$$

where $\hat{E}$ is the limiting field strength for surface emission, perhaps 10^9 volts/m for electrons from a negative particle and 10^{10} volts/m for ions from a positive particle.

The essence of the above relations is simply that one cannot create colloids of the desired low M/q by loading many charge units onto indiscriminately large particles. Rather, one is restricted for reasons of structural strength and surface emission to rather small sizes. For particles having M/q in the range of 10^4 amu/e, for example, the structural limit on particle radius is in the range of 10^{-2} to 10^{-3} μ, and the electron-emission limit even slightly lower. This is at best a few tens of atomic dimensions.

To summarize, then, the concept of colloid acceleration holds promise of extending the useful range of electrostatic propulsion to higher thrust power densities, lower exhaust velocities, and variable specific impulse devices and of utilizing readily available waste materials as propellants. To do so, certain formidable problems in the efficient preparation of sufficiently uniform and intense heavy-ion beams must first be solved, and subsequently the necessary modifications in accelerating electrode and neutralizing arrangements must be implemented.

PROBLEMS

7-1. Show that the current density appearing in Eq. (7-3) is indeed maximized by setting $(dV/dx)_0 = 0$.

7-2. Derive the thrust density relation (7-8) by computing the net force exerted by the electrostatic field on the two electrodes.

7-3. What size power plant will be needed to drive the one-dimensional breakdown-limited cesium thruster discussed in Sec. 7-1 if its accelerating gap is 0.005 m and its beam cross section is 0.01 m^2? If the specific mass of this power plant is 10 kg/kw, what is the maximum attainable acceleration of a spacecraft propelled by this thruster?

7-4. Compute the thrust per unit area, beam power per unit area, exhaust speed, and current density for a one-dimensional accelerator which applies 10,000 volts across a 0.01-m gap to a beam of atomic hydrogen ions. If a power plant of 100 kw is available, of specific mass 10 kg/kw, what maximum acceleration might be achieved by a spacecraft using this thruster?

7-5. Estimate the mean free path for electron-ion momentum-transfer collisions in the exhaust beam of a cesium ion engine operating at thrust density of 100 newtons/m^2 and exhaust speed of 10^5 m/sec.

7-6. Compute the beam power required to achieve a thrust of 10 newtons with a mercury bombardment engine, using a single acceleration potential of 10,000 volts. If a deceleration stage is added, reducing the net potential difference to 3,000 volts, what beam power is needed to produce the same thrust? If electrode gaps can be no less than 0.005 m, what is the maximum current density and minimum jet cross section in each case?

7-7. Estimate the total beam cross section for a mercury bombardment thruster producing 1 newton of thrust at 5,000-sec specific impulse. What would be this area for an 80 percent efficient thruster of the same I_s, consuming 10 megawatts of electric power? Discuss the problems of constructing and operating such a thruster.

7-8. "Design" a 3,000-sec colloid thruster, using ions of 10^5 amu/e. Specifically, what electrode voltages and spacing do you suggest?

7-9. Justify Eqs. (7-33) to (7-35).

REFERENCES

1. Stuhlinger, E.: "Ion Propulsion for Space Flight," McGraw-Hill Book Company, New York, 1964.
2. Langmuir, D. B., E. Stuhlinger, and J. M. Sellen, Jr. (eds.): "Electrostatic Propulsion," vol. 5 of "Progress in Astronautics and Rocketry," Academic Press Inc., New York, 1961.
3. Stuhlinger, E. (ed.): "Electric Propulsion Development," vol. 9 of "Progress in Astronautics and Aeronautics," Academic Press Inc., New York, 1963.
4. Stuhlinger, E.: Electrical Propulsion Systems for Space Ships with Nuclear Power Source, pts. I–III, *J. Astronaut.*, vol. 2, no. 4, pp. 149–152, 1955; vol. 3, no. 1, pp. 11–14, 1956; vol. 3, no. 2, pp. 33–36, 1956.
5. Kaufman, H. R.: Electric Propulsion for Spacecraft, *New Scientist*, vol. 23, p. 263, July, 1964.
6. Mickelsen, W. R.: Electric Propulsion for Space Flight, *Aerospace Eng.*, vol. 19, no. 11, p. 6, November, 1960.
7. Mickelsen, W. R., and H. R. Kaufman: Status of Electrostatic Thrustors for Space Propulsion, *NASA Tech. Note* D-2172, May, 1964.
8. Brewer, G. R.: Current Status of Ion Engine Development, *AIAA Paper* 66-564, 1966.
9. Mickelsen, W. R.: Future Trends in Electric Propulsion, *AIAA Paper* 66-595, 1966.
10. Teem, J., and G. R. Brewer: Current Status and Prospects of Ion Propulsion, *ARS Preprint* 2650-62, 1962.
11. Child, C. D.: Discharge from Hot CaO, *Phys. Rev.*, vol. 32, p. 492, 1911.
12. Schottky, W.: Current between an Incandescent Filament and Coaxial Cylinder, *Phys. Z.*, vol. 15, pp. 526, 624, 1914.
13. Fay, C. E., A. L. Samuel, and W. Shockley: On the Theory of Space Charge between Parallel Plane Electrodes, *Bell System Tech. J.*, vol. 17, p. 49, 1938.
14. Spangenberg, K. R.: "Vacuum Tubes," McGraw-Hill Book Company, New York, 1948.
15. Kaufman, H. R.: One-dimensional Analysis of Ion Rockets, *NASA Tech. Note* D-261, March, 1960.
16. Langmuir, I.: The Effect of Space Charge and Residual Gases on Thermionic Current in High Vacuum, *Phys. Rev.*, vol. 2, p. 450, 1913; Langmuir, I., and K. H. Kingdon: Thermionic Phenomena due to Alkali Vapors, *ibid.*, vol. 21, p. 380, 1923; Langmuir, I., and J. B. Taylor: The Mobility of Cesium Atoms Adsorbed on Tungsten, *ibid.*, vol. 40, p. 463, 1932; Taylor, J. B., and I. Langmuir: The Evaporation of Atoms, Ions and Electrons from Cesium Films on Tungsten, *ibid.*, vol. 44, p. 423, 1933.
17. Datz, S., and E. H. Taylor: Ionization on Platinum and Tungsten Surfaces, I, The Alkali Metals, *J. Chem. Phys.*, vol. 25, p. 389, 1956.
18. Finkelstein, A. T.: A High-efficiency Ion Source, *Rev. Sci. Instr.*, vol. 11, p. 94, 1940.
19. Forrester, A. T., and R. C. Speiser: Cesium-ion Propulsion, *Astronautics*, vol. 4, no. 10, p. 34, October, 1959.
20. Sunderland, R. J., J. R. Radbill, and R. D. Gilpin: Ion Engine Development, I, Diffusion Type Ion Sources, *Advan. Astronaut. Sci.*, vol. 7, p. 148, American Rocket Society, Plenum Press, Inc., New York, 1960.
21. Stavisskii, Yu. Ya., and S. Ya. Lebedev: Surface Ionization of Cesium in the Course of Diffusion through Porous Tungsten, *J. Tech. Phys. U.S.S.R.*, vol. 30, no. 10, p. 1222, 1960.

22. Zuccaro, D., R. C. Speiser, and J. M. Teem: Characteristics of Porous Surface Ionizers, p. 107 in D. B. Langmuir, E. Stuhlinger, and J. M. Sellen, Jr. (eds.), "Electrostatic Propulsion," vol. 5 of "Progress in Astronautics and Rocketry," Academic Press Inc., New York, 1961.
23. Nazarian, G. M., and H. Shelton: Theory of Ion Emission from Porous Media, *ibid.*, p. 91.
24. Worden, D. G.: The Effects of Surface Structure and Adsorption on the Ionization Efficiency of a Surface Ionization Source, *ibid.*, p. 141.
25. Husmann, O. K.: Experimental Evaluation of Porous Materials for Surface Ionization of Cesium and Potassium, p. 195 in E. Stuhlinger (ed.), "Electric Propulsion Development," vol. 9 of "Progress in Astronautics and Aeronautics," Academic Press Inc., New York, 1963.
26. Shelton, H.: Experiments on Atom and Ion Emission from Porous Tungsten, *ibid.*, p. 219.
27. Kuskevics, G., R. M. Worlock, and D. Zuccaro: Ionization, Emission, and Collision Processes in the Cesium Ion Engine, *ibid.*, p. 229.
28. Swanson, L. W., R. W. Strayer, F. M. Charbonnier, and E. C. Cooper: Field Emission Microscope Study of the Kinetics of Cesium Layers on a Tungsten Surface, *ibid.*, p. 165.
29. Dalins, I.: Flow through Porous Media and Its Implications for Ion Rocket Operation, *ARS Preprint* 2362-62, 1962.
30. Marchant, A. B., G. Kuskevics, and A. T. Forrester: Surface Ionization Microscope, *AIAA Paper* 63-018, 1963.
31. Forrester, A. T.: The Ionization of Cesium in Tungsten Capillaries, *Electro-optical Systems Res. Rept.* 20, July, 1964.
32. Kaufman, H. R.: An Ion Rocket with an Electron-bombardment Ion Source, *NASA Tech. Note* D-585, January, 1961; Kaufman, H. R., and P. D. Reader: Experimental Performance of Ion Rockets Employing Electron-bombardment Ion Sources, p. 3 in D. B. Langmuir, E. Stuhlinger, and J. M. Sellen, Jr. (eds.), "Electrostatic Propulsion," vol. 5 of "Progress in Astronautics and Rocketry," Academic Press Inc., New York, 1961; Kaufman, H. R.: Performance Correlation for Electron-bombardment Ion Sources, *NASA Tech. Note* D-3041, 1965.
33. French, P.: Electron Bombardment Ion Source, p. 291 in E. Stuhlinger (ed.), "Electric Propulsion Development," vol. 9 of "Progress in Astronautics and Aeronautics," Academic Press Inc., New York, 1963.
34. Speiser, R. C., and L. K. Branson: Studies of a Gas Discharge Cesium Ion Source, *ARS Paper* 2664-62, 1962; Speiser, R. C., et al.: Cesium Electron Bombardment Ion Engines, *AIAA Paper* 65-373, 1965.
35. Sohl, G., et al.: Life Testing of Electron Bombardment Cesium Ion Engines, *AIAA Paper* 66-233, 1966.
36. Wasserbauer, J. F.: A 5-centimeter-diameter Electron-bombardment Thrustor with Permanent Magnets, *NASA Tech. Note* D-3628, 1966.
37. Miller, N. L., and W. R. Kerslake: Evaluation of Filament Deterioration in Electron-bombardment Ion Sources, *NASA Tech. Note* D-2173, 1964.
38. Eckhardt, W. O., et al.: Liquid-metal Cathode Research, *AIAA Paper* 66-245, 1966; King, H. J., et al.: Electron-bombardment Thrusters Using Liquid-mercury Cathodes, *AIAA Paper* 66-232, 1966.
39. Kerslake, W. R.: Preliminary Operations of Oxide-coated Brush Cathodes in Electron-bombardment Ion Thrustors, *NASA Tech. Mem.* X-1105, 1965.
40. Penning, F. M., and J. H. A. Moubis: Eine Neutronenroehre ohne Pumpvorrichtung, *Physica*, vol. 4, p. 1190, 1937.

41. Meyerand, R. G., Jr., and S. C. Brown: High Current Ion Source, *Rev. Sci. Instr.*, vol. 30, no. 2, p. 110, 1959; Meyerand, R. G., Jr.: The Oscillating-electron Plasma Source, p. 81 in D. B. Langmuir, E. Stuhlinger, and J. M. Sellen, Jr. (eds.), "Electrostatic Propulsion," vol. 5 of "Progress in Astronautics and Rocketry," Academic Press Inc., New York, 1961.
42. Stirling, W. L., J. W. Flowers, and J. S. Luce: Hollow Cathode Arc-ion Sources: Injection Neutralization Possibilities, *ARS Preprint* 1376-60, 1960.
43. MacKenzie, K. R., and R. F. Wuerker: Electron Transfer Discharge Ion Source, p. 299 in E. Stuhlinger (ed.), "Electric Propulsion Development," vol. 9 of "Progress in Astronautics and Aeronautics," Academic Press Inc., New York, 1963.
44. Ardenne, M. von: Über ein Ionenquellen-System mit Massenmonochromator für Neutronengeneratoren, *Phys. Z.*, vol. 42, p. 91, 1942.
45. Heil, H.: Über eine neue Ionenquelle, *Z. Phys.*, January, 1943.
46. Abele, M., and W. Meckbach: Design and Performance of a Hot Cathode Magnetically Collimated Arc Discharge Ion Source, *Rev. Sci. Instr.*, vol. 30, no. 5, p. 335, May, 1959.
47. Burton, B. S., Jr.: The Duoplasmatron: Theoretical Studies and Experimental Observations, p. 21 in D. B. Langmuir, E. Stuhlinger, and J. M. Sellen, Jr. (eds.), "Electrostatic Propulsion," vol. 5 of "Progress in Astronautics and Rocketry," Academic Press Inc., New York, 1961.
48. Braams, C. M., P. Zieske, and M. J. Kofoid: Composition of Noble Gas Ion Beams Produced with a Duoplasmatron, *Boeing Sci. Lab. Rept.* D1-82-0437, 1965.
49. French, P.: Experiments with Arc Ion Sources for Electric Propulsion, *Thompson-Ramo-Wooldridge Rept.* ER-4124, Cleveland, Ohio, June, 1960.
50. Pierce, J. R.: "Theory and Design of Electron Beams," D. Van Nostrand Company, Inc., Princeton, N.J., 1954.
51. Cheever, R. N., and G. E. Bloch: Optical Design Method for High Current Density Ion Engines, *ARS Preprint* 2431-62, 1962.
52. Hamza, V., and E. A. Richley: Numerical Solution of Two Dimensional Poisson Equation: Theory and Application to Electrostatic-ion Engine Analysis, *NASA Tech. Note* D-1323, 1962.
53. Van Duyer, T., and G. R. Brewer: Space-charge Simulation in an Electrolytic Tank, *J. Appl. Phys.*, vol. 30, no. 3, p. 291, 1959.
54. Staggs, J. F.: An Electrolytic Tank Analog for Two dimensional Analysis of Electrostatic-thrustor Optics, *NASA Tech. Note* D-2803, 1965.
55. Ernstene, M. P., et al.: Surface Ionization Engine Development, *J. Spacecraft Rockets*, vol. 3, pp. 744–747, 1966.
56. Pawlik, E. V., P. M. Margosian, and J. F. Staggs: A Technique for Obtaining Plasma-sheath Configurations and Ion Optics for an Electron-bombardment Ion Thruster, *NASA Tech. Note* D-2804, 1965.
57. Mirels, H., and B. M. Rosenbaum: Analysis of One-dimensional Ion Rocket with Grid Neutralization, Lewis Research Center, Cleveland, Ohio, *NASA Tech. Note* D-266, 1960.
58. Sellen, J. M., Jr., and H. Shelton: Transient and Steady State Behavior in Cesium Ion Beams, p. 305 in D. B. Langmuir, E. Stuhlinger, and J. M. Sellen, Jr. (eds.), "Electrostatic Propulsion," vol. 5 of "Progress in Astronautics and Rocketry," Academic Press Inc., New York, 1961.
59. Dunn, D. A., and I. T. Ho: Computer Experiments on Ion-beam Neutralization with Initially Cold Electrons, *AIAA J.*, vol. 1, p. 2770, 1963.
60. Birdsall, C. K., and W. B. Bridges: Space-charge Instabilities in Electron Diode and Plasma Converters, *J. Appl. Phys.*, vol. 32, p. 2611, 1961; Bridges, W. B.,

and C. K. Birdsall: Space-charge Instabilities in Electron Diodes, II, *ibid.*, vol. 34, p. 2946, 1963.
61. Electrostatic Propulsion, *Ramo-Wooldridge Tech. Rept.* RW-RL-155, December, 1959.
62. Dalins, D., R. N. Seitz, and E. W. Urban: Theoretical Study of Ion Beam Neutralization, USAF-NASA Joint Meeting on Electrostatic Propulsion, Beverly Hills, Calif., April, 1961, *AFOSR Rept.* 711.
63. Etter, J. E., et al.: Neutralization of Ion Beams, p. 357 in D. B. Langmuir, E. Stuhlinger, and J. M. Sellen, Jr. (eds.), "Electrostatic Propulsion," vol. 5 of "Progress in Astronautics and Rocketry," Academic Press Inc., New York, 1961.
64. Pearlstein, L. D., M. N. Rosenbluth, and G. W. Stuart: The Neutralization of Ion Beams, p. 379 in E. Stuhlinger (ed.), "Electric Propulsion Development," vol. 9 of "Progress in Astronautics and Rocketry," Academic Press Inc., New York, 1963.
65. Brewer, G. R., M. R. Currie, and R. C. Knecktle: Ionic and Plasma Propulsion for Space Vehicles, *Proc. IRE*, vol. 49, p. 1789, 1961.
66. Kemp, R. F., J. M. Sellen, Jr., and E. V. Pawlik: Neutralizer Tests on a Flight-model Electron-bombardment Ion Thrustor, *NASA Tech. Note* D-1733, 1963.
67. Cybulski, R. J., et al.: Results from SERT I Ion Rocket Flight Test, *NASA Tech. Note* D-2718, 1965.
68. Gold, H., et al.: Description and Operation of Spacecraft in SERT I Ion Thrustor Flight Test, *NASA Tech. Mem.* X-1077, 1965.
69. Nieberding, W. C., and R. R. Lovell: Thrust Measurements of SERT I Ion Thrustors, *NASA Tech. Note* D-3407, 1966.
70. Baldwin, L. V., and V. A. Sandborn: Theory and Application of Hot-wire Calorimeter for Measurement of Ion Beam Power, p. 425 in D. B. Langmuir, E. Stuhlinger, and J. M. Sellen (eds.), "Electrostatic Propulsion," vol. 5 of "Progress in Astronautics and Rocketry," Academic Press Inc., New York, 1961.
71. Krohn, V. E., Jr.: Glycerol Droplets for Electrostatic Propulsion, p. 435 in E. Stuhlinger (ed.), "Electric Propulsion Development," vol. 9 of "Progress in Astronautics and Aeronautics," Academic Press Inc., New York, 1963.
72. Hendricks, C. D., Jr., and R. J. Pfeiffer: Parametric Studies of Electrodynamic Spraying, *AIAA Paper* 66-252, 1966.
73. Cohen, E., C. J. Somol, and D. A. Gordon: A 100-KV, 10-W Heavy-particle Thrustor, *AIAA Paper* 65-377, 1965.
74. Norgren, C. T., and D. S. Goldin: Experimental Analysis of the Exhaust Beam from a Colloidal Thrustor, *AIAA Paper* 64-674, 1964; Goldin, D. S., and G. L. Kvitek: An Analysis of Particle Formation Efficiency in a Colloid Thrustor, *AIAA Paper* 66-253, 1966.
75. Singer, S., N. G. Kim, and M. Farber: An Experimental Study of Colloidal Propulsion Using Sub-micron Solid Particles, *AIAA Paper* 63-052, 1963.
76. Mickelsen, W. R.: Colloid-particle Electrostatic Thrustors, DGRR Sonnenberg Symposium on Electric Propulsion, Braunschweig, West Germany, Feb. 24, 1966.
77. Hunter, R. E., and S. H. Wineland: Exploration of the Feasibility of an Electrodeless Colloid Thrustor Concept, *6th Intern. Symp. Space Tech. Sci.*, Tokyo, November, 1965.
78. Courtney, W. G., and C. Budnik: Colloid Propulsion Using Chemically-formed Particles, *AIAA Paper* 66-254, 1966.
79. Harris, S. P., and M. Farber: Development of a Solid Charged Colloidal Particle Thrustor, *AIAA Paper* 66-255, 1966.

8
Electromagnetic Acceleration—Steady Flow

We come now to the third major class of electric propulsion mechanism: the acceleration of a body of ionized gas by the interaction of currents driven through the gas with magnetic fields established either by those currents or by external means. We have seen how such interactions can constrict a high current arc column in an electrothermal device, how they can contain an ionization discharge in an electron bombardment engine, and how they can direct an exhaust beam in a magnetic expansion thruster. We shall now examine the application of similar magnetic interactions for the direct acceleration of ionized gas flows.

In comparison with pure electrothermal and electrostatic mechanisms, electromagnetic interactions are phenomenologically more complex, analytically less tractable, and technologically more difficult to implement, and hence have lagged in their engineering application. As we shall see, however, they hold promise of providing a combination of high exhaust velocities with high mass flows, and thus justify serious consideration for prime space propulsion.

8-1 CLASSIFICATION OF ELECTROMAGNETIC ACCELERATORS

To illustrate the electromagnetic acceleration concept in its simplest form, consider a flow of ionized gas which is subjected to an electric field **E** and a magnetic field **B**, perpendicular to each other and to the gas velocity **u** (Fig. 8-1). If the gas has a scalar conductivity σ, a current density

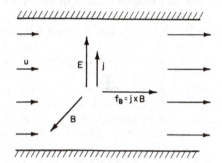

Fig. 8-1 Elementary electromagnetic accelerator.

$\mathbf{j} = \sigma(\mathbf{E} + \mathbf{u} \times \mathbf{B})$ will flow through it, parallel to **E**, and will interact with the magnetic field **B** to provide a distributed body force density $\mathbf{f_B} = \mathbf{j} \times \mathbf{B}$, in the streamwise direction **u**, which will accelerate the gas.

From a particle point of view, the process may be pictured in terms of the mean trajectories of the current-carrying electrons, which, in attempting to follow the applied electric field, are turned in the stream direction by the magnetic field. The streamwise momentum thus acquired by the electrons is transmitted to the bulk of the gas by collisions with the heavy particles or by microscopic polarization fields (Fig. 8-2).

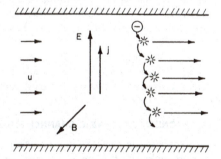

Fig. 8-2 Transfer of magnetic body force to gas stream by electron collisions.

In either representation, note that although it is the electric field which imparts the energy to the gas stream, no macroscopic net space charge is involved in the establishment of the body force, and thus there is no fundamental space-charge limitation on the mass flow density like that arising in pure electrostatic accelerators. Because they use a quasi-neutral ionized gas as a working fluid, devices of this class are often called *plasma thrusters*.

Physically, there are many ways to establish such interacting currents and magnetic fields in an ionized gas flow, in terms of which a subclassification of the family of electromagnetic accelerators may be made. First, one may distinguish between completely steady interactions wherein the current density patterns in the gas, the magnetic field, the flow velocity, and the thermodynamic properties of the gas remain constant in time at every point, and pulsed interactions wherein these elements undergo vigorous pulsations in time. In the former category, one then may further distinguish between the application of external magnetic fields and the use of the fields generated by the current patterns in the gas and its driving circuit. Pulsed acceleration also subdivides further into the series-coupled mode, wherein the discharge current passes directly through the gas between electrodes in contact with the gas, and inductively coupled modes, wherein currents are induced in the gas in response to primary current pulses or oscillations flowing in a closed circuit entirely external to the gas. The extremes of steady and pulsed operation are bridged by a third type of interaction, traveling wave acceleration, wherein an external array of programmed currents generate a continuous electromagnetic wave which propagates through the ionized gas, sweeping it along by interaction of the wave train with currents it induces in the gas.

Superimposed on this temporal subdivision of the phenomena are a variety of practical alternatives for electrode, channel, and field geometries; gas type and density; means of ionization of the gas; and operational details of insulation, switching, gas injection, etc., which distinguish particular devices. Table 8-1 displays some of the possible modes of electromagnetic acceleration which have been studied for possible propulsion application.

In the following sections of this chapter we shall examine the physical characteristics of the steady electromagnetic, or plasma, accelerators. Pulsed and traveling wave interactions will be discussed in Chap. 9.

8-2 MAGNETOGASDYNAMIC CHANNEL FLOW

Electromagnetic acceleration processes embody interlocking aspects of compressible gasdynamics, ionized gas physics, electromagnetic field theory, and particle electrodynamics, the individual analytical complexity of each of which contributes to formidable difficulties in adequate theoretical representation of the composite problems. Analytical progress normally follows only after incisive choice of grossly simplified models which

ELECTROMAGNETIC ACCELERATION—STEADY FLOW

Table 8-1 Classification of electromagnetic accelerators

Time scale of interaction	Steady		Pulsed		Traveling wave
Source of magnetic field	External coils or magnets	Self-induced	Self-induced		Coil sequence or transmission line
Ionization	External	Internal	Internal		External or internal
Primary current source	Direct current supply		Capacitor bank		Radio-frequency supply
Discharge coupling to circuit	Direct		Direct	Inductive	Inductive
Working fluid	Pure or seeded gas		Pure gas; vaporized liquid or solid	Pure gas	Pure gas
Channel geometry	Rectangular or coaxial; constant or variable cross section		Coaxial, pinch, parallel rail, ablating plug	Theta pinch, conical pinch, loop inductor	Rectangular, cylindrical, coaxial; constant or variable cross section
Other distinguishing features	Lorentz or Hall mode		Internal or external switch		Constant or variable phase velocity

retain only the essential physical aspects of the specific situations under study. As examples, the dominant characteristics of various low density steady accelerators can be described on the basis of charged-particle orbit theory like that discussed in Sec. 5-1; for certain pulsed plasma thrusters the entire body of gas to be accelerated may be represented as a constant mass, yet movable, element of a conventional electric circuit (slug model); most traveling wave accelerators are best approached in terms of the behavior of the propagating electromagnetic field patterns; etc. In each case, empirical observation or intuitive logic must precede the mathematical formulation.

Probably the most broadly useful approach to electromagnetic acceleration, however, is that of magnetogasdynamics, wherein one regards the ionized gas medium as a continuum fluid whose physical properties may be adequately described by a set of bulk parameters and whose dynamical behavior may be represented by an appropriate set of continuum conservation relations. Specifically, one adds to the conventional fluid transport parameters of viscosity and thermal conductivity a bulk electrical conductivity, whose relation to the particle properties of the gas under various circumstances has been discussed in Chap. 5. One then includes an electromagnetic body force in the usual gasdynamic equation of motion, allows for electric energy input in the energy relation, and appropriately modifies the equation of state and caloric relation to allow for ionization effects. The set then is closed by statement of Maxwell's equations and suitable constitutive relations, including an Ohm's law. Detailed development of this system of magnetogasdynamic equations is well presented in many texts devoted primarily to this subject [1–6]. In the notation we shall employ, these relations are expressed in the following forms.

Conservation of mass:

$$\frac{\partial \rho}{\partial t} + \nabla \cdot (\rho \mathbf{u}) = 0 \qquad (8\text{-}1)$$

Equation of motion:

$$\rho \left(\frac{\partial \mathbf{u}}{\partial t} + \mathbf{u} \cdot \nabla \mathbf{u} \right) = -\nabla p + (\mathbf{j} \times \mathbf{B}) + \mathbf{f}_v \qquad (8\text{-}2)$$

Energy balance:

$$\rho \left(\frac{\partial}{\partial t} + u \cdot \nabla \right) \left(c_p T + \frac{u^2}{2} \right) = \frac{\partial p}{\partial t} + \mathbf{j} \cdot \mathbf{E} + \phi_t + \phi_v - \phi_r \qquad (8\text{-}3)$$

Here p, ρ, c_p, and T are the pressure, density, specific heat, and absolute temperature of the gas, respectively; $\mathbf{u}$ is the flow velocity; $\mathbf{j}$, $\mathbf{E}$, and $\mathbf{B}$ are the current density and electric and magnetic induction fields. The symbols $\mathbf{f}_v$, ϕ_t, ϕ_v, and ϕ_r represent the net viscous body force density, the net thermal input by conduction processes, the net viscous dissipation, and the net radiant energy loss per unit volume, all of which will be neglected in the example applications to follow.

The appearance of $\mathbf{j} \times \mathbf{B}$ as the electromagnetic body force in the equation of motion is self-evident. Less so, perhaps, is the representation of the rate of total electric energy input as $\mathbf{j} \cdot \mathbf{E}$, with an explicit absence of $\mathbf{B}$ from the energy equation. Actually, $\mathbf{j} \cdot \mathbf{E}$ embodies both a dissipative, or Joule-heating, component and a useful work component, the latter identical with the scalar product of the electromagnetic body force with the stream velocity (Prob. 8-2).

The conservation relations must be supported by an equation of state,

$$p = p(\rho, T) \tag{8-4}$$

a caloric expression relating the specific heat or the specific enthalpy of the medium to its other thermodynamic properties,

$$c_p = c_p(\rho, T) \tag{8-5}$$

$$h = h(\rho, T) \tag{8-6}$$

and the necessary transport coefficients of electrical conductivity, viscosity, thermal conductivity, and radiation,

$$\boldsymbol{\sigma} = \boldsymbol{\sigma}(\rho, T, E, B) \tag{8-7}$$

$$\boldsymbol{\eta} = \boldsymbol{\eta}(\rho, T, E, B) \tag{8-8}$$

$$\cdots \cdots \cdots \cdots$$

where all these may be tensors, but all except $\boldsymbol{\sigma}$ participate only in $\mathbf{f_v}$, ϕ_t, ϕ_v, and ϕ_r, which we are henceforth neglecting.

Only three of Maxwell's relations are strictly needed:

$$\nabla \times \mathbf{E} = -\frac{\partial \mathbf{B}}{\partial t} \tag{8-9}$$

$$\nabla \times \mathbf{H} = \mathbf{j} \tag{8-10}$$

$$\nabla \cdot \mathbf{B} = 0 \tag{8-11}$$

In accordance with the quasi-neutral assumption, the free charge density which is the source of $\mathbf{D}$ does not contribute significantly to the dynamical problem [note that no term $\bar{q}\mathbf{E}$ appears in (8-2)]. If $\bar{q}$ is desired, it may be computed after the dynamical problem has been solved, from the remaining Maxwell relation:

$$\nabla \cdot \mathbf{D} = \bar{q} \tag{8-12}$$

Finally, three electromagnetic constitutive relations must be specified, two of which are almost always assignable their vacuum values:

$$\mathbf{D} = \epsilon_0 \mathbf{E} \tag{8-13}$$

$$\mathbf{B} = \mu_0 \mathbf{H} \tag{8-14}$$

The Ohm's law, however, as shown in Chap. 5, is more cumbersome, and can be written in various forms. Here we display it as a vector equation,

$$\mathbf{j} = \sigma_0 (\mathbf{E} + \mathbf{u} \times \mathbf{B}) + \mathbf{j_H} + \mathbf{j_I} \tag{8-15}$$

where j_H and j_I are possible contributions from Hall current and ion slip, to be discussed later, and all unsteady terms and species gradient terms have been omitted for our applications [1,2].

The system of relations (8-1) to (8-15) can be considerably reduced by eliminating several of the variables (Prob. 8-3). We shall leave them in this form, however, for application to the special problems to follow.

8-3 IDEAL STEADY FLOW ACCELERATION

Armed with the magnetogasdynamic relations, let us approach a particularly simple electromagnetic accelerator—a one-dimensional steady channel flow of an ideal gas with scalar conductivity—for the purpose of exploring some of the characteristic processes and estimating the range of performance attainable by this class of device. As shown in Fig. 8-3, let a channel formed by two conducting walls and two insulating walls contain a stream of ionized gas $u(x)$, across which are applied electric and magnetic fields $E(x)$ and $B(x)$ in the y and z directions, respectively.

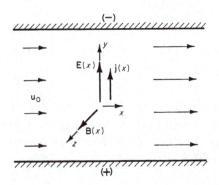

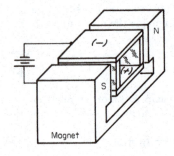

Fig. 8-3 One-dimensional steady electromagnetic accelerator.

We shall specify that the magnetic field generated by the current flowing in the gas is negligible compared with the applied $B(x)$, thereby separating the electromagnetic field relations (8-9) to (8-12) from the dynamics of the problem. Under these assumptions (8-1) to (8-6) reduce to the much simpler set

$$\rho u = F = \text{const} \tag{8-16}$$

$$\rho u \frac{du}{dx} = -\frac{dp}{dx} + jB \tag{8-17}$$

$$\rho u \frac{d}{dx}\left(c_p T + \frac{u^2}{2}\right) = jE \tag{8-18}$$

$$j = \sigma(E - uB) \tag{8-19}$$

$$p = \rho RT \tag{8-20}$$

$$\sigma = \sigma(\rho, T) \tag{8-21}$$

If c_p is presumed constant, consistent with the ideal gas assumption, this set of six relations contains eight variables; hence two must be specified in advance, for example, $E(x)$, $B(x)$ or $u(x)$, $p(x)$, along with the inlet or outlet conditions on the others.

Although no general analytic solutions to this one-dimensional problem have been achieved, several special cases may be reduced to closed form, e.g., adiabatic flows, isothermal flows, constant applied fields, etc. As an illustration, consider the family of isothermal flows, $dT/dx = 0$, for which the equation of motion and energy equation combine to yield

$$Fu\frac{du}{dx} = \xi\left(F\frac{du}{dx} + \frac{dp}{dx}\right) \tag{8-22}$$

where $\xi \equiv E/B$. Noting that

$$\frac{dp}{dx} = RT\frac{d\rho}{dx} = -FRT\frac{1}{u^2}\frac{du}{dx}$$

F and du/dx drop out of (8-22), and we are left with the necessary relation between u and ξ along the channel:

$$\xi = \frac{u^3}{u^2 - RT} \tag{8-23}$$

The quantity $(RT)^{1/2}$ is the isothermal sound speed of the medium, which we hereafter denote by a_T. Note that ξ has a singularity at $u = a_T$, below which it becomes negative. We shall avoid this difficulty by requiring supersonic inlet velocities, $u_0 \geq a_T$.

Returning to the energy equation once again, we write

$$Fu\frac{du}{dx} = \sigma E^2\left(1 - \frac{u}{\xi}\right) = \sigma B^2(\xi^2 - u\xi) \tag{8-24}$$

where the choice of form of the right-hand side depends on the selection of the remaining constraint available to us. Insertion of (8-23) in the first form of (8-24) yields

$$Fu^3\frac{du}{dx} = \sigma E^2 a_T^2 \tag{8-25}$$

which can be integrated immediately if we prescribe σE^2 constant along the channel. The result may be set in dimensionless form:

$$u^* = \left[1 + 4\frac{\sigma E^2 L}{Fu_0^2}\left(\frac{a_T}{u_0}\right)^2 x^*\right]^{1/4} \tag{8-26}$$

where $u^* = u/u_0$, L is the channel length, and $x^* = x/L$ is the dimensionless coordinate. If, instead, we prescribe σB^2 constant along the channel and use (8-23) in the second form of (8-24), the energy equation becomes

$$F\frac{du}{dx} = \sigma B^2 a_T^2 \frac{u^3}{(u^2 - a_T^2)^2} \tag{8-27}$$

Here we must integrate term by term and leave the result in transcendental form:

$$\left(\frac{u_0}{a_T}\right)^2(u^{*2} - 1) - 4\ln u^* + \left(\frac{a_T}{u_0}\right)^2\left[1 - \left(\frac{1}{u^*}\right)^2\right] = 2\frac{\sigma B^2 L}{F}x^* \tag{8-28}$$

The hybrid constraint, $\sigma EB = $ const, leads in similar fashion to the result

$$\left(\frac{u_0}{a_T}\right)^2(u^{*3} - 1) - 3(u^* - 1) = 3\frac{\sigma EBL}{Fu_0}x^* \tag{8-29}$$

A particularly simple solution is available under the constraint of constant power input per unit length of channel, $jE = P$.

$$u^* = \left(1 + 2\frac{PL}{Fu_0^2}x^*\right)^{1/2} \tag{8-30}$$

Note that while a constant electric field, $dE/dx = 0$, is a legitimate special case of (8-26), and likewise $dB/dx = 0$ for (8-28), simultaneous requirement of $dE/dx = dB/dx = 0$ in (8-29), or $dj/dx = dE/dx = 0$ in (8-30), would overspecify the problem in view of the constraint already imposed on u by the isothermal assumption (8-23).

Once $u(x)$ is found from (8-26), (8-28), (8-29), or (8-30), $p(x)$ follows from the gas law and continuity requirement:

$$p = \frac{F a_T^2}{u} \tag{8-31}$$

and $\xi(x)$ may be computed from (8-23).

The dimensionless ratio $\sigma B^2 L/F$ appearing in (8-28) is characteristic of problems of this sort, and will be referred to as the magnetic interaction parameter β. The ratios $\sigma E^2 L/F u_0^2$ in (8-26), $\sigma EBL/F u_0$ in (8-29), and $PL/F u_0^2$ in (8-30) are quite similar to it in the sense that all reflect the ratio of the magnetic body force to the gas inertia, and thus are indicative of the intensity of the electromagnetic interaction. Figure 8-4 displays

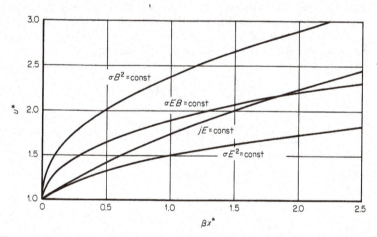

Fig. 8-4 Velocity profiles along one-dimensional isothermal accelerator under various field constraints.

the four solutions (8-26), (8-28), (8-29), and (8-30) as functions of these interaction parameters for an inlet velocity slightly above a_T. The most notable feature of these solutions is the rather weak dependence of u^* on βx^*, which must be a consequence of the isothermal constraint on the constant area flow. To achieve more vigorous acceleration, one or both of these constraints must be relieved.

One might consider replacing the isothermal assumption by an adiabatic requirement, $p/\rho^\gamma = $ const, and indeed special solutions can thereby be obtained [1]. However, this assumption logically also demands an infinite electrical conductivity to preclude electrical heating of the flow, a rather inappropriate description for the modest temperature gas flows which a real channel could tolerate.

If no thermodynamic route is prescribed for the flow, one is free to specify both of the applied field profiles, $E(x)$ and $B(x)$, but this class of problem tends to an algebraic complexity which may obscure the physical behavior. Even the special case of constant fields becomes somewhat ponderous. Persisting with the ideal gas assumption, we make use of the gas constant, $R = c_p - c_v$, and the adiabatic exponent, $\gamma = c_p/c_v$, in the equation of state, thereby eliminating the temperature from the problem:

$$c_p T = \frac{\gamma}{\gamma - 1} \frac{p}{\rho} = \frac{a^2}{\gamma - 1} \tag{8-32}$$

where a is now the adiabatic sound speed $(\gamma R T)^{1/2}$. Using this substitution, the Ohm's law (8-19), and the continuity condition in the form $\rho(du/dx) = -u(d\rho/dx)$, the equation of motion and energy equation become two simultaneous relations in du/dx and dp/dx:

$$F \frac{du}{dx} = -\frac{dp}{dx} + \sigma(E - uB)B \tag{8-33}$$

$$\frac{\gamma}{\gamma - 1}\left(u \frac{dp}{dx} + p \frac{du}{dx}\right) + Fu \frac{du}{dx} = \sigma(E - uB)E \tag{8-34}$$

Eliminating $\sigma(E - uB)$ from these leads directly to an integral condition on the flow,

$$\frac{\gamma}{\gamma - 1} pu + \frac{F}{2} u^2 - \xi(Fu + p) = \mathcal{K} \tag{8-35}$$

where the constant $\mathcal{K}$ is simply the left-hand side evaluated at the inlet, or at the maximum velocity which can be obtained, $u = \xi$:

$$\mathcal{K} = \frac{\gamma}{\gamma - 1} p_0 u_0 + \frac{F}{2} u_0^2 - \xi(Fu_0 + p_0) = \frac{p_\xi \xi}{\gamma - 1} - \frac{F\xi^2}{2} \tag{8-36}$$

To obtain $u(x)$, dp/dx is computed as a function of u from (8-35) and substituted into (8-33), which may then be rearranged and integrated to

ELECTROMAGNETIC ACCELERATION—STEADY FLOW

the imposing form

$$\left[\frac{\gamma(\gamma-1)}{2} - 1 - \gamma(\gamma-1)\frac{u_0}{\xi}\left\{1 + \frac{1}{\gamma M_0^2}\right.\right.$$
$$\left.\left. - \left[\frac{1}{2} + \frac{1}{(\gamma-1)M_0^2}\right]\frac{u_0}{\xi}\right\}\right] \ln\left|\frac{1 - \frac{u_0}{\xi}u^*}{1 - \frac{u_0}{\xi}}\right|$$
$$+ \left[\frac{\gamma^2}{\gamma-1}\frac{u_0}{\xi}\left\{1 + \frac{1}{\gamma M_0^2} - \left[\frac{1}{2} + \frac{1}{(\gamma-1)M_0^2}\right]\frac{u_0}{\xi}\right\} - \frac{\gamma+1}{2}\right]$$
$$\left[\frac{\frac{u_0}{\xi}(u^* - 1)}{\left(1 - \frac{\gamma}{\gamma-1}\frac{u_0}{\xi}u^*\right)\left(1 - \frac{\gamma}{\gamma-1}\frac{u_0}{\xi}\right)}\right.$$
$$\left. + \frac{(\gamma-1)^2}{\gamma}\ln\left|\frac{\frac{\gamma-1}{\gamma} - \frac{u_0}{\xi}u^*}{\frac{\gamma-1}{\gamma} - \frac{u_0}{\xi}}\right|\right] = \beta x^* \quad (8\text{-}37)$$

where M_0 denotes the inlet Mach number u_0/a_0. The profile of u^* tends to be sensitive to the choice of u_0/ξ, and indeed can become double-valued or bounded by an asymptote at $[(\gamma-1)/\gamma]\xi/u_0$ if u_0/ξ is chosen injudiciously. Only very special combinations of inlet conditions and ξ will produce a smooth acceleration to the terminal velocity, $u = \xi$ (Prob. 8-4).

These restrictions on the available acceleration routes will become more evident if we adopt a slightly different approach to the problem which will also permit us to examine certain variable $E(x)$, $B(x)$ solutions [7]. Returning to the momentum and energy relations in the forms of (8-33) and (8-34), let us now eliminate dp/dx and solve for du/dx:

$$\frac{du}{dx} = \frac{\sigma(E - uB)\{E - [\gamma/(\gamma-1)]uB\}}{\rho u^2 + [\gamma/(\gamma-1)](p - \rho u^2)} \quad (8\text{-}38)$$

or in terms of the flow Mach number, $M = u/a$,

$$\frac{du}{dx} = \frac{\sigma B^2}{p}\frac{1}{1 - M^2}(u - \xi)(u - \eta) \quad (8\text{-}39)$$

The quantities $\xi \equiv E/B$ and $\eta = [(\gamma-1)/\gamma]\xi$ appear to be two characteristic velocities whose magnitudes relative to u will determine the sign of du/dx. A similar expression may be derived for the streamwise gradient in the Mach number.

$$\frac{dM}{dx} = \frac{\sigma B^2}{ap} \left\{ \frac{1 + [(\gamma - 1)/2]M^2}{1 - M^2} \right\} (u - \xi)(u - \zeta) \qquad (8\text{-}40)$$

where a is now the conventional sound speed $(\gamma RT)^{\frac{1}{2}}$, and ζ is another characteristic velocity,

$$\zeta = \frac{1 + \gamma M^2}{2 + (\gamma - 1)M^2} \eta \qquad (8\text{-}41)$$

Note that the signs of the derivatives of u and M are determined by the particular ordering of u with respect to a, ξ, η, and ζ, and are not always the same. Table 8-2 displays the signs of du/dx and dM/dx for various regimes of u and M. Note again that this flow is liable to "choking," that is, du/dx can become infinite as M approaches unity. This can be precluded only if u reaches the value of ξ or η as $M \to 1$. On the other hand, if u attains the value of ξ or η at any M other than unity, the flow will cease to accelerate.

Table 8-2 Dependence of flow velocity and Mach number derivatives on characteristic velocities

M	u	$\dfrac{du}{dx}$	$\dfrac{dM}{dx}$
<1	$<\zeta$	+	+
	$\zeta < u < \eta$	+	−
	$\eta < u < \xi$	−	−
	$>\xi$	+	+
>1	$<\eta$	−	−
	$\eta < u < \zeta$	+	−
	$\zeta < u < \xi$	+	+
	$>\xi$	−	−

$\xi = \dfrac{E}{B} \qquad \eta = \dfrac{\gamma - 1}{\gamma} \xi$

$\zeta = \dfrac{1 + \gamma M^2}{2 + (\gamma - 1)M^2} \eta$

SOURCE: E. L. Resler, Jr., and W. R. Sears, The Prospects for Magnetoaerodynamics, *J. Aeron. Sci.*, vol. 25, no. 4, pp. 235–245, April, 1958.

ELECTROMAGNETIC ACCELERATION—STEADY FLOW

This behavior may also be displayed as a graph in the uM plane. In Fig. 8-5, the horizontal lines $u = \xi$ and $u = \eta$ and the vertical line $M = 1$ represent barriers to the flow in the sense that it is not possible to cross them except in very special ways. For example, the only self-consistent continuous acceleration of the gas from subsonic to supersonic

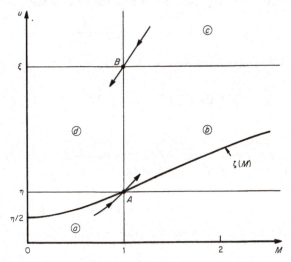

Fig. 8-5 Domains of velocity and Mach number for one-dimensional magnetogasdynamic flow ($\xi = E/B$; $\eta = [(\gamma - 1)/\gamma]\xi$; $\zeta = \{(1 + \gamma M^2)/[2 + (\gamma - 1)M^2]\}\eta$). (*From E. L. Resler, Jr., and W. R. Sears, The Prospects for Magneto-aerodynamics, J. Aeron. Sci., vol. 25, no. 4, pp. 235–245, April, 1958.*)

flow is that passing through $u = \eta$, $M = 1$ from zone ⓐ to zone ⓑ (arrow A). All other routes reach inconsistencies at the barriers that physically must be resolved by shock waves or other dissipative regions not included in this analysis. Likewise, there is only one solution for continuous deceleration from supersonic to subsonic flow, that passing through $u = \xi$, $M = 1$ from zone ⓒ to ⓓ (arrow B). This latter route is of only academic interest for an accelerator since region ⓒ represents a domain of extraction of energy from the gas ($uB > E$). For given assignments of E, B, and $\sigma(T,p)$ there is no guarantee that a flow accelerated from an arbitrary initial state will pass through the tunnel A. To do so $E(x)$, $B(x)$, and the inlet conditions would need to be chosen prudently. Otherwise the barrier problem can be circumvented only by injecting the ionized gas flow into the accelerator at a supersonic velocity between η and ξ.

A constant field accelerator confined to $\eta \le u \le \xi$ would be of little interest for propulsion, since the attainable velocity ratio, $\xi/\eta = \gamma/(\gamma - 1)$, is rather small. However, if one is willing to program the fields along the channel, the local acceleration can be optimized and the attainable velocity increment substantially increased. In particular, it would seem advantageous to retain the product $(u - \xi)(u - \eta)$ at its (negative) maximum throughout the flow, namely, at the value

$$\xi = \frac{2\gamma - 1}{2(\gamma - 1)} u \qquad (8\text{-}42)$$

This produces an optimum local acceleration

$$\frac{\widehat{du}}{dx} = \frac{\sigma B^2}{p}\left(\frac{1}{M^2 - 1}\right)\left[\frac{1}{4\gamma(\gamma - 1)}\right]u^2 = \frac{1}{4(\gamma - 1)}\left(\frac{\sigma B^2}{\rho u}\right)\left(\frac{M^2}{M^2 - 1}\right)u \qquad (8\text{-}43)$$

To integrate this relation along the channel, we shall need an algebraic connection between u and M for the condition (8-42). Dividing (8-39) by (8-40) yields a differential equation,

$$\frac{du}{dM} = \frac{u(u - \eta)}{M(u - \xi)\{1 + [(\gamma - 1)/2]M^2\}} \qquad (8\text{-}44)$$

which, upon use of (8-42), integrates to the form

$$u = \text{const}\left[\frac{M^2}{(2\gamma + 1)/\gamma - M^2}\right]^{1/(2\gamma+1)} \qquad (8\text{-}45)$$

With this, (8-43) may be integrated to the desired optimized velocity profile along the channel:

$$\frac{4\gamma(\gamma - 1)}{2\gamma + 1}\left\{\frac{\gamma + 1}{\gamma}\ln u^* - \frac{\gamma M_0^2 - (2\gamma + 1)}{(2\gamma + 1)\gamma M_0^2}\left[\left(\frac{1}{u^*}\right)^{2\gamma+1} - 1\right]\right\} = \beta x^* \qquad (8\text{-}46)$$

where the interaction parameter here may have the form

$$\beta = \frac{L}{F}\int_0^1 \sigma B^2\, dx^* \qquad (8\text{-}47)$$

Figure 8-6 displays the development of u^* along the channel for various inlet Mach numbers, M_0. Note that, unlike the isothermal profiles derived earlier, we have here a vigorous acceleration which attains a tenfold velocity increase in about two characteristic channel lengths.

Note also that the profiles are relatively insensitive to the inlet Mach number, particularly for large M_0. The corresponding variation

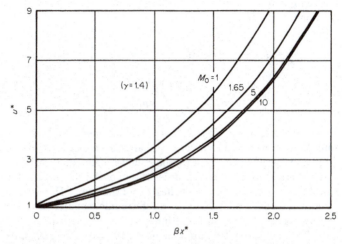

Fig. 8-6 Development of flow velocity along optimized one-dimensional magnetogasdynamic accelerator. *(From E. L. Resler, Jr., and W. R. Sears, The Prospects for Magneto-aerodynamics, J. Aeron. Sci., vol. 25, no. 4, pp. 235–245, April, 1958.)*

in M along the channel follows from (8-45). It is seen that, regardless of the inlet value M_0, the Mach number approaches the same asymptotic value as u^* increases along the channel:

$$M_\infty \to \left(\frac{2\gamma + 1}{\gamma}\right)^{1/2} = \begin{cases} 1.65 & \text{for } \gamma = 1.4 \\ 1.61 & \text{for } \gamma = 1.67 \end{cases} \quad (8\text{-}48)$$

The convergence to this value is very rapid; it is nearly complete for $u^* \approx 3$. The physical implication of this is that our constant-area assumption is constraining much of the energy input to appear as a heating of the gas, with a corresponding increase in the sound speed and lowering of acceleration efficiency (Prob. 8-5). It seems reasonable to expect the performance of the device to improve if the flow is allowed to expand in a diverging channel, thereby recovering some of this thermal energy.

A gently varying channel cross section may be introduced via a "quasi one-dimensional" formulation, wherein one rewrites the continuity relation in the form

$$\rho u A = F \quad (8\text{-}49)$$

and retains the one-dimensional equation of motion and energy equation as statements of the mean flow properties at any given axial position x.

By manipulations identical with those above, one finds (8-39) and (8-40) modified to the forms

$$\frac{du}{dx} = \frac{1}{1-M^2}\left[\frac{\sigma B^2}{p}(u-\xi)(u-\eta) - \frac{u}{A}\frac{dA}{dx}\right] \quad (8\text{-}50)$$

$$\frac{dM}{dx} = \frac{1+[(\gamma-1)/2]M^2}{1-M^2}\left[\frac{\sigma B^2}{ap}(u-\xi)(u-\zeta) - \frac{M}{A}\frac{dA}{dx}\right] \quad (8\text{-}51)$$

The new terms in dA/dx reflect the familiar gasdynamic result that subsonic flows accelerate in converging channels, decelerate in diverging channels, and vice versa for supersonic flows. As suspected, a diverging channel is seen to increase the acceleration of the supersonic flow we have just discussed and to reduce the heating of the flow. However, a rigorous optimization of $A(x)$, $B(x)$, $E(x)$ for given inlet conditions is hardly justified until many other participating effects are included.

Various alternative approaches to the expanding-channel problem can be considered. One may return to the family of isothermal flows and solve the quasi one-dimensional equations for $A(x)$ under other appropriate constraints. For example, the velocity profile may be constrained to the area variation in some convenient functional form, such as a power law [8-10], or it may be directly expressed in powers of the channel length [11]. If large area variation is to be considered, the quasi one-dimensional approximation must be dropped and some form of two- or three-dimensional representation employed. Perhaps the simplest cases of this kind are those which can be represented by two-dimensional source flows, wherein the magnetogasdynamic equations are cast in cylindrical coordinates, and all derivatives in azimuthal angle θ and cross-stream dimension z are neglected. That is, the flow properties are presumed to vary only along radial lines extending from a (fictitious) source axis in the flow. Numerical calculations based on such models confirm the advantages of rapid channel expansion to increase the attainable exhaust speeds for given electrical input [12].

In closing this outline of ideal magnetogasdynamic flows, we might attempt some estimate of the limits imposed on the attainable performance of a steady electromagnetic space thruster by practical operating conditions. From the foregoing examples it appears that the magnetic interaction parameter, $\beta = \sigma B^2 L/\rho u$, is the major index of the intensity of the acceleration process, so it would be instructive to place a limit on the attainable magnitude of this quantity. It is difficult to do this unambiguously because of the several interlocking restrictions that must be placed on the satisfactory operation of such an accelerator. For example:

1. The conductivity of the working fluid is prescribed by its composition, density, and temperature; the latter two, in turn, are limited by the tolerable heat transfer to the channel walls.
2. There are limits to the current density which can be extracted from material electrodes at given temperatures and local electric field strengths.
3. The effective electric field, $E - uB$, must not be so high that the desired uniform discharge breaks down into discrete arc columns. The critical value of this field is a strong function of gas density and composition, and possibly of electrode surface conditions.
4. The strength of the applied magnetic field is limited by the size and type of magnet which could reasonably be carried in the thruster package, and clearly is dependent on the channel dimensions.
5. The gas density cannot become so low that cross-field, or Hall, conduction dominates the conductivity. This in turn depends on the applied B field.
6. The channel size must be reasonable for a spacecraft.
7. The exhaust velocity should be in the desired range for space propulsion.

To construct a numerical estimate we shall need to presume some of the results of following sections and introduce certain empirical data. First, we are interested in exhaust speeds above 10^4 m/sec and in channel lengths of the order of 1 m for a space thruster. Over this dimension we can probably not provide a B field in excess of 1 weber/m². At this field strength, the gas density should nowhere drop below about 0.1 of its STP value, or about 0.1 kg/m³, if Hall currents are not to become dominant. As we shall see, it is doubtful that gas conductivities above 1,000 mhos/m can be sustained short of an arc discharge, a value that is consistent with an observed empirical limit on diffuse discharge current densities of about 10^7 amp/m² $[j = \sigma(E - uB) \approx \sigma uB]$. We are thus led to an upper limit on β of the order of unity, which in the light of our earlier results, such as those shown in Figs. 8-4 and 8-6, is less than optimum for our purpose. For flow injection at 10^3 m/sec, for example, we need velocity ratios from 10 to 50, but these are not readily achievable at $\beta = 1$, even for our optimized accelerator (8-46). Injection at higher speed relieves some of this burden from the electromagnetic stage, but may require a "preaccelerator," such as an arcjet, to provide the high speed inlet flow at the required density. Variable-area accelerators brighten this picture somewhat, but we have not yet considered a variety of gasdynamic and electromagnetic losses which will detract from the performance of the idealized models.

On the positive side, it should be noted that our mythical channel is capable of handling very large mass flows and thus produces very high thrust densities. For the chosen values, ρu^2 has the value 10^7 newtons/m^2, and thus a 10-cm^2 channel awaits only a suitable space power plant to impart some 50 megawatts of thrust power to the stream!

8-4 THERMAL AND VISCOUS LOSSES

For purposes of simple illustration of the scalar crossed-field interaction, we have so far ignored several complicating physical processes, any one of which may influence, or even dominate, the flow in a real accelerator. These consist of certain departures from the assumed patterns of orthogonal electromagnetic fields, currents, and gas flow, which will be discussed in succeeding sections, and of viscous and thermal losses in the body of the flow and at the channel walls. One type of thermal loss, the frozen flow inefficiency, was discussed in the context of electrothermal acceleration in Chap. 6, but is equally appropriate here. To maintain sufficient electrical conductivity to carry the required current density, the gas must be substantially ionized throughout the accelerator duct. To minimize viscous losses, it is desirable to exhaust the flow shortly after the acceleration is completed, but if this is done too abruptly, little of the energy invested in ionization or in the associated store of excitation and thermal energy will be recovered. An exhaust nozzle optimization problem is thus indicated, along with an evident need to employ gases of high molecular weight, which also exhibit good conductivity at modest temperatures.

Even more serious, however, are the viscous and thermal losses experienced by the hot gas stream at the duct walls and the associated distortions of the dynamic and electrical properties of the flow in the nearby regions. Conventional high speed gas flows in channels are well known to develop viscous and thermal boundary layers on the walls which can subtract considerable momentum and energy from the stream and cause severe departures from the inviscid velocity profile over significant portions of the channel cross section. In magnetogasdynamic flows of the type we are considering, such disturbances in the dynamical properties of the flow will be transcribed into corresponding distortions of the electrical properties of the fluid, and thus into further dynamical disturbances. For example, the temperature gradient through a thermal boundary layer will predicate a gradient in the electrical conductivity of the gas, with corresponding variations in the resistive energy deposition and in the local electric field, which in turn will reinfluence the dynamical processes in this region. Near electrode surfaces there are superimposed the additional disturbances of anode or cathode sheaths characteristic of any gaseous discharge. Even on insulator surfaces, boundary conditions on electric and magnetic field components must be met (Prob. 2-1). From all these interacting processes there results a net heat transfer rate which determines surface erosion, and thereby the maximum free stream flow temperature, and a viscous skin friction which is the dominant dynamical loss in most accelerators of this type.

Detailed analysis of the magnetogasdynamic boundary layer problem would take us too far afield. Certain limited solutions to it have been found [13,14] which confirm various intuitively reasonable effects and evaluate their relative importance for specific circumstances. For example, it is found that significant amounts of heat may be liberated within the boundary layer by resistive dissipation over regions of low conductivity, as well as by conventional viscous dissipation. From these sources, and from the high temperature free stream, heat is transported to the wall by the usual fluid conduction and convection processes, and also by free electrons, both in conduction and in field-enhanced diffusion. There is found to be some increase in electric field in the boundary layer to maintain current continuity over the lower conductivity region, but this effect is somewhat ameliorated by the reduction in the back emf, $\mathbf{u} \times \mathbf{B}$, in the same region.

Quantitative calculations for such boundary layers based on relatively simple free stream flows predict levels of viscous drag and wall heat flux which are discouraging to the propulsion application of the scalar crossed-field accelerator. In the range of operation proposed earlier, many kilowatts per square centimeter would be transferred to the electrodes, which thus would have to be vigorously cooled to retain their structual integrity. Unless this could be accomplished regeneratively, a large fraction of the electrical energy input would be lost from the stream, and the efficiency of the device would be correspondingly low. Experimental studies have tended to confirm this pessimism. Devices of this class display propensities to serious electrode and insulator erosion which not only lower their dynamical performance, but complicate their operation in the sense of contaminating the flow, changing the channel geometry, and limiting their operational lifetime (Fig. 8-7).

It appears that steady high enthalpy magnetogasdynamic acceleration, already of marginal propulsion performance in the inviscid idealization, probably cannot function effectively in channels of small width-to-length aspect ratio. For propulsion purposes, we must turn to major modifications of the original concept. These might include operation in large aspect ratio, two-dimensional configurations wherein electrode surface is minimized; establishment of some form of magnetic protection for the solid surfaces; use of highly seeded gases and/or nonequilibrium ionization, which would provide adequate electrical conductivity at lower temperatures; conduction in some tensor mode, in which the electrodes

Fig. 8-7 Experimental crossed-field accelerator, using arcjet source. Note electrode erosion and separation of luminous gases from electrode surfaces. *(AVCO, Space Systems Division, Wilmington, Mass.)*

bear less current density, or unsteady operation, whereby the mean value of the intermittent wall losses would be more tolerable. Several of these possibilities will be examined in the following sections.

8-5 FIELD GEOMETRY CONSIDERATIONS

The second idealization of the simple accelerator of Sec. 8-3 lies in its particular assumptions of field and current conduction patterns. Several possible fundamental contradictions are ignored in that model, any one of which may become important in a given real situation. Specifically, we have violated Maxwell's relations, ignored channel end effects, inconsistently neglected tensor conductivity components, and improperly dropped self-induced magnetic fields! This and the two following sections will attempt to repair some of this distortion.

Early in Sec. 8-2 it was stated that the neglect of self-induced fields uncoupled Maxwell's relations from the magnetogasdynamic equations of motion. In point of fact this is only partially true; it still remains to satisfy the field curl relations

$$\frac{\partial E_x}{\partial y} = \frac{\partial E_y}{\partial x} \tag{8-52}$$

$$\frac{\partial B_x}{\partial z} = \frac{\partial B_z}{\partial x} + \mu j_y \tag{8-53}$$

With reference to Fig. 8-3, relation (8-52) implies that the assumption of an electric field solely in the y direction is inconsistent with a variation of that field along the channel, $\partial E_y/\partial x$, such as that utilized in our optimized accelerator [(8-42)]. Although we are free to set $E_x = 0$ at any one position, say, on the centerline of the flow, a finite component must arise at other transverse locations, for example, $E_x = (A/2)(\partial E_y/\partial x)$, at the walls of a channel of width A. It then follows from the assumption of scalar conductivity that corresponding axial components of current exist off axis, becoming largest at the channel walls. Since electric field and current density vectors must be normal to highly conducting surfaces, however, we are led to the necessity of a converging or diverging electrode geometry (Fig. 8-8). This in turn is inconsistent with a strictly one-

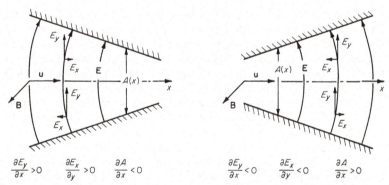

Fig. 8-8 Electric field curvature and corresponding channel area variation necessitated by irrotationality of **E**.

dimensional flow assumption, $dA/dx = 0$. Relation (8-53) implies a similar difficulty with the assumption of B solely in the z direction, even if the current density j_y is regarded as a negligible source. The import of our inconsistencies in this regard is to restrict the validity of the previous calculations to narrow channels and weak axial gradients in the fields (Prob. 8-6). Other cases must be approached with the inclusion of (8-52) and (8-53) and the appropriate electrode boundary conditions in the formulation of the magnetogasdynamic problem.

Distortion of the electric and magnetic field patterns also may occur near the channel inlet and exit as a result of the discontinuities in the electrode and magnet geometry in these regions. For example, a pair of plane parallel electrodes terminating abruptly at a given axial position cannot sustain a uniform parallel electric field of similarly abrupt termination. Rather, consistent with the solution to Laplace's equation for this boundary configuration, the field will balloon out of the aperture and weaken somewhat inside (Fig. 8-9). The magnetic field likewise will have a fringe region near the termination of its source.

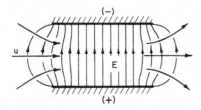

Fig. 8-9 Field fringing effects in magnetogasdynamic accelerator.

In the inlet and exit field fringing regions one must thus expect lower current densities with significant axial components near the electrodes and corresponding components of the $\mathbf{j} \times \mathbf{B}$ body force directed normal to the axis. At the channel entrance, for example, these forces will tend to constrict the stream toward the axis, while at the exit they will tend to expand the stream. If the magnetic field should end more abruptly than the electric field, there could be some regions where current flows outside of any significant B field, thereby dissipating energy in Joule heating without accomplishing useful acceleration.

The seriousness of these end effects clearly depends on the aspect ratio of the channel and on the specific details of the electrode and magnet geometry. In principle, they could be reduced by the use of electrode and magnet guard-surface arrangements to sustain uniform fields beyond the current flow region, but this sophistication might be too ponderous for a space thruster. On the other hand, there are cases where fringing field gradients are beneficial to accelerator operation, as in the magnetic expansion thrusters discussed earlier and in the self-field accelerators to follow. Indeed, since free charges can acquire organized drift motions in nonuniform magnetic fields (Prob. 5-2), field-gradient acceleration at the exit may enhance the performance of a variety of accelerators.

8-6 TENSOR CONDUCTIVITY: HALL EFFECT AND ION SLIP

The hypothetical accelerators of Sec. 8-3 presumed that the electrical conductivity of the medium remained scalar in the presence of the applied magnetic field. However, we have previously seen (Sec. 5-4) that a biasing field of this sort will curve the trajectories of the conduction electrons and thereby give rise to a component of current mutually orthogonal to the crossed **E** and **B** fields. The relative importance of this Hall current component to that along the **E** field depends on the ratio of the electron gyro frequency to its collision frequency, ω_B/ν_c. Specifically, for the geometry of Fig. 8-3, we must allow two components of current:

$$j_y = \sigma_{yy}(E - uB) = \frac{1}{(\omega_B/\nu_c)^2 + 1}\sigma_0(E - uB) \tag{8-54}$$

$$j_x = \sigma_{xy}(E - uB) = -\frac{\omega_B/\nu_c}{(\omega_B/\nu_c)^2 + 1}\sigma_0(E - uB) \tag{8-55}$$

where again $\sigma_0 = ne^2/m\nu_c$. Recall that the ratio ω_B/ν_c, or its equivalent, $\omega_B\tau$, where τ is the mean free time between electron collisions, is frequently called the Hall parameter, Ω (Sec. 5-2). If $\Omega \ll 1$, the gas conducts in the simple scalar fashion assumed earlier. If $\Omega \approx 1$, the Hall current component j_x becomes important, and the parallel current j_y is reduced. If $\Omega \gg 1$, the Hall current dominates the problem, but attains only $1/\Omega$ the magnitude of the scalar current for $\Omega \ll 1$.

It is thus imperative to ascertain the magnitude of Ω for any particular domain of operation of the crossed-field accelerator. The gyro frequency follows directly from its single-particle definition:

$$\omega_B = \frac{eB}{m} = 1.76 \times 10^{11}B \quad \text{sec}^{-1} \text{ for } B \text{ in webers/m}^2 \tag{8-56}$$

The electron collision frequency is less precisely definable, however, since it is an average over the entire electron swarm and since the relevant atomic cross sections are not exactly known. As originally introduced (Sec. 5-2), ν_c was an effective collision frequency for transport processes, and hence only approximately related to the isolated-particle collision cross sections [Eq. (5-16)]:

$$\nu_c \approx \sum_j n_j Q_{ej}{}^{(p)} \bar{v}_j \tag{8-57}$$

where $\bar{v}_j$ is the mean speed of the electrons with respect to various species j particles with which they collide. Except for very strong electric fields, very low densities, or other pathological situations, the free electrons may be expected to retain an essentially isotropic maxwellian distribution, with mean thermal velocity defined by their temperature:

$$\bar{v} = \left(\frac{8kT}{\pi m}\right)^{1/2} = 6.2 \times 10^3 T^{1/2} \quad \text{m/sec} \quad (8\text{-}58)$$

This is normally so much higher than the speeds of the heavy particles with which they collide, that it serves as the v_j required in (8-57). The relative effectiveness of various species in interrupting the electron orbits clearly depends on the composition of the gas, particularly on its ionization level. If the gas is so slightly ionized that the electrons collide most frequently with neutral particles, (8-57) becomes, simply,

$$\nu_c \approx n_0 Q_{e0} \bar{v} \approx 6.2 \times 10^3 T_e^{1/2} n_0 Q_{e0} \quad (8\text{-}59)$$

As we saw in Sec. 4-2, electron-neutral elastic cross sections for all but the alkali atoms are in the range of 10^{-20} to 10^{-19} m², and thus we might write a rough index of the Hall parameter for a mildly ionized common gas in the form

$$\Omega \approx \frac{B}{\tilde{\rho} \tilde{Q}_0 \tilde{T}_e^{1/2}} \quad (8\text{-}60)$$

where $\tilde{\rho}$ = density relative to STP
$\tilde{Q}_0$ = cross section, in units of 10^{-20} m²
$\tilde{T}_e$ = electron temperature, in units of electron volts

The Hall parameter tends to lower as the ionization level of the gas increases because of the disproportionately large influence of electron-ion coulomb encounters on the effective collision frequency. Well below 1 percent ionization, these interactions completely control the electron migration, and (8-57) effectively becomes

$$\nu_c \approx n_+ Q_{e+}^{(p)} \bar{v} \quad (8\text{-}61)$$

where the energy-dependent coulomb cross section $Q_{e+}^{(p)}$ has been discussed in Sec. 4-3:

$$Q_{e+}^{(p)} = \frac{\pi e^4}{(4\pi\epsilon_0)^2 \mathcal{E}^2} \ln \Lambda \approx \frac{5.4 \times 10^{-9}}{T_e^2} \quad \text{mks} \quad (8\text{-}62)$$

Here we have again taken $\ln \Lambda \approx 10$ and set $\mathcal{E} = \frac{1}{2} m \bar{v}^2$. Using the same mean velocity, the electron-ion collision frequency may be written

$$\nu_c \approx 3 \times 10^{-5} \frac{n_+}{T_e^{3/2}} \quad (8\text{-}63)$$

which, in comparison with (8-59), provides a criterion for the relative importance of the two classes of collision (Prob. 8-8). If (8-63) prevails, the Hall parameter becomes

$$\Omega \approx 6 \times 10^{15} \frac{BT_e^{3/2}}{n_+} \approx 3 \times 10^{-4} \frac{B\tilde{T}_e^{3/2}}{\alpha\tilde{\rho}} \qquad (8\text{-}64)$$

where $\alpha \equiv n_+/(n_+ + n_A)$ is the fractional ionization, and must be independently related to T_e and ρ by a Saha equation or other appropriate statement (Prob. 8-8).

Regardless of the particular collision process which dominates, we see from (8-60) or (8-64) that attempts to increase the magnetic interaction parameter β by reducing the density level of the gas or by increasing the magnetic field will drive the Hall parameter upward. In those domains of accelerator operation where it becomes significantly large, one then has the alternatives of allowing the Hall current j_x to flow uninhibited and accepting the added component of $\mathbf{j} \times \mathbf{B}$ body force, or of suppressing j_x by application of an axial electric field component E_x. To accomplish the latter, we return to the conductivity tensor formulation and require

$$j_x = \sigma_{xx}E_x + \sigma_{xy}(E_y - uB) = 0 \qquad (8\text{-}65)$$

which yields the necessary axial field component

$$E_x = -\left(\frac{\sigma_{xy}}{\sigma_{xx}}\right)(E_y - uB) = +\frac{\omega_B}{\nu_c}(E_y - uB) \qquad (8\text{-}66)$$

Note that by applying this axial component, the transverse current j_y is just restored to its scalar value,

$$j_y = \sigma_{yx}E_x + \sigma_{yy}(E_y - uB) = \sigma_0(E_y - uB) \qquad (8\text{-}67)$$

The practical problem of applying such an axial field to the flow may be handled by segmenting the electrodes into many small sections, each insulated from its neighbors so that it may be maintained at a slightly different potential (Fig. 8-10). Such segmentation will introduce various

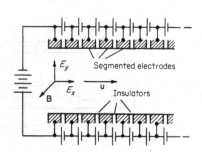

Fig. 8-10 Segmented electrode structure to provide axial electric field component.

new disturbances into the flow associated with the field and current concentrations at the electrode-insulator interfaces, which may be particularly troublesome if other two-dimensional effects, such as boundary layers and fringing fields, are also significantly present.

Instead of striving to eliminate the Hall current, one may attempt to utilize it for stream acceleration. Low density, high **B**-field devices

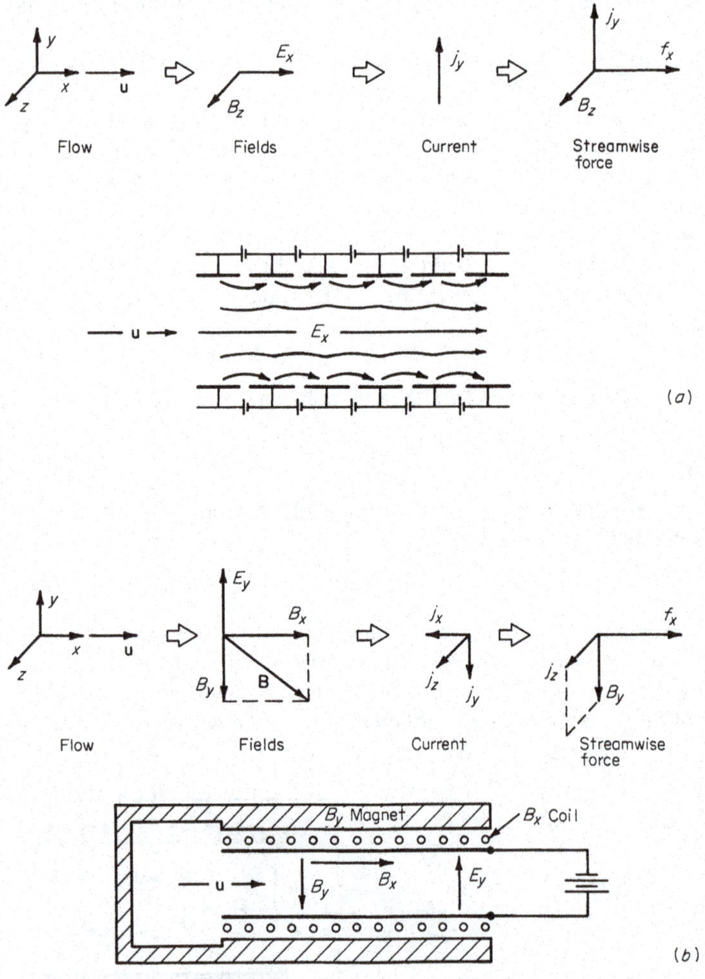

Fig. 8-11 Schematic representation of various Hall accelerator configurations. *(a)* Rectangular type I; *(b)* rectangular type II; *(c)* coaxial type I; *(d)* coaxial type II.

ELECTROMAGNETIC ACCELERATION—STEADY FLOW

which invoke the interaction between a Hall current and the applied magnetic field as the dominant body force on the gas—so-called *Hall accelerators*—have been constructed in various geometries. Basically, there are two possible modes of implementation, one requiring a streamwise applied electric field, the other a streamwise component of applied magnetic field. In the former (Fig. 8-11a), the transverse current j_y is now a Hall current whose interaction with the orthogonal B_z field

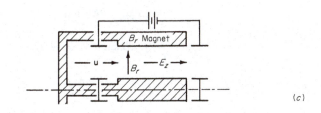

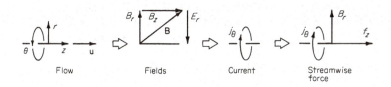

(c)

(d)

yields the desired axial body force:

$$j_y = \sigma_{yx}E + \sigma_{yy}(-uB) = \frac{\Omega}{\Omega^2 + 1}\sigma_0 E - \frac{1}{\Omega^2 + 1}\sigma_0 uB$$

$$\approx \frac{\sigma_0 E}{\Omega}\left(1 - \frac{uB/E}{\Omega}\right) \qquad (8\text{-}68)$$

There is also an unavoidable axial Hall current component, driven by the transverse back emf, uB, producing an unwanted transverse force which gives rise to a balancing pressure gradient.

In the second class of Hall accelerator (Fig. 8-11b), the applied electric field may be normal to the flow, as in the scalar accelerator, but the magnetic field is applied obliquely in the plane of **E** and **u**. There thus arise three components of current (Prob. 8-9):

$$j_x = \frac{\Omega \sigma_0}{1 + \Omega^2}(uB \cos^2 \alpha - \Omega E \sin \alpha \cos \alpha) \qquad (8\text{-}69)$$

$$j_y = \sigma_0 E + \frac{\Omega \sigma_0}{1 + \Omega^2}(uB \sin \alpha \cos \alpha - \Omega E \sin^2 \alpha) \qquad (8\text{-}70)$$

$$j_z = \frac{\sigma_0}{1 + \Omega^2}(\Omega E \sin \alpha - uB \cos \alpha) \qquad (8\text{-}71)$$

the z component of which interacts with B_y to produce the desired axial body force:

$$f_x = -j_z B_y = \frac{\sigma_0 B}{1 + \Omega^2}(\Omega E \sin \alpha \cos \alpha - uB \cos^2 \alpha) \qquad (8\text{-}72)$$

Note that there is an optimum inclination of **B**; in fact, $\Omega E/uB$ must exceed $\cot \alpha$ for any acceleration at all. Optimization of (8-72) automatically sets j_x from (8-69) at a negative maximum, suggesting that performance could be improved by adding an axial component of electric field just sufficient to reduce j_x to zero. This done, j_z becomes

$$j_z = \frac{\sigma_0}{1 + \Omega^2 \sin^2 \alpha}(\Omega E_y \sin \alpha - uB \cos \alpha) \qquad (8\text{-}73)$$

which yields a correspondingly larger axial body force (Prob. 8-9).

Both of these Hall accelerator concepts are more readily implemented in coaxial geometries, where the Hall currents may close on themselves without passing to electrode surfaces. For the first type of accelerator one would apply an axial **E** field and a radial **B** field, which would yield an azimuthal Hall current to interact with B (Fig. 8-11c). For the second type, the electric field would be radial and the magnetic field would have both radial and axial components. Again, the Hall currents would be predominantly azimuthal and interact with B_r to produce the desired axial acceleration (Fig. 8-11d). Devices of this type are discussed more fully in Sec. 8-9.

The sequence of cause and effect implicit in the above descriptions of Hall accelerators, e.g., a component of **E** interacting with a component of **B** to yield a component of **j**, which in turn interacts with another component of **B** to yield a particular component of force, etc., is perhaps too artificial. The composite process is internally self-consistent from first principles if formulated in adequate generality. That is, the current density vector is uniquely prescribed by the pattern of applied fields and the state of the medium in accordance with a generalized Ohm's law, readily assembled from the conductivity tensor components used heretofore (Prob. 8-10).

$$\mathbf{j} = \sigma_0(\mathbf{E} + \mathbf{u} \times \mathbf{B}) - \frac{\Omega}{B}(\mathbf{j} \times \mathbf{B}) \qquad (8\text{-}74)$$

The cross product of this total current with the magnetic field vector yields the total body force on the gas without further rationalization.

It is frequently illuminating, however, to take advantage of the essential low density character of the medium and to consider the individual particle trajectories in the collisionless limit ($\Omega \rightarrow \infty$). In the first class of Hall accelerator, for example, we find the free electrons executing a cycloidal **E** × **B** drift in the negative y direction, while the heavy ions, presumably having gyro radii many times larger than the channel dimensions, closely follow the applied axial electric field. These ions transmit the momentum they thus acquire to the neutrals by collision (Fig. 8-12a). In the annular version of the device, the electrons would be completely trapped in an endless azimuthal drift, with the ions again free to follow the axial **E** field (Fig. 8-12b). In this sense, the Hall accelerator is an electrostatic device which is relieved of space-charge limitations by the neutralizing effect of the trapped electrons. Note that, so long as there are adequate ion-neutral collisions, this direct ion acceleration does not produce a current relative to the axial gas flow; rather, the electric body force may be regarded as applied to the entire flowing mass.

The particle orbit image of the second class of Hall accelerator is more complex because the fields are not necessarily orthogonal. If $E_x = 0$, the electrons execute parabolically displaced helices whose axes lie in the **B**, **E** × **B** plane, advancing in the streamwise direction (Sec. 5-1 and Fig. 8-12c). The ions will again feel the electric field, but since this is purely transverse, it cannot directly supply the thrust. The streamwise momentum gain here is acquired first by the electrons, which, in actual operation, will transmit it to the heavy-particle component of the gas only by the occasional collisions or by charge separation in regions of

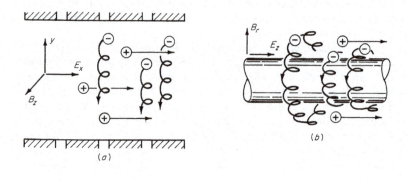

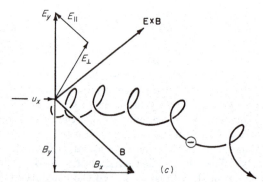

Fig. 8-12 Electron and ion trajectories in various Hall accelerators. *(a)* Rectangular type I; *(b)* coaxial type I; *(c)* rectangular type II.

field nonuniformity, such as at the channel exit. If the axial electron motion is suppressed by an applied E_x, however, the ions will again feel a streamwise electric force.

In the usual range of Hall accelerator operation, the trajectories of the individual ions are little affected by the magnetic field, and the mean ion motion is essentially the same as the bulk streaming of the gas as a whole. Under extreme conditions of very low densities and very high magnetic fields, however, ion-neutral collisions may become sufficiently rare that the ions are able to transverse the accelerator channel or to achieve a cycloidal drift motion of their own, somewhat independently of the neutral gas flow. This condition is known as *ion slip;* if severe, it implies an uncoupling of the electromagnetic processes from the gasdynamics and a corresponding loss of accelerator efficiency. Inclusion of the contribution of ion drifts to the current density is a straightforward but somewhat detailed calculation, yielding a more general form of the generalized Ohm's law [1],

$$\mathbf{j} = \sigma_0(\mathbf{E} + \mathbf{u} \times \mathbf{B}) - \frac{\Omega}{B}(\mathbf{j} \times \mathbf{B}) + (1 - \alpha)^2 \frac{\Omega\Omega_+}{B^2}[(\mathbf{j} \times \mathbf{B}) \times \mathbf{B}] \quad (8\text{-}75)$$

where $\Omega_+ = \omega_+/\nu_c^+$ is the ion Hall parameter, and α is the fractional degree of ionization, $n_+/(n_A + n_+)$. For a fully ionized gas, the ion slip term disappears since the mean ion motion is then essentially identical with the mean mass motion. In a partially ionized gas, the relevant ion collision frequency ν_c^+ is that against neutrals. Ion-electron collisions have little effect on ion trajectories; ion-ion collisions do not affect the mean ion motion, to first order. Thus, while the ratio of ion to electron Hall parameters is depressed by the particle mass ratio, it may still become significant in cases where coulomb interactions determine the electron collision frequency (Prob. 8-11).

The admission of Hall current and ion slip effects clearly elevates the crossed-field accelerator to a full-blown two-dimensional or even three-dimensional problem. Simultaneous solution of the magnetogasdynamic conservation equations with the generalized Ohm's law is found to yield gas flow and current density patterns with cross-stream variations comparable with the streamwise gradients. These solutions typically display a concentration of current density on the upstream edge of the cathode and on the downstream edge of the anode, consistent with the Hall effect for the prevailing conditions, which predicates transverse pressure gradients and corresponding thermodynamic variations of comparable importance (Fig. 8-13). Such skewed conduction patterns have been well confirmed experimentally [15,16].

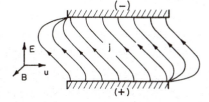

Fig. 8-13 Typical current density distribution in crossed-field accelerator with free stream Hall parameter of order unity.

8-7 SELF-INDUCED FIELDS

A substantial nonlinearity was removed from the scalar accelerator problem by requiring that the magnetic field generated by the current flowing across the gas be everywhere insignificant in comparison with the externally applied field. This restriction is seldom completely justified and totally excludes from consideration a very interesting class of high current accelerators. To establish the domain of importance of self-field effects, return for a moment to the simple orthogonal geometry of Fig. 8-2 and the y component of the curl **B** equation (8-53). The transverse field gradient $\partial B_x/\partial z$ can now consistently be set to zero, leaving

$$\frac{dB}{dx} = -\mu j \qquad (8\text{-}76)$$

or the equivalent integral requirement,

$$\Delta B = B_0 - B_L = \mu \int_0^L j(x)\, dx = \mu J \qquad (8\text{-}77)$$

where J is the total current crossing the channel, per unit breadth z. If we insert the limiting conditions discussed in Sec. 8-3, that is, $L = 1$ m, $j = 10^7$ amp/m², we find $\Delta B = 4\pi$ weber/m², an order of magnitude greater than the largest external field considered. The effect diminishes for shorter channels, but as a rough rule of thumb, significant self-fields can be anticipated whenever current densities reach 10^5 or 10^6 amp/m².

A useful index of the relative importance of the self-induced field is the magnitude, relative to unity, of a dimensionless group of parameters called the magnetic Reynolds number,

$$R_B = \mu\sigma uL \qquad (8\text{-}78)$$

The relevance of this group can be demonstrated by various crude arguments. For example, approximating J in (8-77) by $\sigma(E - u_0 B_0)L \approx \sigma u_0 B_0 L$, we see that

$$\frac{\Delta B}{B_0} \approx R_{B_0} \qquad (8\text{-}79)$$

where the subscript zero again refers to the inlet values. Thus, if $R_B \ll 1$, the induced field can reasonably be neglected; if $R_B \gg 1$, it dominates the problem.

When R_B is large, the essential relation between the induced field and its source current substantially changes the character of the magnetogasdynamic accelerator problem. To illustrate this, let us now assume that the external magnetic field is totally negligible in comparison with the self-induced field, and return again to the strictly one-dimensional formulations (8-16) to (8-21). We may rewrite the momentum equation in the form

$$F\frac{du}{dx} + \frac{dp}{dx} = jB = -\frac{B}{\mu}\frac{dB}{dx} \qquad (8\text{-}80)$$

ELECTROMAGNETIC ACCELERATION—STEADY FLOW

which has an immediate first integral:

$$Fu + p + \frac{B^2}{2\mu} = Fu_0 + p_0 + \frac{B_0^2}{2\mu} = \mathcal{P} \tag{8-81}$$

The quantity $B^2/2\mu$ is often called the magnetic pressure, in terms of which (8-81) expresses the conservation of the sum of total pressure plus momentum flux. Ohm's law now may be expressed as a differential relation among B, E, and u:

$$\frac{dB}{dx} = -\mu\sigma(E - uB) \tag{8-82}$$

and the current may likewise be eliminated from the energy equation:

$$F\left(c_p \frac{dT}{dx} + u\frac{du}{dx}\right) = -\frac{E}{\mu}\frac{dB}{dx} \tag{8-83}$$

For a given conductivity function $\sigma(T,p)$, Eqs. (8-81) to (8-83) and an equation of state relate five variables, p, T, u, E, B, and we may again search for special solutions. For constant E field, for example, the energy equation also has a simple integral:

$$F\left(c_p T + \frac{u^2}{2}\right) + \frac{EB}{\mu} = \text{const} = \mathcal{E} \tag{8-84}$$

Using the perfect-gas relation in the form $Fc_p T = [\gamma/(\gamma - 1)]\, pu$ and comparing (8-84) with (8-81) provides a single algebraic relation between u and B along the channel:

$$\frac{\gamma+1}{2(\gamma-1)} Fu^2 + \frac{\gamma}{\gamma-1}\left(\frac{B^2}{2\mu} - \mathcal{P}\right)u + \left(\mathcal{E} - \frac{EB}{\mu}\right) = 0 \tag{8-85}$$

A corresponding relation between p and B may be similarly obtained, and the conductivity function $\sigma(p,u)$ may then be expressed solely in terms of B and the initial conditions. Insertion of this $\sigma(B)$ and the $u(B)$ solution from (8-85) into the Ohm's law relation (8-82) provides a differential equation for $B(x)$, from whose solution $u(x)$ and $p(x)$ can also be evaluated via (8-85) and (8-81).

Analytic results can also be found for the constant pressure accelerator. Here we may write the momentum and energy relations:

$$F\frac{du}{dx} = -\frac{B}{\mu}\frac{dB}{dx} \tag{8-86}$$

$$\frac{\gamma}{\gamma-1} p_0 \frac{du}{dx} + Fu\frac{du}{dx} = -\frac{E}{\mu}\frac{dB}{dx} = \frac{1}{\mu^2\sigma}\left(\frac{dB}{dx}\right)^2 - \frac{uB}{\mu}\frac{dB}{dx} \tag{8-87}$$

Eliminating du/dx from these fortuitously also removes u and the non-linearity in dB/dx, leaving only a linear equation for $B(x)$:

$$\frac{dB}{dx} + \frac{\gamma}{\gamma - 1} \frac{\mu \sigma p_0}{F} B = 0 \qquad (8\text{-}88)$$

In general, $\sigma(u,p_0)$ can be expressed as $\sigma(B,p_0)$ via the simple integral of (8-86), and (8-88) may then be solved for $B(x)$. Note that $E(x)$ is by no means constant here; rather, it must be evaluated from (8-82) after $B(x)$ and $u(x)$ are found.

This class of problem displays somewhat troublesome exit conditions, which can be illustrated quickly by the special case of constant σ. Here the familiar exponential solution of (8-88) can be incorporated into (8-86) to yield the velocity profile:

$$u - u_0 = \frac{B_0^2 - B^2}{2\mu F} = \frac{B_0^2}{2\mu F}(1 - e^{-2\nu x}) \qquad (8\text{-}89)$$

where $\nu = [\gamma/(\gamma - 1)]\mu\sigma p_0/F$. Note that this form of solution requires a field at the exit plane, $x = L$, which is not strictly zero but rather has the value $B_L = B_0 e^{-\nu L}$, and in this sense the model is not totally commensurate with a pure self-field accelerator. For cases of large νL the distinction is negligible, but is nonetheless reflected in the evaluation of B_0 in terms of the total current per unit channel breadth:

$$B_0 = \frac{\mu J}{1 - e^{-\nu L}} \qquad (8\text{-}90)$$

Other special cases of one-dimensional self-field acceleration, such as adiabatic and isothermal flows, can be similarly evaluated (Prob. 8-12). Each of these displays the distinguishing characteristics of all self-field accelerators—a quadratic dependence of the flow velocity increase on the discharge current and a shifting of the centrode of the body force profile somewhat upstream compared with the corresponding external field accelerators. The latter property, which follows from the inherent monotonic decay of the self-field along the channel, helps to relieve the mismatch between E and uB which detracts from the efficiency of constant B devices. That is, the steady rise in u along the channel combines with the steady decay in B to provide a more uniform profile of back emf, uB.

A subclass of self-field accelerators of particular interest and analytical tractability consists of those wherein the current conduction zone is sufficiently narrow that one can afford to surrender detailed information about its interior structure in favor of discrete jump conditions, much in the spirit of classical gasdynamic shock waves. For one-dimensional "sheet" accelerators of this class we may write conservation requirements directly in terms of total current and a single electric field value, or input power per unit channel width y and breadth z:

$$F(u_L - u_0) + (p_L - p_0) = \frac{\mu J^2}{2} \qquad (8\text{-}91)$$

$$\frac{\gamma}{\gamma - 1}(p_L u_L - p_0 u_0) + \frac{F}{2}(u_L^2 - u_0^2) = JE \qquad (8\text{-}92)$$

Since the interior of the sheet is not to be examined, Ohm's law in the differential form (8-82) is clearly inapplicable, but we do need a functional relation between the total current J and the applied potential per unit channel width E to complete specification of the problem. In practice this is difficult to specify analytically because of the very complex current conduction mechanisms which prevail in narrow intense sheets of this type (Sec. 9-4). However, J and E are readily measured experimentally, and an empirical relation may be assigned on this basis. Alternatively, one may replace the Ohm's law statement by a thermodynamic constraint and compute the corresponding $J(E)$ relation a posteriori. For example, if the gas pressure is kept the same on both sides of the sheet, the velocity jump depends only on the total current and mass flow:

$$u_L - u_0 = \frac{\mu J^2}{2F} \qquad (8\text{-}93)$$

and the corresponding discharge characteristic is found to be

$$E = \left(\frac{\gamma}{\gamma - 1}\frac{p_0}{2F} + \frac{u_0}{2}\right)\mu J + \frac{\mu^2 J^3}{8F} \qquad (8\text{-}94)$$

In most accelerators of this type, both p_0 and Fu_0 are negligible compared to μJ^2, in which limit $u_L \to \mu J^2/2F = 4(E/B_0)$, in contrast to the familiar E/B limit for external field accelerators. Note also that in this idealization the efficiency $\eta = (Fu_L^2/2)/EJ \to 1$. Other special thermodynamic jump conditions can be similarly imposed on the sheet, such as the isothermal jump, $p_0 u_0 = p_L u_L$, or the adiabatic jump, $p_0 u_0^\gamma = p_L u_L^\gamma$, with similar results. The constant velocity "pump," $u_0 = u_L$, of course displays substantially different behavior (Prob. 8-13).

The self-field mode of magnetogasdynamic acceleration is particularly attractive for space-thruster applications. The ability of the discharge to supply its own magnetic field relieves the burden of a large external magnet and its auxiliary power supply from the thruster package and allows significantly more freedom in channel geometry. In particular, very short channels, with correspondingly lower wall losses, are now possible. Also, at the high current density levels implicit in self-field operation, the conductivity of the gas is maintained at a high level by the discharge itself, without the need for a preionization stage in the channel. The evident drawback is in the extremely high gas temperatures which must be sustained in the channel and exhaust plume without deterioration of the solid surfaces or excessive conduction and radiation losses. In addition, some problems of discharge pattern stability may arise at these current levels. A leading example of this class of thruster, the magnetoplasmadynamic arc, will be described in Sec. 8-9.

8-8 SOURCES OF THE CONDUCTING GAS

In much of the preceding discussion it has been presumed that the electrical conductivity of the working fluid is largely at our disposal, to be assigned in accordance with the optimum dynamical requirements of the particular problem. Actually, the satisfactory preparation of the ionized gas for a magnetogasdynamic channel accelerator is one of the most troublesome aspects of its operation. At temperatures which can reasonably be tolerated by material walls, certainly less than 3000°K, the equilibrium ionization of the common gases is far too low to provide adequate electrical conductivities for efficient, vigorous $\mathbf{j} \times \mathbf{B}$ acceleration. One must then look to one of three less simple alternatives to provide the desired medium: (1) "seed" the common gas with a fraction of an easily ionized impurity; (2) establish some type of nonequilibrium ionization to elevate the electron density; or (3) contain much hotter gases away from the material surfaces by appropriate magnetic fields. Each alternative has several possible methods of implementation, most of which raise additional analytical and practical complications. Here we can discuss them only briefly.

SEEDED GASES

Two advantages are obtained by the addition of small amounts of an alkali metal, such as cesium, rubidium, or potassium, to the gas stream. First, in the temperature range of interest, the ionization level and electrical conductivity become orders of magnitude higher, and second, the conductivity becomes relatively insensitive to temperature. For example, in the range from 1000 to 3000°K, the ionization of any pure common gas is in a low-level exponential range [Eq. (3-28)]:

$$n \approx G n_0^{1/2} T^{3/4} e^{-\epsilon_i/2kT} \qquad (8\text{-}95)$$

Since the prevailing electron collisions are with neutrals, $\nu_c \propto n_0 T^{1/2}$, and the scalar conductivity is dominated by the same exponential dependence:

$$\sigma_0 = \frac{ne^2}{m\nu_c} \propto n_0^{-1/2} T^{1/4} e^{-\epsilon_i/2kT} \qquad (8\text{-}96)$$

A small fraction of alkali seed material added to the gas will be more substantially ionized at this temperature, and the conductivity proportionally increased, up to the point where electron-ion coulomb encounters assume control of the collision frequency. Beyond this level, further creation of free electrons will be accompanied by corresponding increase in ν_c, thereby flattening the dependence of conductivity on temperature for such a mixture (Fig. 8-14).

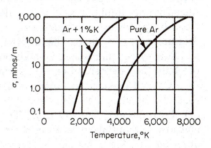

Fig. 8-14 Comparison of electrical conductivity of pure argon and argon seeded with 1 percent potassium, both at 1 atm. (*From R. J. Rosa, Phys. Fluids, vol. 4, p. 182, 1961, and A. B. Cambel, "Plasma Physics and Magnetofluidmechanics," McGraw-Hill Book Company, New York, 1963.*)

The optimum seeding ratio may be estimated by differentiating the conductivity expressed in terms of the participating species densities and cross sections:

$$\sigma = \frac{ne^2}{m\bar{\nu}_c} = A \frac{n}{n_s Q_s + n_0 Q_0 + n Q_+} \tag{8-97}$$

where $A = e^2(\pi/8mkT)^{1/2}$ Invariably, the seed atom cross section Q_s will be much larger than that of the carrier gas Q_0, because of the large dimension of the valence electron orbit, and the coulomb cross section of the seed ion Q_+ will be even larger at these temperatures, so that, while $n \ll n_s \ll n_0$, none of the three products in the denominator of (8-97) may be neglected. However, if we assume that only seed material is ionized, and this only slightly, so that a relation of the form (8-95) remains appropriate, a useful simplification develops. Inverting (8-97), differentiating with respect to n_s, and setting equal to zero, we find

$$\left(\frac{n}{dn/ds} - n_s\right) Q_s = n_0 Q_0 \tag{8-98}$$

But invoking (8-95) in the form $n \propto n_s^{1/2}$, it follows that $n/(dn/ds) = 2n_s$, whereby we obtain a maximum conductivity with the seed ratio

$$\left(\frac{n_s}{n_0}\right)_{opt} = \frac{Q_0}{Q_s} \tag{8-99}$$

a ratio typically in the range of 10^{-3} to 10^{-2}. Note that this result is independent of the value of Q_+.

The choice of a specific seeding material is based primarily on its ionization potential, since this enters exponentially into the available electron density, and thence into the conductivity for a given gas temperature. Cesium (3.87 ev), rubidium (4.16 ev), and potassium (4.32 ev) have the lowest ionization potentials among the elements and are the most commonly employed. Unfortunately, the same electronic structure which gives these substances their low ionization potentials also makes them extremely active chemically, and thus difficult to handle. Further, since they are liquid or solid at standard temperature and pressure, they must be preheated to temperatures sufficiently high that their vapor pressure corresponds to the desired seeding ratio, and the entire gas handling system must then be maintained at an elevated temperature to preclude their deposition on cold surfaces.

Certain chemical compounds may eventually prove more suitable than the alkali elements as seeding materials. For example, Ba_2O (4.0 ev), LaO (4.8 ev), ZrO (5.5 ev), UO (4.7 ev), and UO_2 (4.3 ev), all have ionization potentials lower than their constituent elements, and raise the question whether other, more exotic compounds might not have lower potentials than even the alkali elements [17].

NONEQUILIBRIUM IONIZATION

If the free-electron component of an ionized gas is maintained at a temperature above that of the heavy-particle components, the ionization level and the electrical conductivity are found also to exceed the equilibrium values for the prevailing atomic temperature. This result may be rationalized on the basis that, since the electron-atom ionization processes and most of the recombination processes depend heavily on the energy of the incoming electron, it is reasonable to expect that the ionization level and conductivity are primarily determined by the electron temperature, rather than by the heavy-particle temperature. In fact, there appears to be some justification for computing the ionization level of such a gas from a Saha relation, using the electron temperature [18,19].

Nonequilibrium gaseous conductors of this sort are quite familiar in low-density low-current discharges of many types, most notably the simple glow discharges (Sec. 6-4) and the rf inductive discharges (Sec. 6-7), where the distinguishing characteristic is good electrical conduction with a minimum of heat transfer to the surroundings. In a magnetogasdynamic accelerator, the gas would need to be brought to the desired nonequilibrium state by a discharge at the channel entrance, or prior to it, and this state would need to be maintained along the complete channel length, while yet sustaining the desired level of current conduction. This places an upper limit on the gas density which can be accommodated at a given electric field strength, above which the interspecies collisions will drive the two temperatures together (Prob. 8-14). Low densities, in turn, favor tensor-conductivity effects, with their associated practical complications for scalar accelerators. It thus seems that the Hall current accelerator stands to benefit most from the use of nonequilibrium gases of high electron temperature.

MAGNETIC CONFINEMENT

We have seen that the electrothermal arcjet benefits from the confinement of the hotter portions of its chamber flow to the center of the channel, away from the constrictor and nozzle walls, whereby an average gas temperature far above the softening point of the wall material can be tolerated. To create a similar temperature profile across a magnetogasdynamic channel by magnetic means might seem less attractive since a corresponding profile in electrical conductivity and Hall parameter would be created which would tend to complicate the flow. Certain experimental results, however, suggest that this may indeed be a profitable approach. For example, it has been found empirically that it is possible to withdraw electrodes slightly beyond the luminous edges of an ionized gas stream without drastically reducing the current which can be driven through it at a given voltage [20]. In this mode of operation, the electrode erosion is substantially reduced for a given mass flow and rate of acceleration. This phenomenon of remote conduction presumably implies a large peripheral range for the current-carrying electrons, an effect that has also been observed in certain Langmuir probe studies of plasma jet flows [21], but no satisfactory theoretical explanation yet exists. A particularly evident example of magnetic confinement that permits elevated gas temperatures arises in the magnetoplasmadynamic arc, discussed in the following section.

8-9 THE MAGNETOPLASMADYNAMIC ARC

Prior to 1964, no device remotely resembling a practical space thruster of the steady flow electromagnetic class yet existed. Although the essential elements of the magnetogasdynamic interactions discussed above had been well confirmed by various laboratory experiments, efforts to implement them into an efficient gas accelerator were mired in the complications of producing and handling appropriately ionized seeded-gas flows, of preserving the electrode and insulator surfaces against the thermal and chemical attack of such flows, of providing satisfactory magnetic field geometries and strengths, and of maintaining electrical discharge patterns of optimum intensity and uniformity. A few experimental crossed-field accelerator channels had been made to function, but only when surrounded by massive preionizing equipment and electromagnet coils, which hardly seemed adaptable to spacecraft engines. Although these cumbersome devices displayed impressive combinations of several newtons of thrust at 20,000- to 30,000-m/sec exhaust speeds [20,22], it was generally felt that the need for external magnets and the serious viscous and thermal wall losses precluded use of this mode of acceleration for space flight. Accelerators using the Hall current mode had also been operated, but these too were rather ponderous systems of low efficiency and less thrust density than the scalar conduction devices [23–25]. While interest in magnetogasdynamic channels for ground-based power generation or wind-tunnel type of test facilities remained active [16,26], their application to steady electromagnetic thrusters seemed unpromising, at least until very high space power levels became available.

At this time, a remarkable and significant development arose from the somewhat prosaic electrothermal field which served to rejuvenate abruptly the interest in magnetogasdynamic thrusters. In an empirical series of experiments with a conventional short arcjet device, it was found that by drastically reducing the propellant mass flow, and thereby the pressure in the arc chamber into the millimeter range, arc currents as high as 3,000 amp could be drawn across the electrodes without serious erosion. Under these conditions, the exhaust velocity of the hydrogen flow could be increased to values of the order of 100,000 m/sec, and the overall efficiency reached 50 percent [27].

A rapid succession of similar experiments at various laboratories quickly confirmed this finding and extended the performance to even more impressive levels. With some care, this mode of operation was also established in devices of the Hall accelerator class, as well as in a variety of conventional arcjet configurations. In retrospect, some hints of this high-performance capability were recovered from much earlier experiments with both classes of accelerator [23,28].

As the number of experimental demonstrations of this extraordinary effect increased, the urgency of its physical interpretation also rose, but this was not to be a trivial problem. The reasonable supposition was that the high current densities in the arc were generating self-magnetic fields within the chamber sufficiently intense to produce substantial electromagnetic acceleration of the flow, but the details of the mechanism were far from clear. The practical implication was, however, entirely evident: despite the discouraging analytical prognoses and the unhappy experimental heritage of steady electromagnetic accelerators, here was a valid example operating in an interesting thrust and impulse range at reasonable efficiencies, with no external magnet, no preionization gear, and minimal electrode erosion. And to complete the humiliation of the scientific approach, the device had originally been designed for a totally different, electrothermal, class of operation!

In lieu of any adequate analytical models of this accelerator, there now followed a flurry of rather empirical experimental activity at several laboratories, wherein various gases, electrode geometries, mass flow rates, voltage and current levels, and external biasing magnetic fields were tried, and the resulting performance cataloged. Although these improved on the original operation only slightly and provided little basic insight into the detailed mechanisms of the acceleration, a second remarkable property was uncovered. It was possible in some circumstances to produce significant thrust and to develop an extended luminous exhaust plume with no propellant mass flow whatsoever into the arc chamber [29,30]!

This latter idiosyncrasy understandably cast some doubts on the validity of the originally indicated high performance levels and demanded immediate explanation. Other studies of the device momentarily deferred to an intensive attack on this paradox, results of which indicated that under some circumstances the "flowless" operation proceeded by an ingestive recirculation of the residual gas in the vacuum tank, in other cases by a gradual vaporization and expulsion of electrode material, and in some cases by a magnetic interaction with the tank walls. With all such effects accounted for, however, the original high-performance results with low but finite mass flow through the arc were largely sustained, and the search for identification and optimization of the mechanism was resumed.

Unfortunately, this early phase of discovery and excitement was followed by an extended period of frustration in attempting to comprehend and implement this promising mechanism and in adequately testing devices of this type. Indeed, their ultimate utility for space propulsion is even now not entirely established. However, they provide such an interesting basis for application of many of the electromagnetic and particle processes we have been discussing in the abstract, that a more detailed description and analytical examination will be instructive within our context.

PHENOMENOLOGICAL OBSERVATION

Early in its career, this type of accelerator was dubbed the *magnetoplasmadynamic* (MPD) *arc*, although other titles, such as *thermoionic accelerator* and *high-impulse arcjet*, appear in the literature. Figure 8-15 shows a schematic drawing and photograph of one MPD arc assembly, and Fig. 8-16 shows the device in vacuum tank operation. Table 8-3 lists typical terminal parameters and performance data for the device shown. In operation, this arc emits an intensely luminous exhaust plume which extends many orifice diameters downstream. In many cases a coaxial structure is visible in this plume, consisting of an intense axial column extending from the cathode tip (cathode jet) and a somewhat less intense coaxial shell extending from the lip of the anode orifice (anode jet), with an even less intense region separating the two.

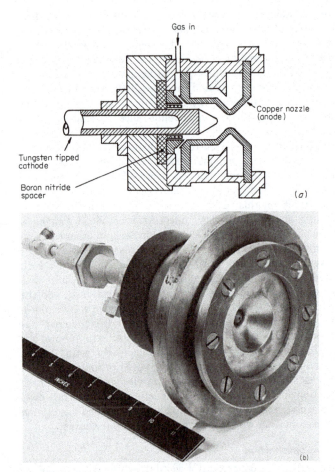

Fig. 8-15 50- to 250-kw MPD arc accelerator. (a) Section; (b) photograph. (*AVCO, Space Systems Division, Wilmington, Mass.*)

Although the arc will operate without the external field from the coaxial solenoid, it tends to be more stable with such a field applied. The outer edge of the luminous plume is observed to contract toward the axis somewhat as this field is intensified, and the performance data are influenced somewhat by its magnitude and axial location.

Velocity measurements in the exhaust jet confirm the high specific impulses inferred from the mass flow and thrust data. Significant fractions of the arc current are found to flow out into the plume and back

Fig. 8-16 MPD arc in vacuum tank operation. (*AVCO, Space Systems Division, Wilmington, Mass.*)

Table 8-3 Typical terminal performance characteristics of magnetoplasmadynamic arc shown in Fig. 8-15

Propellant	Hydrogen			Ammonia				Argon	
Current, amps	1,500	2,500	1,500	2,000		2,500		1,500	2,100
Mass flow rate, gms/sec	0.050	0.020	0.050	0.030	0.030	0.015	0.015	0.032	0.032
Voltage	71	68	78	36	39	37	35	25	30
Thrust, newtons	1.21	0.98	2.22	0.88	1.13	0.90	0.95	0.54	0.68
Specific impulse, sec	2,460	5,000	4,520	3,000	3,840	6,140	6,500	1,720	2,160
Efficiency, %	13.7	23.6	25.4	24.0	27.0	36.6	34.4	12.1	11.5

SOURCE: Arcjet Technology Research and Development, *AVCO Space Systems Division RAD-TR*-65-37, *NASA CR*-54867, December, 1965.

again, apparently along the anode and cathode jets, for many orifice diameters downstream [31,32]. Electron density and electron temperature profiles, both axially and radially in the jet, are found to be rather flat and in the range of 10^{14} cm^{-3} and 10^{4}°K, respectively. Significant electron densities are found beyond the luminous boundaries of the plume. Gas temperatures in excess of 10^{4}°K in some parts of the plume are indicated by various measurements, and total enthalpy levels are correspondingly high [33].

Interpretation of these observations, leading to comprehension of the overall acceleration process, probably is closely tied to three basic questions: (1) How can current densities of the order of 10^9 amp/m^2 be drawn from the cathode without its immediate destruction? (2) What is the detailed pattern of current flow in the arc, its associated magnetic field, and the resulting body force distribution? (3) What is the role of particle collisions in the process; i.e., is the interaction truly magnetogasdynamic or is it more nearly a collisionless particle stream acceleration? Definitive responses to these questions can be provided only by much more extensive diagnostic studies of the interior of the arc and the exhaust plume, a most complex and hostile environment for precise work. Lacking these, the only recourse is to postulate various empirical models and to compare their gross performance predictions with less difficult test results. Several such models have been proposed for the MPD arc. One treats it as a coaxial self-field accelerator with scalar conductivity; a second regards it as a coaxial Hall accelerator of the second class; a third considers it as a collisionless free-particle accelerator; another proceeds from finite collision orbit theory; etc. The main distinction among all these is between the continuum or magnetogasdynamic point of view and the particle approach. Although the actual process probably involves some aspects of both these mechanics, it is instructive to adopt each approach separately and compare their predictions.

MAGNETOGASDYNAMIC DESCRIPTION

In the continuum representation, several components of magnetogasdynamic interaction may be proposed (Fig. 8-17). First, a streamwise (axial) acceleration is provided by the crossing of the radial arc current with the self-generated azimuthal magnetic field. This we shall refer to as the electromagnetic "blowing," which is essentially the scalar crossed-field interaction discussed in detail earlier. Second, there may be an electromagnetic "pumping" process wherein axial components of arc current cross with the azimuthal B field to establish a radial gradient in the gasdynamic pressure which provides a reaction force on the cathode surface. Third, if an external magnetic field is applied, having components in the axial and radial directions, an electromagnetic "swirling" may be generated by $j_r B_z$ or by $j_z B_r$. Finally, if tensor conductivity effects are admitted, various Hall interactions may arise, such as the crossing of an azimuthal Hall current component with the radial component of the external field.

For the present we shall ignore the external field, and consider only the blowing and pumping mechanisms. For simple illustration we first construct two idealized models which display these interactions separately;

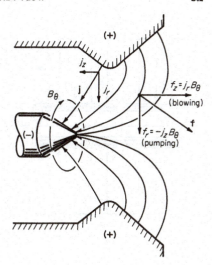

Fig. 8-17 Magnetogasdynamic model of the MPD accelerator.

then a more realistic model which allows both mechanisms to act simultaneously is examined.

The blowing mechanism may be evaluated in a model that allows only radial current flow, azimuthally and axially uniform between coaxial cylindrical electrodes (Fig. 8-18a). From Maxwell's $\nabla \times \mathbf{B}$ relation, the self-induced field is seen to be purely azimuthal, linear in z, and inversely proportional to r over the gap:

$$B_\theta(r,z) = \frac{\mu J}{2\pi r}\left(1 - \frac{z}{z_0}\right) \qquad (8\text{-}100)$$

where $J = 2\pi r z_0 j_r$ is the total arc current. The body force density is thus purely axial and proportional to z/r^2:

$$f_z(r,z) = j_r B_\theta = \frac{\mu J^2}{4\pi^2 r^2 z_0^2}(z_0 - z) \qquad (8\text{-}101)$$

The total axial force applied to the gas stream is the integral of f_z over the gap volume:

$$F_z = \frac{\mu J^2}{4\pi^2 z_0^2}\int_0^{z_0}\int_0^{2\pi}\int_{r_c}^{r_a}\frac{z_0 - z}{r^2}\, r\, dr\, d\theta\, dz = \frac{\mu J^2}{4\pi}\ln\frac{r_a}{r_c} \qquad (8\text{-}102)$$

The coefficient $\mu/4\pi$ has the value 10^{-7} in the mks system; hence, for reasonable anode-to-cathode radius ratio, significant thrust from this mechanism arises quadratically in the 1,000-amp range. The result expressed by Eq. (8-102) may readily be shown to be appropriate for arbitrary distribution of current density along the z axis, provided the

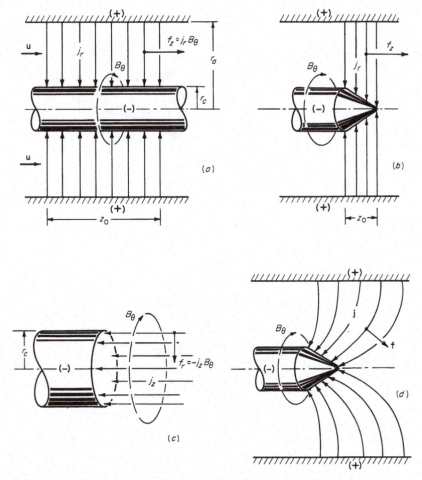

Fig. 8-18 Idealized models for analysis of MPD accelerator. (a) Uniform radial current; (b) radial current into conical cathode; (c) uniform axial current; (d) hybrid model.

azimuthal symmetry is maintained (Prob. 8-15). It does, however, need to be amended if the radial current enters the end rather than the side surface of the cathode. Consider another idealized case involving a conical-tip cathode, as sketched in Fig. 8-18b. Relations (8-100) and (8-101) remain valid for this case, but now the lower limit of the inner integral in (8-102) must be extended to the value $r_c(1 - z/z_0)$, with the consequent appearance of a second term in the result, representing the blowing contribution within r_c:

$$F_z = \frac{\mu J^2}{4\pi}\left(\ln\frac{r_a}{r_c} + \frac{1}{2}\right) \tag{8-103}$$

The value of the second term is sensitive to the particular current density distribution over the cathode surface. In the case just computed, $j(r)$ varies as $1/r$; if it were prescribed uniform over the cathode surface (with the corresponding $1/z$ variation over the anode surface), the contribution would be less:

$$F_z = \frac{\mu J^2}{4\pi}\left(\ln\frac{r_a}{r_c} + \frac{1}{4}\right). \tag{8-104}$$

The electromagnetic pumping contribution may be illustrated in another idealized arc model, wherein current enters the end surface of a cylindrical cathode in a uniform normal beam (Fig. 8-18c). The magnetic field within the arc is again azimuthal, but now proportional to radius:

$$B_\theta(r) = \frac{\mu J r}{2\pi r_c^2} \tag{8-105}$$

The body force density is therefore radial, and in equilibrium must be balanced by a radial gas pressure gradient:

$$f_r = j_z B_\theta = \frac{\mu J^2 r}{2\pi^2 r_c^4} = -\frac{dp}{dr} \tag{8-106}$$

The profile of gas pressure over the cathode face is thus parabolic.

$$p(r) = p_0 + \frac{\mu J^2}{4\pi^2 r_c^2}\left[1 - \left(\frac{r}{r_c}\right)^2\right] \tag{8-107}$$

where p_0 is the pressure outside of r_c. Since this gap pressure on the cathode is not balanced at the anode end of the discharge, the integral of $p - p_0$ over the cathode surface imparts an additional increment of thrust to the arc chamber:

$$F_c = 2\pi \int_0^{r_c}(p - p_0) r\, dr = \frac{\mu J^2}{8\pi} \tag{8-108}$$

Note that this result is again quadratic in total arc current and is independent of cathode radius. Indeed, it may also be shown to be independent of the radial distribution of current density over the cathode surface (Prob. 8-16).

If we now were to construct an idealized hybrid model involving all the elements of interaction discussed separately above (Fig. 8-18d), we should find the total electromagnetic acceleration to be just the sum of the blowing and pumping contributions derived separately, e.g.,

$$F = \frac{\mu J^2}{4\pi}\left(\ln\frac{r_a}{r_c} + \frac{3}{4}\right) \tag{8-109}$$

for uniform current density over the cathode end surface. This result is independent of the specific path of the arc current between the anode and cathode, and thus is appropriate to the curved diffuse arc patterns actually observed in the MPD arcs (Fig. 8-17). This proof may be more elegantly formulated in terms of the magnetic stress tensor, a powerful analytic technique which replaces detailed volume integration of the $\mathbf{j} \times \mathbf{B}$ pattern by simpler surface integrals [33]. Consider a second-rank tensor quantity $\mathfrak{B}$ constructed to satisfy the vector identity

$$\mathbf{j} \times \mathbf{B} = \frac{1}{\mu}(\nabla \times \mathbf{B}) \times \mathbf{B} = \nabla \cdot \mathfrak{B} - \frac{1}{\mu}\mathbf{B}\nabla \cdot \mathbf{B} \tag{8-110}$$

Since $\nabla \cdot \mathbf{B}$ is identically zero, application of the divergence theorem to (8-110) permits expression of the volume integral of the magnetic body force density in terms of a more readily evaluated surface integral:

$$\int_V \mathbf{j} \times \mathbf{B}\, dV = \int_V \nabla \cdot \mathfrak{B}\, dV = \int_A \mathfrak{B} \cdot \mathbf{n}\, dA \tag{8-111}$$

where $\mathbf{n}$ is a unit outward normal vector to the element dA of any surface A surrounding the volume V. To satisfy the relation (8-110), the tensor $\mathfrak{B}$ must have the elements $\mathfrak{B}_{ij} = (1/\mu)(B_i B_j - \tfrac{1}{2}\delta_{ij}B^2)$ in rectangular coordinates. In the cylindrical coordinates more convenient to the geometry of our MPD arc, the tensor takes the doubly contravariant form

$$\mathfrak{B} = \frac{1}{\mu}\begin{Bmatrix} B_r^2 - \dfrac{B^2}{2} & \dfrac{B_r B_\theta}{r} & B_r B_z \\ \dfrac{B_\theta B_r}{r} & \dfrac{B_\theta^2 - B^2/2}{r^2} & \dfrac{B_\theta B_z}{r} \\ B_z B_r & \dfrac{B_z B_\theta}{r} & B_z^2 - \dfrac{B^2}{2} \end{Bmatrix} \tag{8-112}$$

In the problem at hand, $B_r = B_z = 0$; so the tensor is diagonal, with elements

$$\mathfrak{B}_{rr} = \mathfrak{B}_{zz} = -\frac{B_\theta^2}{2\mu}$$

$$\mathfrak{B}_{\theta\theta} = +\frac{B_\theta^2}{2\mu r^2} \tag{8-113}$$

$$\mathfrak{B}_{\theta\theta} = +\frac{B_\theta^2}{2\mu r^2}$$

ELECTROMAGNETIC ACCELERATION—STEADY FLOW

The most convenient surface for integration is the cylindrical shell totally enclosing the arc current region, as shown in Fig. 8-19, which may be

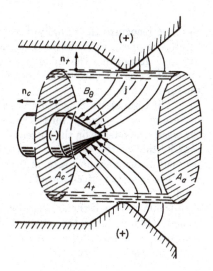

Fig. 8-19 Control surface for integration of the magnetic stress tensor.

treated in three sections: A_a, the circular surface at the anode end; A_c, the circular surface at the cathode end; and A_t, the cylindrical side element. Since B is zero everywhere on A_a, there is no contribution to the surface integral here. A_c is described by a unit normal vector $\mathbf{n}_c = 0, 0, -1$; hence

$$\mathfrak{B} \cdot \mathbf{n}_c = 0, 0, \frac{B_\theta^2}{2\mu} \tag{8-114}$$

Again assuming uniform current density over the cathode surface, we have the B_θ profile

$$B_\theta = \begin{cases} \dfrac{\mu J}{2\pi r_c^2} r & r < r_c \\ \dfrac{\mu J}{2\pi r} & r > r_c \end{cases} \tag{8-115}$$

For this profile

$$\int_{A_c} \mathfrak{B} \cdot \mathbf{n}\, dA = \frac{\mu J^2}{4\pi}\left(\ln\frac{r_a}{r_c} + \frac{1}{4}\right) \tag{8-116}$$

directed along the positive z axis, i.e., a blowing component in agreement with Eq. (8-104).

On the cylindrical side surface, $n_t = 1, 0, 0$; hence

$$\mathcal{B} \cdot \mathbf{n}_t = -\frac{B_\theta^2}{2\mu}, 0, 0 \qquad (8\text{-}117)$$

Here we recognize the radially directed pumping component, which, as before, we presume establishes a balancing gas pressure gradient within the arc. Recall that the unbalanced axial component of this hydrostatic force acting on the cathode is dependent only on the total arc current, in the form $\mu J^2/8\pi$.

The total electromagnetic thrust thus is independent of the detailed pattern of the arc, except for the ratio of anode to cathode radius and for the weak dependence on the current distribution over the cathode surface contributing to the sweeping term. Note also that the thrust is independent of the mass flow rate, provided the latter does not alter the total arc current or the effective radii of arc attachment on the cathode and anode. To this extent, then, the exhaust velocity should scale inversely with the mass flow if the acceleration conforms to the postulated mechanisms. (It was during the course of efforts to exploit this low mass flow, high specific impulse range that the previously mentioned zero mass flow operation was encountered.)

PARTICLE DESCRIPTION

Although the magnetogasdynamic model of the MPD arc makes no reference to gas properties, it presumes a continuum interaction in the manner in which it invokes the $\mathbf{j} \times \mathbf{B}$ body force density. However, since the large arc currents and high exhaust velocities are favored by low gas pressures in the arc chamber, some particle mean free paths and gyro radii may become comparable with the chamber dimensions, and orbit considerations may become relevant in its optimum range of operation. A second proposed model of the MPD device considers the extreme case of completely collisionless current conduction and particle acceleration in the arc [35]. As shown in Fig. 8-20, the model assumes that collisions are important only near the anode where ions are created. These ions are accelerated without further collision radially inward by the electric field and are deflected axially outward by the azimuthal self-magnetic field until they join an axial stream of electrons emitted from the cathode. The device is thus, basically, an electrostatic ion accelerator, with space-charge neutralization provided by the cathode electron beam.

This model yields certain correlations of its terminal properties which differ from the continuum case, and thus may serve to test its relative validity. First, since the radial current is carried by ions, it is limited by the available mass flow $\dot{m}$:

$$J \leq \frac{\dot{m}e}{m_+} \qquad (8\text{-}118)$$

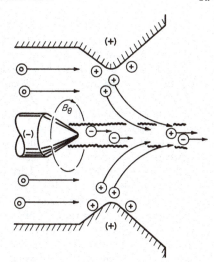

Fig. 8-20 Collisionless model of MPD arc.

where m_+ is the ion mass, and single ionization is presumed. Second, the ions must be able to negotiate the radial-to-axial turn; i.e., their local gyro radii must be of the order of their radial position in the arc or less:

$$r_B^+(r) = \frac{m_+ v_+}{eB} = \frac{2\pi r m_+ v_+}{\mu e J} \leq r \qquad (8\text{-}119)$$

Strictly, J here should be the enclosed current, rather than the total current; i.e., the outermost ions feel a weaker B field. Since these will be less deflected, however, they will soon penetrate farther into the field, etc., and the criterion should be qualitatively appropriate. In similar spirit, we can replace the variable ion velocity $v_+(r)$ in (8-119) by the full exhaust speed derived from the field, u, which then is limited to the approximate value

$$u \leq \frac{\mu e J}{2\pi m_+} \qquad (8\text{-}120)$$

The exhaust velocity is also limited by the available potential drop across the electrodes:

$$u \leq \left(\frac{2Ve}{m}\right)^{1/2} \qquad (8\text{-}121)$$

If we use the equalities throughout, comparison of (8-118), (8-120), and (8-121) predicts an arc voltage that varies as J^2, a thrust that varies as J^2, and an exhaust speed proportional to J and inversely proportional to m_+. The essential distinctions between these predictions and those of the magnetogasdynamic model thus stem from the specific coupling of arc current to mass flow [Eq. (8-118)] and to exhaust speed [Eq. (8-120)], which are not demanded by the latter.

The validity of the particle model clearly hinges on the collisionless assumption, and this may be unreasonable for much of the domain over which the MPD arc is observed to function. However, two features of it might be more generally relevant, and provide some insight into two prominent aspects of the arc's operation. The first relates to the ability of the cathode to sustain the extremely high current densities. Normally, concentrations of current of this intensity would rapidly erode a cathode surface by ion bombardment, but here we may speculate that the strong magnetic field near the cathode tip deflects the incoming ions sufficiently to protect it from excessive heating. This magnetic protection of the cathode may be expected to function over that portion of the arc near the cathode tip, where $\omega_B^+/\nu_c^+ + \gg 1$ and where the ion gyro radius is small compared with the cathode dimension, both of which can prevail at far higher densities than could be allowed for use of the collisionless theory over the remainder of the arc chamber.

The second instructive contribution of the collisionless approach concerns the coupling of arc current to mass flow rate, which may be relevant to the "zero mass flow" and spuriously high exhaust velocity observations mentioned earlier. If the device is operated in a regime where the equality of Eq. (8-118) applies and the mass flow is then reduced, the arc will be starved for current carriers, and the current will tend to lower. If the current is forcibly maintained at its previous level by the external circuit, however, the arc may elect to obtain current-carrying particles from other than the inlet mass flow, e.g., from the electrode surface material or by recirculating a fraction of its exhaust plume. Thus, although Eq. (8-118) may not accurately describe the higher mass flow domain of MPD arc operation, it may establish the lower mass flow limit before recirculation or electrode consumption begins.

HEURISTIC COLLISIONAL DESCRIPTION

The gap between the magnetogasdynamic theory and the collisionless approach can be bridged somewhat by inclusion of collisional effects in the particle orbit mechanics. To do this accurately and thoroughly would be a formidable task, but qualitative patterns of current and thrust densities can be predicted from examination of the electron and ion Hall parameters, Ω and Ω_+, and the respective gyro radii, r_B and r_B^+ (Sec. 5-2), over the arc chamber and plume regions [36]. The method proceeds from some initial estimates, presumably based on experimental data, of the patterns of the various gas properties which determine the Ω's and

r_B's, namely, the degree of ionization, the electron and heavy-particle temperatures, the electron density, and the magnetic field. As an example of the technique, consider a hydrogen MPD arc of sufficient current density that the gas may be regarded as fully dissociated and ionized throughout the current-carrying domain. Since only coulomb collisions occur in such an environment, the electron Hall parameter is given by (8-64) in the convenient form

$$\Omega \approx 6 \times 10^{15} \frac{BT_e^{3/2}}{n} \quad \text{mks} \tag{8-122}$$

The corresponding ion Hall parameter is largely irrelevant in this case, since the ion collision frequency is ill-defined. Ion-ion collisions serve mainly to maintain isotropy and equilibrium in the ion random thermal motion, rather than to exert any damping on the ion current; ion-electron collisions detract no significant momentum from the ion motion; and ion-neutral collisions are excluded by the complete ionization assumption. We thus may approximate the mean ion motion as a free particle drift. The gyro radii of both electrons and ions strongly affect the current conduction pattern, particularly when they become comparable with the arc dimensions. These may be written

$$r_B = 3.5 \times 10^{-8} \frac{T_e^{1/2}}{B} \quad \text{mks} \tag{8-123}$$

$$r_B^+ = \begin{cases} 1.5 \times 10^{-6} \frac{(T_+)^{1/2}}{B} & \text{mks} \\ 1.0 \times 10^{-8} \frac{u}{B} & \text{mks} \end{cases} \tag{8-124}$$

where it is presumed that the electron thermal velocity everywhere exceeds the drift speed, but the latter value of r_B^+ should be used wherever the streaming component of ion motion u exceeds the random thermal ion speed.

The technique now proceeds by insertion of measured or estimated patterns of n, T_e, T_+, B, and, where necessary, u, in (8-122) to (8-124), to provide corresponding maps of Ω, r_B, and r_B^+ over the arc and plume regions. From these, for a given electric field distribution in the same regions, individual particle trajectories can be sketched in the spirit outlined in Secs. 5-2 and 8-6. These in turn provide clues to the overall current conduction pattern, and thence to the regions where thrust is imparted to the gas stream. Figure 8-21 sketches typical results of one such calculation of the electron and ion mean motion trajectories. In this case a large electron Hall parameter at the edge of the cathode jet constrains the electrons emitted from the cathode surface to a cross-field

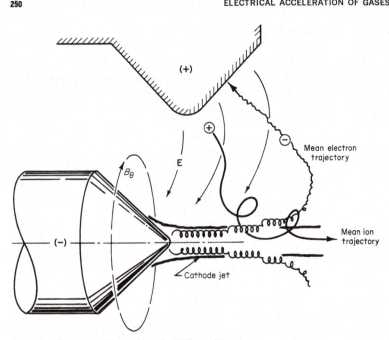

Fig. 8-21 Electron and ion motions in MPD accelerator.

axial drift until some downstream position where Ω has relaxed sufficiently to permit a collisional radial motion to develop. Only beyond this point can electrons migrate back toward the anode. In the absence of significant radial electron current near the electrodes, the ion current becomes an important factor. In the case sketched, the ion gyro radius is sufficiently small near the cathode jet to permit the ions also to lapse into axial cross-field drift there. The overall picture of this arc is thus qualitatively similar to the collisionless model discussed earlier.

The dominant thrust mechanisms now follow from the particle orbit patterns. In the outer regions of the plume we find a collision-dominated scalar conduction region providing a $\mathbf{j} \times \mathbf{B}$ body force like that in the elementary crossed-field accelerator. The steep radial decay of the magnitude of the self-generated magnetic field in this region reduces the relative importance of this thrust domain, and the curvature of the electric field lines here causes the thrust density vector to tilt radially outward near the anode surface. At intermediate radial and axial positions, we find a regime of predominantly electron Hall current and ion scalar current, yielding a radial component of thrust associated with the direct acceleration of the ions by the electric field. This we expect to manifest itself in a gasdynamic pressure over the cathode surface. Within the cathode jet, we find a region where both ions and electrons are constrained to an axial drift at a speed given by the local ratio of E/B, contributing a thrust determined by the mass flow concentrated in this jet. Since the self–magnetic field decreases monotonically with radius inside the jet, the charged-particle trajectories here are complex figures, which are, however, effectively bounded by the maximum field at the jet edge (Prob. 8-17).

The effect of an external magnetic field on the operation of the arc can also be seen by this approach. Starting from an empirical representation of the total magnetic field pattern, presumably based upon an experimental determination of the external field with the arc off, one again constructs maps of the Hall parameters and gyro radii over the region of interest. From these, and from the presumed electric and magnetic field distributions, electron and ion trajectories may again be sketched. Figure 8-22 shows the results of such a construction for a predominantly axial external **B** field which is everywhere much stronger than the self-induced field. For the example chosen, the electrons emitted from the cathode tip again find themselves in a region of large Ω, but now their **E** × **B** drift is in the azimuthal direction. In addition, since **E** and **B** are not orthogo-

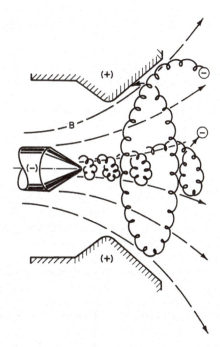

Fig. 8-22 Electron motion in MPD accelerator with large external magnetic field.

nal here, a component of motion is permitted along the **B** lines. The composite electron trajectory in this region is thus a cycloidally embossed spiral about the z axis, emanating from the cathode and confined to the funnel-shaped surface of which the local **B** line is an element. This type of motion continues until the electron progresses downstream to a lower-Ω collisional regime, where it can migrate across the **B** lines back toward the anode. As it approaches the anode, however, it again enters a high-Ω region, where it must lapse into another spiraling trajectory like that found near the cathode.

In this particular calculation, the ion gyro radii near the anode and cathode are found to be much smaller than the electrode gap, predicating spiraling ion trajectories in these regions very much like those of the electrons. Again, it is not until they reach some position farther downstream that the ions can cross the **B** lines. Since the direction of the azimuthal **E** $\times$ **B** drift is the same for ions and electrons, this model predicts no net Hall current, but only a swirling motion of the gas as a whole in these anode and cathode plasma funnels.

Although it is inappropriate to protract discussion of orbital analysis of the MPD arc by further illustrations, it should be noted that most of the phenomenological traits displayed in the experimental operation of these devices can be qualitatively rationalized by this technique. Plume protraction, ambient-gas ingestion, arc swirling, cathode protection, and the gross aspects of the current density distribution, all follow logically from its self-consistent application under the prevailing conditions. The extent to which it can provide quantitative representation of these phenomena clearly rests on the accuracy with which the participating parameters can be determined over the region of interest.

One point which has been neglected by all the foregoing analyses is the possibility of an electrothermal contribution to the performance of the MPD arc. Bearing in mind that the devices in which the MPD effect was first identified had originally been designed as electrothermal thrusters, it is reasonable to suspect that resistive plasma heating processes continue to participate in the overall gas acceleration mechanism in some way. At first glance, the electrothermal contribution might seem to be minor in an arc producing exhaust velocities up to 10^5 m/sec, because of the excessive gas temperatures which would be required to yield comparable speeds via nozzled expansion. We have just seen, however, that one effect of the large self-induced or external **B** fields is to displace the collisional plasma region downstream from the electrodes. By so doing, the arc may now be able to sustain much higher currents and gas temperatures in this "ohmic" region than if it were in direct contact with the material electrodes. The extreme point of view would be to regard the device as still an electrothermal accelerator, with the hottest portion of the gas kept away from the electrodes, and the exhaust stream focused by the "magnetic nozzle" established by the field pattern. We should thus again have a prime example of a hybrid thruster wherein both electrothermal and electromagnetic processes contributed substantially to the gas acceleration.

In the analysis of the arc as a whole, however, the distinction between electrothermal and electromagnetic acceleration mechanisms becomes somewhat vague and academic. On an atomic scale, the distinction relates to the extent to which the charged particles can first randomize and then reorganize the energy imparted to them by the electromagnetic fields, over the dimensions available in the device. Randomization is evident in the collisional low-Ω regions of the arc, but in the free-orbit drifting regimes and in the effectively collisionless motion of the ions, the criteria for thermalization are less clear. For example, does one regard a charged particle constrained to cycloidal helical drift by the magnetic field pattern to be in organized streaming or random thermal motion?

Effective electrothermal conversion is important, however, from the standpoint of the overall efficiency of the MPD device as a thruster. Were only electromagnetic effects producing the gas acceleration, the efficiency would need to be assigned on the basis of the ratio of some appropriate integration of $(\mathbf{j} \times \mathbf{B}) \cdot \mathbf{u}$ over the entire volume to a similar integration of $\mathbf{j} \cdot \mathbf{E}$. That is, all Joule heating of the gas would be regarded as an energy loss. If, on the other hand, the device retained some of its electrothermal heritage and could redirect a significant portion of this randomized energy into streaming motion, the overall efficiency would benefit accordingly. In the limit, only frozen flow radiation and electrode losses would detract from its performance, and its efficiency could be significantly higher.

Regardless of the particular physical models invoked to describe the details of MPD arc operation, it is clear that this is a device which fortuitously embodies intensities and geometries of current and electric and magnetic fields which circumvent several of the classical limitations of the conventional electromagnetic and electrothermal accelerators. Indeed, in comparison with the idealized uniform channel flow concepts with which this chapter began, or with the elementary core-flow arcjets discussed in Chap. 6, the MPD arc is a renegade, violating our estimates of reasonable values for current density from an electrode, gas density, gas temperature, channel length, channel profile, gas conductivity, and magnetic interaction parameter (Prob. 8-18). Yet this very irregularity of its domain of operation may be indicative of the situation prevailing in the entire field of plasma propulsion, namely, that the acceleration of a high temperature, highly ionized, nonuniform working fluid by the application of interwoven electric and magnetic fields simply will not prosper under conventional gasdynamic implementation, but rather demands a new technology yet to be developed.

PROBLEMS

8-1. What electromagnetic plasma accelerators are analogous to the following conventional electric motors: (a) series-wound dc motor; (b) shunt-wound dc motor; (c) polyphase induction motor?

8-2. Show that $\mathbf{j} \cdot \mathbf{E}$ is the proper expression of the rate of total electric energy input to a magnetogasdynamic channel flow. Express $\mathbf{j} \cdot \mathbf{E}$ as a sum of useful work and dissipative components.

8-3. Reduce the system of magnetogasdynamic equations (8-1) to (8-15) by assuming f_V, ϕ_v, ϕ_t, ϕ_r, j_H, $j_I = 0$ and eliminating $\mathbf{j}$ and $\mathbf{E}$. Interpret the resulting relations.

8-4. Derive relations between u_0 and p_0 and u_0 and M_0 which guarantee smooth passage of the flow through $M = 1$ for a constant field, constant area channel.

8-5. There is some ambiguity in the definition of an efficiency for a magnetogasdynamic accelerator. One might use the ratio of the increase in kinetic energy of the flow along the channel to the electric input power,

$$\eta_1 = \frac{(F/2)(u_L{}^2 - u_0{}^2)}{\int_0^L \mathbf{j} \cdot \mathbf{E}\, dx}$$

or alternatively, the ratio of kinetic energy of the exhaust flow to the sum of inlet kinetic energy, inlet enthalpy, and electric input power,

$$\eta_2 = \frac{\tfrac{1}{2} F u_L{}^2}{\tfrac{1}{2} F u_0{}^2 + F c_p T_0 + \int_0^L \mathbf{j} \cdot \mathbf{E}\, dx}$$

Compute η_1 and η_2 for a uniform field, constant area accelerator, with sonic inlet velocity and maximum exhaust velocity, as functions of u_0/ξ. Examine the limiting cases for maximum and minimum possible u_0/ξ. For what values of u_0/ξ are η_1 and η_2 a maximum? Explain why η_1 may exceed unity.

8-6. To what extent are the limiting performance estimates at the close of Sec. 8-3 compromised by the neglect of streamwise components of electric field?

8-7. Why are electron-electron collisions not involved in the calculation of the electron Hall parameter for a substantially ionized gas?

8-8. Express the ratio of electron-ion to electron-neutral collision frequency in terms of electron temperature and degree of ionization. Given a gas in thermal equilibrium at 0.1 STP density which satisfies a small ionization Saha relation of the form (3-28), where $G' = 2 \times 10^{11}$ for n_0 in m^{-3} and $\varepsilon_i = 15$ ev, and assuming $Q_{e0} = 10^{-16}$ cm^2, compute the temperature at which electron-ion collisions take over control of the Hall parameter Ω. What is the corresponding value of Ω?

8-9. Derive the components of current density in a rectangular Hall accelerator of the second class for the two cases $E_x = 0$ and $j_x = 0$. Derive the optimum angle of inclination of B and the corresponding optimum axial force density f_x, in each case.

8-10. Derive the generalized Ohm's law in the form of Eq. (8-74).

8-11. Express the ratio of ion to electron Hall parameters for an equilibrium partially ionized gas in terms of mass ratio, degree of ionization, and ratio of electron-ion to ion-neutral cross section. Under what conditions might this Hall parameter ratio approach or exceed unity?

8-12. Explore possible solutions for a one-dimensional induced-field accelerator under constraints of (a) isothermal flow, (b) adiabatic flow, (c) constant current density, and (d) any others permitting closed-form solutions.

8-13. Explore possible solutions of a self-field sheet accelerator under constraints of

(a) isothermal flow, (b) adiabatic flow, (c) zero acceleration (i.e., a pump), and (d) any others permitting simple solutions.

8-14. Estimate the maximum gas density which will allow an electron temperature of 4000°K to be sustained in a nitrogen stream at 2000°K for an applied electric field of 10^3 volts/m.

8-15. Show that the axial thrust generated by an azimuthally uniform radial discharge between two concentric cylinders is properly given by Eq. (8-102), irrespective of the axial distribution of current density.

8-16. Show that total force on the cathode in Fig. 8-18c is independent of the radial distribution of axial current density over the cathode surface.

8-17. Sketch typical trajectories for electrons within the cathode jet of an MPD arc, assuming $B_\theta(r) \propto 1/r$, $E_r \propto 1/r$, and no collisions.

8-18. Estimate a range of magnetic interaction parameter for the MPD arc operation of Table 8-3.

REFERENCES

1. Sutton, G. W., and A. Sherman: "Engineering Magnetohydrodynamics," McGraw-Hill Book Company, New York, 1965.
2. Spitzer, L., Jr.: "Physics of Fully-ionized Gases," Interscience Publishers, Inc., New York, 1956.
3. Cowling, T. G.: "Magnetohydrodynamics," Interscience Publishers, Inc., New York, 1957.
4. Ferraro, V. C. A., and C. Plumpton: "An Introduction to Magneto-fluid Mechanics," Oxford University Press, Fair Lawn, N.J., 1961.
5. Pai, S.: "Magnetogasdynamics and Plasma Dynamics, Prentice Hall, Inc., Englewood Cliffs, N.J., 1962.
6. Cambel, A. B.: "Plasma Physics and Magnetofluidmechanics," McGraw-Hill Book Company, New York, 1963.
7. Resler, E. L., Jr., and W. R. Sears: The Prospects for Magneto-aerodynamics, *J. Aeron. Sci.*, vol. 25, no. 4, pp. 235–245, April, 1958.
8. Rosa, R. J.: Engineering Magnetohydrodynamics, Ph.D. thesis, Cornell University, Ithaca, N.Y., 1956.
9. Mirels, H.: Analytical Solution for Constant Enthalpy MHD Accelerator, *AIAA J.*, vol. 2, no. 1, p. 145, January, 1964.
10. Ring, L. E.: Status of MHD Accelerators for Test Facilities, AIAA and Northwestern University Sixth Biennial Gas Dynamics Symposium, Evanston, Ill., Aug. 25–27, 1965, *AIAA Paper* 65-631.
11. Kerrebrock, J. L., and F. E. Marble: Constant-temperature Magneto-gasdynamic Channel Flow, *J. Aerospace Sci.*, vol. 27, p. 78, 1960.
12. Sherman, A.: Theoretical Performance of a Crossed Field MHD Accelerator, *ARS J.*, vol. 32, p. 414, 1962.
13. Kerrebrock, J. L.: Electrode Boundary Layers in Direct Current Plasma Accelerators, *J. Aerospace Sci.*, vol. 28, p. 631, 1961.
14. Bush, W. B.: Compressible Flat Plate Boundary Layer Flow with an Applied Magnetic Field, *J. Aerospace Sci.*, vol. 27, p. 49, 1960.
15. Demetriades, S. T., and P. D. Lenn: Electrical Discharge across a Supersonic Jet of Plasma in Transverse Magnetic Field, *AIAA J.*, vol. 1, no. 1, p. 234; no. 8, p. 1967, 1963.
16. Duclos, D. P., et al.: Physical Property Distributions in a Low Pressure Crossed-field Plasma Accelerator, *AIAA J.*, vol. 3, no. 11, p. 2026, November, 1965.
17. Foley, R. T., and T. A. Vanderslice: Seeding of Gases for Magnetohydrodynamic Applications, *ARS J.*, vol. 32, no. 10, p. 1573, October, 1962.

18. Kerrebrock, J. L.: Nonequilibrium Ionization due to Electron Heating, I, Theory, *AIAA J.*, vol. 2, no. 6, p. 1072, June, 1964; Kerrebrock, J. L., and M. A. Hoffman: Nonequilibrium Ionization due to Electron Heating, II, Experiments, *AIAA J.*, vol. 2, no. 6, p. 1080, June, 1964.
19. Zukoski, E. E., et al.: Experiments Concerning Non-equilibrium Conductivity in a Seeded Plasma, *AIAA J.*, vol. 2, no. 8, p. 1410, August, 1964; Cool, T. A., and E. E. Zukoski: Recombination, Ionization, and Nonequilibrium Electrical Conductivity in Seeded Plasmas, *Phys. Fluids*, vol. 9, p. 780, 1966.
20. Demetriades, S. T., et al.: Three-fluid Non-equilibrium Plasma Accelerators, p. 461 in E. Stuhlinger (ed.), "Electric Propulsion Development," vol. 9 of "Progress in Astronautics and Aeronautics," Academic Press Inc., New York, 1963; Demetriades, S. T.: Momentum Transfer to Plasmas by Lorentz Forces, p. 297 in T. P. Anderson et al. (eds.), "Physico-chemical Diagnostics of Plasmas," Northwestern University Press, Evanston, Ill., 1964.
21. Jahn, R. G., and W. von Jaskowsky: Langmuir Probe Measurements in the Plasma Jet, *Plasmadyne Corp. Rept.* PLR-34, August, 1959; Shielded Probe Measurements in the Plasma Jet, *Plasmadyne Corp. Rept.* PLR-72, June, 1960.
22. Blackman, V. H., and R. J. Sunderland: Experimental Performance of a Crossed-field Plasma Accelerator, *AIAA J.*, vol. 1, no. 9, p. 2047, September, 1963.
23. Powers, W. E., and R. M. Patrick: A Magnetic Annular Arc, *Phys. Fluids*, vol. 5, p. 1196, 1962; Patrick, R. M., and A. M. Schneiderman: Performance Characteristics of a Magnetic Annular Arc, *AIAA J.*, vol. 4, no. 2, p. 283, February, 1966.
24. Cann, G. L., and G. L. Marlotte: Hall Current Plasma Accelerator, *AIAA J.*, vol. 2, no. 7, p. 1234, July, 1964.
25. Sevier, J. R., et al.: Coaxial Current Accelerator Operation at Forces and Efficiencies Comparable to Conventional Crossed-field Accelerators, *ARS J.*, vol. 32, no. 1, p. 78, January, 1962.
26. Carter, A. F., et al.: Research on a Linear Direct-current Plasma Accelerator, *AIAA J.*, vol. 3, no. 6, p. 1040, June, 1965.
27. Ducati, A. C., et al.: Experimental Results in High Specific Impulse Thermoionic Acceleration, *AIAA J.*, vol. 2, no. 8, p. 1452, August, 1964.
28. Hess, R. V.: Experiments and Theory for Continuous Steady Acceleration of Low Density Plasmas, *Proc. 11th Intern. Astronaut. Federation Congr.*, Springer-Verlag OHG, Berlin, 1960.
29. Ducati, A. C., et al.: Recent Progress in High Specific Impulse Thermo-ionic Acceleration, AIAA 2d Aerospace Sciences Meeting, New York, January, 1965, *AIAA Paper* 65-96.
30. Bennett, S., et al.: MPD Arc Jet Engine Performance, AIAA 2d Annual Meeting, San Francisco, July, 1965, *AIAA Paper* 65-296.
31. Cann, G. L., et al.: Hall Current Accelerator, *Electro-optical Systems, Inc., Tech. Rept.* 5470-Final, February, 1966.
32. Powers, W. E.: Measurements of the Current Density Distribution in the Exhaust of an MPD Arc Jet, AIAA 3d Aerospace Sciences Meeting, New York, January, 1966, *AIAA Paper* 66-116.
33. Kelly, A. J., et al.: Electron Density and Temperature Measurements in the Exhaust of an MPD Source, *AIAA J.*, vol. 4, no. 2, p. 291, February, 1966.
34. Jackson, J. D.: "Classical Electrodynamics," John Wiley & Sons, Inc., New York, 1962.
35. Stratton, T. F.: High Current Steady State Coaxial Plasma Accelerators, *AIAA J.*, vol. 3, no. 10, p. 1961, October, 1965.
36. Jahn, R. G.: An Electron's View of an MPD Arc, *Giannini Scientific Corp. Tech. Rept.* 5QS085-968, August, 1965.

9
Unsteady Electromagnetic Acceleration

9-1 MOTIVATION AND CLASSIFICATION

Considering the attractive performance range and operational simplicity of the steady magnetoplasmadynamic arc discussed in the preceding chapter, there might seem to be little motivation for extending the study of electromagnetic acceleration into the inherently more complex unsteady domain. Faced with the substantial analytical and technological development needed to refine the steady MPD accelerator into a satisfactory space propulsion unit, one may understandably require clear-cut incentives to introduce the additional complications of unsteady fields and flows into electromagnetic propulsion devices. However, such incentives do exist, in the form of certain physical characteristics of unsteady discharges which presage possibilities for even higher plasma acceleration performance than the steady devices attain. Specifically, three potential advantages of unsteady operation merit consideration:

1. The thrust efficiency of the steady MPD accelerator tends to improve with increasing discharge current density. In a space thruster of

this type the maximum current will be limited either by the tolerable electrode erosion or by the available power supply. Since the total thrust of a self-field device scales quadratically with the arc current, there would seem to be some advantage in passing this current through it in intermittent intense pulses, thereby attaining higher average thrust, probably with less electrode erosion, than by a steady consumption of the same total power.

2. Operation of the arc in brief intense pulses may preclude the development of certain undesirable kinetic equilibrations within the accelerating plasma. For example, if the entire acceleration event is accomplished rapidly enough, it may be possible to keep the free-electron component of the gas from thermalizing with the high energy streaming ions, and subsequently squandering this energy via inelastic collisions and radiation.

3. Unsteady electromagnetic field effects may produce beneficial spatial nonuniformities in the discharge pattern. The most striking example of this is the development of a "skin effect," in transient discharges, whereby the entire current may be constrained to flow in a narrow region at a favored location in the discharge channel, thereby providing a more effective coupling between the accelerating field and the ambient gas.

The task of evaluating the validity and applicability of such unsteady acceleration benefits is lightened somewhat by two factors. First, it is found that, under some circumstances, introduction of unsteady effects simplifies, rather than complicates, the theoretical formulation. For example, if the current is indeed confined to a narrow region, finite-jump, or *snowplow*, models, akin to classical gasdynamic shock wave treatments, which make no reference to interior structure of the current zone, may provide useful information (Sec. 8-7). Even in those cases where details of the discharge structure are important, extremely high current levels assure essentially full ionization; hence a two-fluid domain of predominantly electrodynamic, rather than gasdynamic, interaction will prevail.

As a second aid, there exists a substantial amount of previous experimental experience with closely related acceleration devices. Most relevant are the family of plasma injectors, or "guns," developed in connection with the extensive program on controlled thermonuclear fusion [1]. These are supplemented by a few smaller scale devices for the production of "plasmoids" for various interaction experiments [2], some electromagnetic shock tubes for high enthalpy gasdynamic and gaskinetic studies [3], and certain exotic spectroscopic light sources [4]. In a rather different class of unsteady interaction, traveling-wave electron accelerators have been in operation for many years [5], and provide valuable background for continuous-wave-train plasma acceleration.

UNSTEADY ELECTROMAGNETIC ACCELERATION

The last reference brings us to a necessary subdivision of the concept of unsteady electromagnetic acceleration. In the following discussion we shall distinguish between pulsed devices, wherein the entire plasma production-acceleration-ejection event is accomplished within a single burst of electrical discharge, and traveling-wave accelerators, wherein a continuous electromagnetic wave train convects a periodically modulated, but otherwise steady, gas flow. This distinction will diffuse somewhat as a pulsed device is brought to more and more rapid repetition rates and the successive discharge events interact with each other, both in the plasma and in the driving circuit. Perhaps a more basic distinction is between those devices which have electrodes in direct contact with the gas, and thus require a zone of conducting plasma to complete the circuit, and those which induce currents in the gas flow in response to a changing current in a complete external circuit. The former are probably most applicable to propulsion, and we shall allow them the designation *pulsed plasma accelerators*, with no other qualification. Much of the concept and mechanics of these devices can be directly extended to the *induction accelerators*, although the limiting case of traveling-wave machines yields more naturally to a somewhat different analytical formalism.

9-2 PULSED PLASMA ACCELERATION

Doubtless the first thoughts of pulsed plasma acceleration were inspired by phenomenological observation, and the most effective introduction to this topic is still a phenomenological one. The basic process can be demonstrated by any of several simple and familiar laboratory devices. Consider the pair of parallel-rail electrodes sketched in Fig. 9-1a. If a large capacitor, charged to sufficiently high voltage, is abruptly switched across these rails, an arc will initiate between them at the closest accessible position to the capacitor circuit, and then will propagate down the rails, away from its original position. A similar effect can be observed in a "button gun" (Fig. 9-1b), where a blob of plasma created between the electrode tips is ejected normally outward; in a "T tube" (Fig. 9-1c), where discharge plasma impels a gasdynamic shock wave down a glass pipe; in a "coaxial gun" (Fig. 9-1d), where an annular sheet of plasma sweeps axially between coaxial conductors; and in a "dynamical pinch" discharge (Fig. 9-1e), where a cylindrical plasma sheet implodes radially inward from the electrode periphery.

Each of these experiments displays the same sequence of three physical events: (1) the discharge initiates in the geometrical configuration which provides the minimum inductance for the electrical circuit it completes; (2) the discharge intensifies into a narrow current-conducting zone; and (3) this current zone accelerates toward a configuration of maximum inductance. Detailed analytic description of any of these three processes

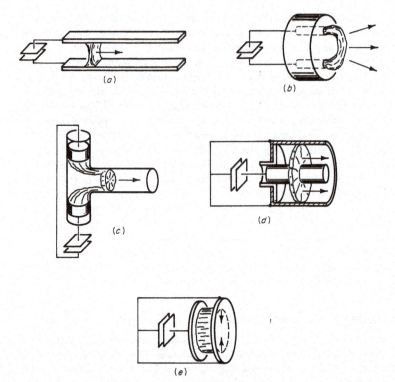

Fig. 9-1 Various pulsed plasma accelerators (schematic). *(a)* Parallel-rail accelerator; *(b)* button gun; *(c)* T tube; *(d)* coaxial gun; *(e)* linear pinch.

has not been achieved, but their general relation to more familiar electrodynamic phenomena seems plausible. The initiation of the arc in the minimum inductance configuration may be associated with a high-frequency skin effect wherein a rapidly changing current tends to be excluded from the interior of a conductor by the back emf of its own rapidly changing magnetic field (Prob. 9-1). In the situations depicted here, one must first postulate a transient low-level discharge over the entire electrode gap, which provides an initial gas conductivity, but then rapidly intensifies preferentially at the low inductance location.

The maintenance of a narrow arc sheet or filament is most probably associated with the initial localization of the arc; with the strong nonlinearity in gas conductivity as a function of current density, i.e., in the arc current-voltage characteristic; and with the action of the self–magnetic forces over the brief time scale of the event. That is, the entire acceleration process is completed before the sharp current density profile created at the outset has time to diffuse into the ambient gas ahead of it or into the magnetic field behind it.

The acceleration itself is mainly provided by the interaction of the arc current with the self-magnetic field produced by the complete circuit. Regardless of the sign of the current, this force always acts to expand the area of the circuit loop, and hence drives the plasma away from the external conductors. The microscopic details of the $\mathbf{j} \times \mathbf{B}$ interaction in the gas are difficult to identify precisely, but may involve substantial interior electric fields, as well as the primary magnetic forces on the charged particles. Some electrothermal contribution to the acceleration process may be provided by a preferential expansion of the heated electron component of the discharge plasma away from the inception boundary. As in the magnetic expansion thruster (Sec. 6-6), the force associated with this expansion must be transmitted to the heavy particles via a streamwise electric field established by a slight separation of the electron and ion component distributions. An index of the relative importance of this contribution is the ratio of the electron pressure to the magnetic pressure at the back of the current sheet.

For a device of this type to function effectively as a pulsed plasma thruster, it must also properly attend to a sequence of gasdynamic processes, including the initial filling of the electrode gap with propellant gas, a channeling of the flow during the acceleration phase, and the ejection of the accelerated mass at the end of the channel. The complete cycle of a pulsed plasma thruster thus involves the following series of events:

1. The appropriate amount of electrical energy must first be accumulated in the external circuit. Since we desire pulses of very large current (10^4 to 10^6 amp), at sufficient voltage to break down a few centimeters of cold gas (10^3 to 10^5 volts), an electrical storage unit such as a capacitor, or possibly an inductor (Prob. 9-2), will be required, unless a prodigious power plant is available. When switched across the electrode gap, this storage unit must deliver its electrical energy with a minimum of internal dissipation or inductive delay.

2. The propellant gas must then be admitted into the electrode gap in a manner that establishes the desired mass density distribution prior to discharge. Since the exit of the accelerator channel opens onto free space, this gas injection must be accomplished rapidly, and in precise correlation with the discharge initiation, lest too much gas leak out the channel before breakdown or too much enter after the event is over. For single pulse or slow repetitive operation this requires an extremely rapid gas valve. For very high frequency

pulse repetition, it may be possible to operate with a steady inlet flow.
3. Discharge breakdown must occur at the proper time in the gas-filling cycle and in the proper geometrical location in the channel. The latter is normally accomplished automatically by the aforementioned skin effect for a reasonably designed channel; the former will require either a switch in the external circuit or proper utilization of the breakdown potential vs. pressure characteristic of the propellant gas in the gap (Sec. 6-4). An external switch permits precise correlation of the discharge with the gas-filling cycle, but will inevitably consume some of the electrical energy and add some inductance to the circuit. Self-triggering of the discharge when the injected gas pressure builds up to the breakdown level eliminates the switch, but restricts operation to specific combinations of gap spacing, applied voltage, and gas density and may itself involve significant dissipation of electrical energy in the poorly conducting initial breakdown phase.
4. Once a breakdown is established, it must intensify rapidly to a current layer that is stable, that covers the entire channel cross section, that is of sufficiently high conductivity to make resistive losses negligible, and that is sufficiently dense to be impermeable to the ambient gas which it must accelerate. These requirements bear, first, on the characteristics of the external circuit, which must deliver sufficient total current at sufficiently rapid rise time that the skin effect operates effectively at the attainable gas conductivity; second, on the ambient gas density, which must be high enough to sustain the arc, but not so high as to overload it with excessive mass; and third, on the channel geometry, which must lend itself to a transverse current zone across the entire section.
5. The current layer thus formed must now accelerate down the channel under its own magnetic field, effectively accumulating the ambient gas either by pushing it like a piston or by entraining it within itself, until an appropriate velocity is attained. During this process the current layer must not become porous to the ambient gas ahead of it, nor should it develop dynamic instabilities which distort or disrupt its profile. In addition, the gas accumulation process should involve a minimum of thermalization and dissipation in the body of accelerated gas. Relation of these requirements to the external electrical parameters, channel geometry, and gas properties is perhaps the most challenging basic problem of this field.
6. When the accelerated mass of gas and plasma reaches the end of the channel, it must be ejected with a minimum of thermal and electromagnetic loss. On the one hand, it is desirable that the magnetic

field prevailing near the exhaust aperture contribute to a focusing of the exhaust plume into a well-collimated stream and provide a region where substantial recovery of thermal energy in the plasma may occur. On the other hand, this exit field must not impede the ejection of the plasma in a dynamical sense; i.e., the plasma must not become tied to the field lines. Finally, the field pattern at exhaust must not itself involve substantial electromagnetic energy, which later will relapse into the external circuit, to be dissipated in useless secondary breakdowns of the gap. Here again, precise relation of these requirements to the controllable parameters of the problem is poorly understood.

7. Finally, the entire electromagnetic and gasdynamic system must return to its initial situation, prior to the initiation of a new cycle. Involved here are many diverse processes, such as the ringdown pattern and recharging profile of the electrical storage unit, the relaxation of any residual ionization in the channel, the thermal and chemical relaxation of the electrode surfaces, the completion of the gas valve cycle, the opening and rearming of the switch, etc., any one of which could limit the repetition rate of the accelerator.

The grand sum of this list of requirements, of course, is simply the efficient conversion of electrical energy into the kinetic energy of the exhaust stream. The tools available for accomplishing these specifications are, as before, ad hoc analytical models, empirical testing, and—perhaps more essential here than anywhere else in the entire electric propulsion field—basic experiments on the component physical processes. In the balance of this chapter, we shall attempt some illustration of each of these methods of attack.

9-3 CIRCUIT ANALYSES OF PULSED ACCELERATORS

The phenomenological observation of narrow arc current layers in accelerators of this class suggests a theoretical idealization wherein these layers are represented by discrete, albeit movable, elements of a simple inductance-capacitance-resistance (LCR) series circuit [6]. Consider, for example, the parallel-plate accelerator sketched in Fig. 9-2, across which is switched a capacitance C, initially charged to a voltage V_0. Breakdown of the gas completes a circuit which has a certain initial inductance L_0, determined by the external circuit geometry and by the inherent inductance of the capacitors, and a certain initial resistance R_0 contributed partly by the external circuit, including the capacitors, and partly by the discharge itself. As the discharge intensifies and propagates down the channel, the total circuit inductance will increase and the resistance of the

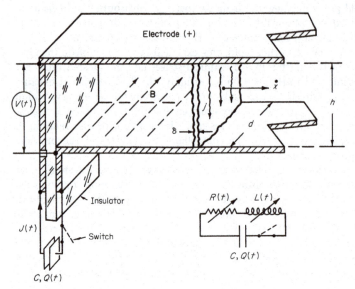

Fig. 9-2 Parallel-plate accelerator.

arc will change in time. The waveforms of current flowing through the circuit $J(t)$ and voltage appearing across the electrodes $V(t)$ will reflect these changes in circuit parameters, $L(t)$, $R(t)$. For example, application of Maxwell's $\nabla \times \mathbf{E}$ equation around the changing circuit yields the voltage-current characteristic

$$V = JR + \dot{\phi} = JR + \frac{d}{dt}(LJ) = JR + L\dot{J} + J\dot{L} \qquad (9\text{-}1)$$

where dots denote time derivatives, and ϕ is the magnetic flux defining the inductance L (Sec. 2-3). The power delivered by the capacitance to the circuit is thus

$$P = JV = J^2R + LJ\dot{J} + J^2\dot{L} = J^2R + \frac{d}{dt}(\tfrac{1}{2}LJ^2) + \tfrac{1}{2}J^2\dot{L} \qquad (9\text{-}2)$$

The three terms on the right represent, respectively, rate of resistive heat generation, rate of change of energy stored in magnetic fields, and rate of work done on the moving arc sheet. Presumably, it is the last term we wish to optimize.

Let us now assume that all the energy initially stored in the capacitance, $W_0 = \frac{1}{2}CV_0^2$, is delivered to the circuit in a single pulse of current lasting a time τ. We may write

$$W_0 = \int_0^\tau P\, dt = \int_0^\tau (J^2 R + \tfrac{1}{2} J^2 \dot{L})\, dt \tag{9-3}$$

where the magnetic field energy term has disappeared in the integration since the current is zero at $t = 0$ and $t = \tau$ by definition. Note that $\dot{L}$ has the dimensions of a real impedance, like R, and that its magnitude relative to R will be one index of the effectiveness of the accelerator. For example, we may define an electrical efficiency

$$\eta_e = \frac{1}{2} \frac{\int_0^\tau J^2 \dot{L}\, dt}{W_0} \tag{9-4}$$

which regards the $J^2 R$ integral, although partially a thermal energy input to the gas, as a total loss, and the $J^2 \dot{L}$ integral as completely convertible to kinetic energy of the plasma layer; i.e., it neglects dynamical losses.

To estimate the magnitude of η_e, let us define a fictitious reference current J_0 equal to the amplitude of the sinusoidal current which would have arisen in this circuit had R been zero and L constant at the initial value L_0; that is, $W_0 = \frac{1}{2} L_0 J_0^2$. Since in the real circuit L monotonically increases from L_0 at breakdown and R is finite, we may presume that $J < J_0$ throughout the pulse (Prob. 9-3), and thus (9-4) yields the useful inequality

$$\eta_e = \frac{\int_0^\tau J^2 \dot{L}\, dt}{J_0^2 L_0} < \frac{\int_0^\tau J_0^2 \dot{L}\, dt}{J_0^2 L_0} = \frac{\Delta L}{L_0} \tag{9-5}$$

where ΔL is the total increase in circuit inductance achieved during the current pulse. Relation (9-5) implies that efficient single-pulse accelerators will be characterized by inductance changes during acceleration at least comparable with the initial inductance. Violation of this criterion will be indicative either of excessive resistive losses or of an uncoupling of the plasma element from the circuit before completion of the current pulse, with consequent energy residue left in the capacitance or in the electromagnetic field.

The $\Delta L/L_0 \approx 1$ criterion relates primarily to the geometry and scale of the accelerator, and might best be illustrated by an order-of-magnitude examination of a specific case. If the plates of our linear accelerator are sufficiently broad so that edge effects of the fields and flow may be neglected, say, $h/d \approx 0.1$, the inductance they contribute to the circuit is simply $\mu_0(h/d)x$, where x is the displacement of the current sheet from its initial position. Since this is the only time-varying contribution to the total circuit inductance, we have

$$\dot{L} = \mu_0 \frac{h}{d} \dot{x} \qquad \text{henry/sec} \qquad (9\text{-}6)$$

If we center our interest on the exhaust speed range $\dot{x}_e \approx 5 \times 10^4$ m/sec, $\dot{L}$ will not exceed about 0.005 ohm. Therefore all resistances in the circuit, including the discharge itself, must be small compared with this value if the bulk of the applied power is not to be dissipated in Joule heating.

The available inductance change is similarly very small. Even if we allow the channel to be 0.50 m long, a rather generous size as we shall see later, ΔL will still be only about 5×10^{-8} henry. To assemble an external circuit with fixed inductance less than this value requires considerable care.

The required capacitance C and initial voltage V_0 are controlled by a number of conflicting factors, most notably by the current sheet opacity requirement discussed above, by the ringdown pattern of the complete LCR circuit, and by the specific mass of the available capacitors. For example, it is found empirically that a linear current density of at least 10^5 amp/m must pass through the current sheet for it to accelerate and sweep ambient gas effectively; hence we require some 10^4 amp for a 0.10-m-wide accelerator. Using the time scale $\tau = l/\dot{x} \approx 10^{-5}$ sec, we require an initial charge storage $Q_0 \approx J\tau > 0.1$ coul to sustain the pulse over the complete acceleration, and hence a capacitance $C = Q_0/V_0 > 0.1/V_0$ farad. The initial voltage, in turn, is bounded by the requirements that it be large enough to accomplish the breakdown and to drive the current through the total impedance of the circuit, and not so large that it is troublesome from an insulation or power supply standpoint, say, $1{,}000 < V < 20{,}000$ volts in this example. The indicated minimum capacitance is thus in the range of 5 to 100 μf.

The second constraint on the capacitance can be illustrated by reference to an LCR circuit of fixed elements, where relation (9-1) can be written as a differential equation in Q by recognizing that $J = -\dot{Q}$ and $V = Q/C$ in this circuit:

$$L_0 \ddot{Q} + R_0 \dot{Q} + \frac{Q}{C} = 0 \qquad (9\text{-}7)$$

For initial conditions $Q = Q_0$, $J_0 = -\dot{Q}(0) = 0$ at $t = 0$, the solution has two forms. For $C < 4L_0/R_0^2$ there is a damped oscillatory behavior (Fig. 9-3a):

UNSTEADY ELECTROMAGNETIC ACCELERATION

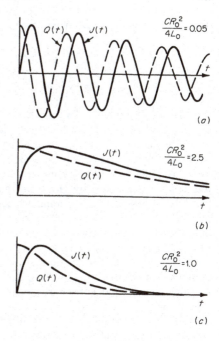

Fig. 9-3 Current waveforms for simple LCR circuits. (a) Underdamped; (b) overdamped; (c) critically damped.

$$Q = \frac{Q_0}{\omega\sqrt{L_0 C}}\, e^{-[(R_0/2L_0)t]} \sin(\omega t + \delta)$$

$$J = -\dot{Q} = \frac{Q_0}{\omega L_0 C}\, e^{-[(R/2L_0)t]} \sin \omega t$$

where
$$\omega = \left(\frac{1}{L_0 C} - \frac{R_0^2}{4L_0^2}\right)^{1/2}$$

and
$$\delta = \tan^{-1}\left(\frac{4L_0}{R_0^2 C} - 1\right)^{1/2}$$

(9-8)

while for $C > 4L_0/R_0^2$, the current pulse is overdamped (Fig. 9-3b):

$$Q = \frac{Q_0}{\omega'\sqrt{L_0 C}}\, e^{-[(R_0/2L_0)t]} \sinh(\omega' t + \delta')$$

$$J = \frac{Q_0}{\omega' L_0 C}\, e^{-[(R_0/2L_0)t]} \sinh \omega' t$$

where
$$\omega' = \left(\frac{R_0^2}{4L_0^2} - \frac{1}{L_0 C}\right)^{1/2} = i\omega$$

and
$$\delta' = \tanh^{-1}\left(1 - \frac{4L_0}{R_0^2 C}\right)^{1/2}$$

(9-9)

The special case $C = 4L_0/R^2$ is critically damped (Fig. 9-3c):

$$Q = Q_0 \left(1 + \frac{R_0}{2L_0} t\right) e^{-[(R_0/2L_0)t]}$$

$$J = \frac{Q_0}{L_0 C} t e^{-[(R_0/2L_0)t]} \qquad (9\text{-}10)$$

In the example accelerator chosen, we have committed ourselves to a ratio $4L_0/R_0^2 \approx 4\,\Delta L/R_0^2 \approx 10^{-2}$. Thus some 10^4 μf of capacitance would be needed to critically damp the current pattern. The foregoing analysis strictly applies only to a circuit of constant parameters, but it serves to indicate that any comparable match for our dynamic circuit is put out of reach both by the total mass and volume which would be involved in assembling this immense capacitance and by the intolerably slow current rise time which it would predicate. We can probably handle only 1 percent of this value at best; hence we must be content with a nearly free-ringing circuit [Eq. (9-8) and Fig. 9-3a].

If we use the bare minimum of capacitance, say 5 μf, the period of oscillation of this circuit, $T = 2\pi \sqrt{L_0 C}$, is shorter than the desired acceleration time, so that current reversals will occur while the plasma is in motion down the channel. Ideally, such reversal might not seem detrimental since current and self-B-field reverse in phase and the body force remains positive, albeit pulsating. In practice, however, it is observed under most conditions that current will not reverse in the initial current layer, but rather a secondary "crowbar" breakdown will occur back at the initial breakdown position, at a time slightly prior to circuit current reversal, thereby effectively short-circuiting the accelerating initial sheet from the external supply. Although such secondary discharges may also propagate down the channel, they tend to be less manageable than the primary sheet, since they have only the residue of gas left behind by the initial sheet in which to form and accelerate.

To preclude secondary breakdowns, one might consider the addition of capacitance until the half-period of the circuit, with dynamic distortion included, exceeds the acceleration time of the channel. However, this technique would protract the current rise time to nearly one-half the total pulse length, thus reducing the skin-effect current sheet concentration and lowering its gas-sweeping effectiveness. To eliminate oscillatory behavior of the circuit without excessive capacitance and associated slow rise time, a more complex external configuration, such as a pulse-forming network or lumped-element transmission line, is needed, which can deliver a more appropriately tailored pulse of current to the discharge. The attractiveness of this approach is enhanced by consideration of the dynamic efficiency of the gas-sweeping process, discussed in the following sections.

UNSTEADY ELECTROMAGNETIC ACCELERATION

The foregoing sequence of rationalization has been presented for a particular accelerator geometry and size, but the results are quite characteristic of most devices of this class. In general, they call for very large currents, while presenting extremely low dynamical impedance to the driving circuit, and thus they place severe demands on it for very low external inductance and resistance and very high capacitance and voltage. More specific design criteria cannot be developed until the gasdynamical aspects are included in the problem, since these determine the rate of increase of inductance, which must be the dominant circuit load in an efficient accelerator.

9-4 DYNAMICAL MODELS

The essential function of the devices and circuits described above is to accelerate mass by the self-applied magnetic forces. Since the discharge current which provides the force in each case is itself determined by the position and velocity of the movable circuit element, i.e., the plasma under acceleration, the general dynamical problem is nonlinear. Furthermore, the microscopic details of the acceleration process, wherein ambient gas particles are ionized and accumulated by the propagating current sheet and given streamwise momentum, are complex and obscure. Consequently, analytic approach to this class of problem is usually based on one of several heuristic models which may adequately represent the gross dynamical features in particular domains of operation.

SLUG MODEL

In the simplest idealization, one presumes that the entire mass of gas to be accelerated is enveloped in the initial breakdown event at the minimum inductance configuration, and thereafter no mass is accumulated or lost as the accelerating plasma traverses the channel. This *slug model* is probably relevant only to a few special situations, such as the purposeful injection of a highly localized ambient mass distribution near the channel inlet, or the creation of the plasma by vaporization of a metallic film initially closing the circuit in this region [7], but it provides some insight into the scaling properties of accelerators of this class. Historically, this model was developed in connection with a proposed "electromagnetic cannon," wherein a solid conducting element was to be accelerated along parallel rails to a formidable muzzle velocity [8].

Returning to the quasi-infinite parallel-plate geometry of the previous section, Newton's law for this system may be written

$$m\ddot{x} = hd \int_0^\delta jB\, dx = \frac{\mu_0 h}{2d} J^2 = \tfrac{1}{2} L_1 J^2 \qquad (9\text{-}11)$$

where m = total mass accelerated
δ = current zone thickness
L_1 = channel inductance per unit length

The circuit equation (9-1) can be written with the capacitor voltage included in the form

$$\frac{d}{dt}[J(L_0 + L_1 x)] + JR + \frac{Q}{C} = (L_0 + L_1 x)\dot{J}$$

$$+ JL_1 \dot{x} + JR - V_0 + \frac{1}{C}\int_0^t J\, dt = 0 \quad (9\text{-}12)$$

Relations (9-11) and (9-12) are simultaneous equations in $J(t)$ and $x(t)$ in terms of the parameters m, L_0, L_1, V_0, C, and R, if R is presumed a constant throughout the process. These may be nondimensionalized in various ways to generalize numerical solutions and to establish the scaling laws. For example, the choice

$$x^* = \frac{L_1}{L_0} x$$

$$t^* = \frac{1}{\sqrt{L_0 C}} t \quad (9\text{-}13)$$

$$J^* = \frac{1}{V_0}\sqrt{\frac{L_0}{C}} J$$

reduces Eqs. (9-11) and (9-12) to the forms

$$(x^*)'' = \alpha (J^*)^2 \quad (9\text{-}14)$$

$$(1 + x^*)(J^*)' + J^*(x^*)' + \beta J^* = 1 - \int_0^{t^*} J^*\, dt^* \quad (9\text{-}15)$$

with dimensionless initial conditions

$$\begin{aligned} x^*(0) &= 0 \\ (x^*)'(0) &= 0 \\ J^*(0) &= 0 \\ (J^*)'(0) &= 1 \end{aligned} \quad (9\text{-}16)$$

where primes indicate differentiation with respect to t^*, and α and β are the relevant scaling parameters:

$$\alpha = \frac{C^2 V_0^2 L_1^2}{2 m L_0} \quad (9\text{-}17)$$

$$\beta = R\sqrt{\frac{C}{L_0}} \quad (9\text{-}18)$$

In β we recognize our earlier index of the damping characteristic of a simple RCL circuit. If $\beta \gg 1$, the complete circuit is predisposed to an overdamped waveform; if $\beta \ll 1$, the tendency is toward an oscillatory waveform, provided the dynamic contribution to the impedance, $\dot{L}$, is not excessive. The relative importance of the dynamic impedance is latent in the other parameter, α. If we cast it in the forms

$$\alpha = \frac{C^2 V_0^2 L_1^2}{2mL_0} = \frac{1}{2} \frac{(\tfrac{1}{2}CV_0^2)(L_1 \dot{x})^2}{(\tfrac{1}{2}m\dot{x}^2)(L_0/C)}$$

$$= \frac{1}{2}\left(\frac{\tfrac{1}{2}CV_0^2}{\tfrac{1}{2}m\dot{x}^2}\right)\left(\frac{\dot{L}}{R}\right)^2 \beta^2 = \frac{1}{8\pi^2}\left(\frac{\tfrac{1}{2}CV_0^2}{\tfrac{1}{2}m\dot{x}^2}\right)\left(\frac{2\pi\sqrt{L_0 C}}{L_0/\dot{L}}\right)^2 \quad (9\text{-}19)$$

we can identify four ratios of interest: the ratio of energy initially stored in the capacitance to the kinetic energy of the accelerated mass, which should be of order unity in an efficient accelerator; the ratio of dynamic impedance to circuit resistance, $\dot{L}/R$; the static circuit damping parameter β; and the ratio of the resonant period of the initial circuit, $2\pi\sqrt{L_0 C}$, to the time required for the inductance to increase by one unit of L_0. We thus expect that systems of large α will display current waveforms which are severely damped and protracted in wavelength as time progresses (Prob. 9-4).

Numerical calculations based on (9-14) to (9-16) confirm these interpretations of the scaling parameters. For $\alpha \gg 1$, one finds distorted damped current waveforms $J^*(t^*)$ and associated monotonically accelerating slug trajectories $x^*(t^*)$ which achieve some terminal velocity at current completion or exhaust, determined by the initial energy storage and resistive losses throughout the pulse (Fig. 9-4a). For $\alpha \ll 1$, $\beta \ll 1$, $J^*(t^*)$ describes slowly damped, slightly protracted sinusoidal oscillations (Fig. 9-4b). The corresponding $x^*(t^*)$ trajectories clearly do not oscillate, since the body force does not reverse on current reversal but may display comparatively minor fluctuations in their derivatives in approaching a terminal velocity.

In most propulsion devices for which this model is at all relevant, practical requirements on the component parameters place α and β well in the latter, oscillatory domain. In applying this analysis to these cases, it is presumed that the plasma slug remains coupled to the circuit after current reversal, i.e., that the plasma slug is the only element closing the external circuit at all times. As mentioned earlier, many devices have a tendency to initiate a new crowbar discharge, back at the initial low inductance position in the channel, at or even prior to the time of current reversal. This will divert a major portion of the circuit current from the primary sheet and clearly invalidate the analysis from this time on. To preclude current reversal before plasma ejection, it is normally necessary

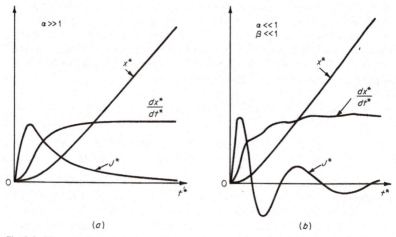

Fig. 9-4 Dimensionless current waveforms and current sheet trajectories from one-dimensional slug model. (a) $\alpha \gg 1$; (b) $\alpha \ll 1$, $\beta \ll 1$.

to enforce a protracted nonreversing current waveform with a more elaborate external circuit. Although the dynamical equation remains valid for this case, the circuit statement now needs to be much more detailed [9].

SNOWPLOW MODEL

The obvious deficiency in the slug model is the neglect of entrainment of additional ambient mass as the current zone propagates down the channel. This sweeping process is characteristic of the majority of pulsed plasma accelerators, and the effectiveness with which it is accomplished bears heavily on the overall efficiency of a given device. The simplest analytical model which incorporates the sweeping feature regards the current zone, driven by its own magnetic field, as an impermeable, completely absorbing surface propagating into the ambient gas [10]. The mass overtaken by this advancing surface is assumed to be accumulated within it and subsequently to travel along with it; hence the nomen *snowplow*. As we shall see later, this model has implicit limitations on the attainable dynamic efficiency.

The circuit equation remains the same as for the slug model (9-12), but the statement of Newton's law now must contain another inertial term to allow for the rate of mass accumulation:

$$m\ddot{x} + \dot{m}\dot{x} = \tfrac{1}{2} L_1 J^2 \qquad (9\text{-}20)$$

It will be assumed that when initially formed, the current sheet embodies a certain small but finite mass m_0. This conforms with experimental experience and precludes a troublesome singularity in the acceleration of initially zero mass. After inception, the current sheet accumulates mass at a rate determined by the sheet velocity, the channel dimensions, and the ambient mass density distribution $\rho(x)$ which we shall first assume to be uniform:

$$m(x) = m_0 + hd \int_0^{t(x)} \rho(x)\dot{x}\, dt = m_0 + hd\rho x \qquad (9\text{-}21)$$

The dynamical equation thus may be written

$$(m_0 + hd\rho x)\ddot{x} + hd\rho \dot{x}^2 = \tfrac{1}{2} L_1 J^2 \qquad (9\text{-}22)$$

A nondimensionalization similar to the slug model serves here also:

$$x^* = \frac{L_1}{L_0} x$$
$$t^* = \frac{1}{\sqrt{L_0 C}} t \qquad (9\text{-}23)$$
$$J^* = \frac{1}{V_0}\sqrt{\frac{L_0}{C}}\, J$$

The dynamical and circuit equations then take the forms

$$(m^* + x^*)(x^*)'' + [(x^*)']^2 = \alpha(J^*)^2 \qquad (9\text{-}24)$$

$$(1 + x^*)(J^*)' + J^*(x^*)' + \beta J^* = 1 - \int_0^{t^*} J^*\, dt^* \qquad (9\text{-}25)$$

where
$$\alpha = \frac{C^2 V_0^2 L_1^2}{2\mu L_0}$$
$$\mu = hd\rho \frac{L_0}{L_1} \qquad (9\text{-}26)$$
$$\beta = R\sqrt{\frac{C}{L_0}}$$
$$m^* = \frac{m_0}{\mu} = \frac{L_1 m_0}{hd\rho L_0}$$

Computations based on this model display two anticipated differences from those of the slug model (Fig. 9-5). First, the current sheet need not always accelerate monotonically along the channel. In low current portions of the circuit cycle it may accumulate mass more rapidly than the driving body force can sustain at the prevailing velocity, and hence be obliged to decelerate momentarily. Second, the acceleration

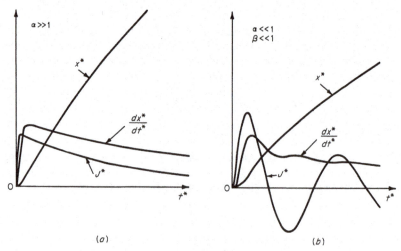

Fig. 9-5 Dimensionless current waveforms and current sheet trajectories from one-dimensional snowplow model. (a) $\alpha \gg 1$; (b) $\alpha \ll 1$, $\beta \ll 1$.

tends to be most vigorous at early times where the mass accumulated is yet small, and to weaken later when the bulk of the mass has become entrained. Otherwise, the scaling parameters play the same role in damping and distorting the waveform as before. If the initial mass parameter m^* is small, its effect on the current sheet trajectory tends to disappear after a small fraction of the channel has been traversed.

Despite the severity of its assumptions, the snowplow model is quite successful in predicting current sheet trajectories for a variety of pulsed plasma accelerators over a broad range of operation. As cast above, the model ignores possible short-circuiting effects of secondary breakdowns at current reversal, and must be amended if these are known to prevail in a given device. Again, the dynamical equation can be coupled with more sophisticated external circuits, provided they can be reasonably represented analytically. In some cases it is possible to specify the current waveform in advance, leaving only one equation [(9-24)] to be solved for the trajectory $x^*(t)$. For example, if the dynamical development of the discharge is only a small portion of the total circuit impedance, i.e., if the discharge and energy source are weakly coupled, (9-25) may be solved with $x^* = (x^*)' = 0$, to provide a first approximation to the current waveform. Weakly coupled systems of this sort are common in laboratory experiments where effectiveness of energy transfer from source to plasma is not a prime consideration. For propulsion application, however, tight coupling, i.e., matching of the dynamical impedance of the discharge to the output impedance of the source, is clearly essential.

GASDYNAMIC MODELS

The first refinement of the snowplow model which suggests itself is the inclusion of more realistic gasdynamic properties, most notably the finite compressibility of the entrained gas. That is, while still treating the current sheet as an impermeable surface, it will now be regarded as a piston which drives a gasdynamic shock wave ahead of it at somewhat higher velocity (Fig. 9-6). This shock front abruptly accelerates the

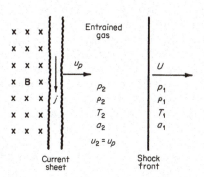

Fig. 9-6 One-dimensional constant speed shock wave model.

overrun gas and compresses it to a high but finite density and pressure in a small but finite layer between itself and the current sheet. The familiar normal shock wave relations [11] can be applied to this model to relate the state of the entrained gas and shock speed to the ambient conditions and piston speed, provided equilibrium is achieved in the overrun gas, the piston speed is constant, and an effective ratio of specific heats, γ, can be identified for the process. For example, for the very high piston speeds u_p of interest here, the shock speed U and the density, pressure, and temperature ratios across the shock front are closely approximated by the simple expressions

$$U = \frac{\gamma + 1}{2} u_p \qquad (9\text{-}27)$$

$$\frac{\rho_2}{\rho_1} = \frac{\gamma + 1}{\gamma - 1} \qquad (9\text{-}28)$$

$$\frac{p_2}{p_1} = \frac{\gamma(\gamma + 1)}{2} \left(\frac{u_p}{a_1}\right)^2 \qquad (9\text{-}29)$$

$$\frac{T_2}{T_1} = \frac{\gamma(\gamma - 1)}{2} \left(\frac{u_p}{a_1}\right)^2 \qquad (9\text{-}30)$$

where a_1 is the sound speed in the ambient gas. The width of the shocked gas zone, its density and temperature, are seen to depend heavily on γ, and this is a strongly temperature dependent quantity. For the velocity range of interest, the associated shocked gas temperatures are so high (10^4 to $10^{5}°K$) that most of the internal degrees of freedom are highly excited and γ closely approaches unity (Sec. 6-2). Small uncertainties in γ are thus magnified in the computation of the shocked gas density and all the collisional processes which depend on it. In a sense, the snowplow model may be regarded as the special case of a shock wave model for a $\gamma = 1.00$ process, for in this limit the gas is completely compressible and the piston and shock front velocities coincide. Beyond the sensitivity to γ, there are theoretical uncertainties in the rates of attainment of equilibrium behind the shock, particularly in the excitation, ionization, and radiation processes. Consequently, application of this model normally requires initial knowledge of an additional parameter beyond the piston velocity, e.g., the piston-shock separation.

Even if equilibrium is completely established at the shock front, the gas captured between it and the piston will be uniform only in the special case of constant piston speed, constant duct area, and uniform ambient conditions. If the piston accelerates, or if the channel area or ambient gas distribution is nonuniform, the entrained gas zone will reflect a history of the accumulation process, and streamwise gradients in its properties will exist. Methods of characteristics can be applied to such situations, but these normally require tedious iterative evaluation [12].

More ambitious gasdynamic models might attempt to allow for additional observed nonidealities, such as a finite width and conductivity of the current sheet "piston," with consequent diffusion of itself and its magnetic field into the shocked gas zone; leakage of shocked gas through the current piston; preionization of the ambient gas by radiation from the shocked gas and current zone; finite shock front thickness; etc. The complexities of such models largely vitiate the essential advantages of coupling gasdynamic formulation with simple circuit analysis, and still lack certain observed details of pulsed current zone structure, such as interior streamwise electric fields and different density distributions for the three participating species. To incorporate these elements one must turn to a multicomponent gasdynamic analysis or to a complete gas-kinetic representation.

MULTIFLUID MODELS

To retain the advantages of continuum representation while yet incorporating interspecies interactions, one may regard the working medium as a mixture of three artificial "fluids," composed of neutral particles, positive single ions, and electrons, respectively. Equations of mass conservation, dynamics, energy, state, etc., may be written for each of the fluids, and these will be coupled by a variety of physical processes and constraints. For example, the mass conservation relation for the neutral fluid will involve at least two source terms representing loss by ionization processes and gain by recombination processes. Corresponding terms will appear in the ion-fluid and electron-fluid mass relations, subject to the constraint on conservation of total mass of the complete system. In the dynamical relations, each fluid must be allowed to exert a "viscous" drag on the other; for example, if the ion fluid is moving faster than the neutral fluid at a given location, the former will tend to accelerate the latter, and the latter decelerate the former, at a rate dependent on the relative velocity and the effectiveness of the collisional coupling between them. In similar fashion, terms representing interspecies energy transfer, driven by differences in species temperatures, will appear in the energy relations, and disparities in electron and ion fluxes will appear as current densities in Ohm's law.

The dynamical relations for the charged fluids will also contain body force terms in the applied and interior electric and magnetic fields, and these in turn will be constrained by Maxwell's relations and by appropriate boundary conditions. It should be noted that any local disparity between ion and electron densities provides a source term for $\nabla \cdot \mathbf{E}$, a feature which has not arisen in any of our previous magnetogasdynamic analyses.

Little would be gained by displaying the imposing array of equations which result from this approach, particularly since the specific form of many of the coupling terms depends on the available means for evaluating them. It is clear that the spirit of the method is magnetogasdynamic, with Hall effects, ion slip, and charge separation specifically included. With initial and boundary conditions properly posed, the system in principle may be solved to yield complete profiles in space and time of all species densities, velocities, temperatures, etc., and of all field patterns and currents arising in the given pulsed plasma acceleration event. In practice, such solutions have been obtained only in a few special cases, and then only by elaborate machine computations [13].

GAS-KINETIC MODELS

In certain situations, particularly in low density environments, it may be more instructive to approach the analysis from a particle, rather than a continuum, point of view. In the same spirit of orbital mechanics outlined in Sec. 8-9, individual electron, ion, and neutral trajectories may be examined under the prevailing circumstances of Hall parameters, gyro radii, and mean free paths, and thereby a pattern for the current conduction and mass acceleration processes may emerge [14]. As a rudimentary example of the application of this technique to the problem under study, let us idealize the current sheet as a simple discontinuity in magnetic and electric field which propagates at velocity u_s into a neutral gas at rest (Fig. 9-7a). Assume that as each ambient neutral particle is overrun by this sheet, it is ionized by some unspecified agent and that the electron and ion thus formed then execute collisionless motions in the prevailing fields. Finally, assume that the gyro radius of the electron is much smaller than the sheet and channel dimensions, and that of the ion is much larger.

If the applied electric field E_y were the only component present, the electron would undertake an $E_y \times B_z$ drift purely in the x direction, parallel to the sheet motion, and thus would contribute no conduction current j_y. The much heavier ion would also attempt to respond to the fields by a streamwise drift, but, because of its large gyro radius and inertia, would accomplish only a negligible acceleration along the applied field E_y. Neither process can be long sustained, however, for the resultant streamwise separation of the electron and ion will rapidly give rise to, and be opposed by, an interior electric field component E_x, that is, a polarization field. The particle trajectories thus will actually develop in accordance with the resultant **E** field, which has both transverse (applied) and streamwise (polarization) components, as shown in Fig. 9-7a. For example, the electron now will have a streamwise component of drift which permits it to keep pace with the propagating front, and a transverse component which constitutes the bulk of the conduction current j_y. The polarization field E_x also serves the essential function of accelerating the ion from rest to its stream speed. Although the initial direction of ion motion will be along the resultant **E**, a transverse magnetic force arises with the increasing velocity and eventually cancels the applied field E_y, leaving an axial streaming as needed.

The same process can be developed in a frame of reference fixed on the current sheet discontinuity (Fig. 9-7b). In this system the ambient neutral particle approaches the sheet at velocity u_s, which is momentarily retained by the ion-electron pair formed shortly after entrance into the sheet. Within the sheet the applied electric field, $E' = E - u_s B$ (Sec. 2-4), is nearly zero because of the high plasma conductivity; hence each particle attempts to execute a circular orbit from its point of creation in the B field. The electron with small gyro radius accomplishes this readily, but the heavy ion is little affected by the same magnetic field, and would continue in the streamwise direction of its parent neutral, were it not for the polarization electric field arising from its separation from the electron. This streamwise field E_x serves, on the one hand, to decelerate the ion to rest in the sheet and, on the other, to induce a transverse $E_x \times B_z$ drift of the electron which provides the necessary current

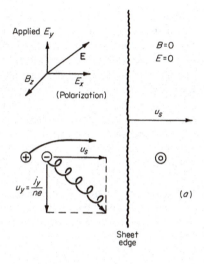

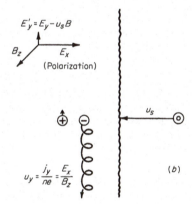

Fig. 9-7 Gas-kinetic model of one-dimensional acceleration. *(a)* Laboratory frame; *(b)* current sheet frame.

conduction j_y. In this framework, the current sheet may in a sense be regarded as a miniature self–Hall accelerator, with the electrons constrained by the self–magnetic field to drifting motion, and the electric body force, derived from an inertially generated streamwise field, applied directly to the ions.

Extension of the single particle trajectories thus deduced to the bulk behavior of a collisionless swarm is straightforward. Clearly, the discrete separation of a specific electron-ion pair now transcribes into a slight displacement of the centrodes of electron and ion density distributions, and the polarization field profile $E_x(x)$ is consistent with a one-dimensional Poisson's equation. It should perhaps be noted that even for very modest charge densities, a minute charge separation will provide sufficiently large fields to accelerate the ions over the available distance, and for many purposes the quasi-neutrality assumption is still tenable.

The model used above is, however, greatly oversimplified in its field geometry and in its total neglect of collisions. In some aspects it is even inconsistent, as in the neglect of the finite width of the current zone and corresponding B-field transition, in ignoring the streamwise current associated with the charge separation, in the neglect of ion rebound in the conservative polarization field, and in the lack of a specific mechanism for the ionization event. Proper inclusion of collisional effects ameliorates most of these difficulties without basically altering the fundamental conduction and acceleration mechanisms displayed in the simple model [15]. Indeed, analogies to them can be identified in the full-blown multifluid models cast in continuum nomenclature.

9-5 DYNAMICAL EFFICIENCY OF PULSED ACCELERATION

In any electric propulsion concept, considerable premium must be placed on the efficiency with which input power can be converted to useful streaming motion of the gas. As mentioned earlier, pulsed plasma accelerators have certain peculiar advantages for minimizing relative losses in ionization and heat transfer in their high enthalpy domain of operation. However, they are also inherently prone to a special type of dynamic inefficiency arising from the impulsive character of the gas acceleration process. Returning to the simple snowplow image, this loss may be visualized as the sum effect of all the inelastic collisions made by the ambient particles on the "sticky" piston which accelerates them. Alternatively, in the shock wave representation of the process, one recognizes that strong shocks propagating at constant velocity into gases at rest convert only about one-half of the available energy into streaming motion of the shocked gas and one-half into internal energy. Or finally, in the gas-kinetic models, one observes that any orbital or collisional process which traps particles in the current sheet inevitably produces transverse particle velocity components comparable with the sheet speed.

The one evident exception to this impulsive acceleration loss is the slug model, wherein all the mass to be accelerated resides in the current layer at the start, whence it may then be accelerated as a rigid body with perfect dynamical efficiency. This distinction suggests that the dynamical efficiency of the sweeping process in the general case will be dependent upon the detailed profile of the acceleration in time and on the ambient mass distribution. In particular, we expect that accelerators wherein most of the mass is collected early in the sheet motion, at relatively low velocity, and then is accelerated to terminal speed will have higher dynamical efficiency than those wherein most of the mass is entrained at high velocity.

To explore this argument more rigorously [16], return again to the conventional snowplow idealization of the process, wherein a thin current sheet is driven by its own magnetic field into a gas at rest and accumulates within a negligibly narrow region on its surface all the gas that it overtakes. Starting from a simple newtonian statement of the overall dynamics of the situation, one can identify two components of the rate of energy deposition:

$$uF = u\frac{d}{dt}mu = \frac{d}{dt}\frac{1}{2}mu^2 + \frac{1}{2}\dot{m}u^2 \tag{9-31}$$

where F is the instantaneous magnetic driving force, m the mass accumulated on the piston, and u its velocity. The first term on the right is the rate of increase of streaming kinetic energy of the swept gas immediately useful for propulsion. The second term represents the dissipation associated with the inelastic collision of the incoming particles with the snowplow piston.

Imagine that the preceding process starts with some initial velocity u_0 and continues over a period of time t_f, at the end of which the piston has progressed a distance x_f and has accumulated a total mass m_f, which is now moving at a velocity u_f. The ratio of interest is that of the integrals of the foregoing two terms over this period,

$$\alpha = \frac{\int_0^{t_f} \frac{1}{2}\dot{m}u^2 \, dt}{\frac{1}{2}m_f u_f^2} \tag{9-32}$$

which determines the dynamical efficiency of the process,

$$\eta_d = \frac{1}{1 + \alpha} \tag{9-33}$$

It is useful now to convert to dimensionless distance, velocity, and time variables:

$$x^* = \frac{x}{x_f} \qquad u^* = \frac{u}{u_0} \qquad t^* = \frac{tu_0}{x_f} \tag{9-34}$$

The geometry of a specific accelerator and the ambient gas density distribution may be embodied in the functional dependence of accumulated mass on piston position:

$$m^*(x^*) = \frac{m(x^*)}{m_f} \tag{9-35}$$

In terms of these dimensionless quantities, the energy-division ratio α becomes

$$\alpha = \frac{\int_0^{t_f^*} \left(\frac{dx^*}{dt^*}\right)^3 \frac{dm^*}{dx^*} dt^*}{\left(\frac{dx^*}{dt^*}\right)_f^2} = \frac{\int_0^1 (u^*)^2 \frac{dm^*}{dx^*} dx^*}{(u_t^*)^2} \tag{9-36}$$

The development of the dimensionless velocity profile $u^*(x^*)$ is determined by the magnitude and waveform of the discharge current, as well as by the geometry and ambient gas distribution. The point discussed here, however, may be demonstrated by examining the behavior of α for various assumed velocity profiles. For example, any constant velocity accelerator is seen to subscribe to the previously mentioned equal partition criterion, since here u^* has the value 1 throughout, and hence

$$\alpha = \int_0^{m_f^*} dm^* = 1 \tag{9-37}$$

Note that this result is independent of accelerator geometry and initial gas distribution.

Other velocity profiles must be examined in the light of specified initial mass distributions $m^*(x^*)$. For all uniform cross-section accelerators (coaxial guns, parallel-plate accelerators, T tubes, etc.) having uniform initial gas fills, $m^*(x^*) = x^*$ and dm^*/dx^* disappears from the α integral. As one illustration of these cases, consider the family of exponential velocity profiles $u^* = e^{\kappa x^*}$, which yield

$$\alpha = \frac{1}{2\kappa}(1 - e^{-2\kappa}) \tag{9-38}$$

and the corresponding range of efficiencies shown in Fig. 9-8a.

In a similar way, the general power law profiles $u^* = 1 + b(x^*)^n$ yield

$$\alpha = \frac{1}{(1+b)^2}\left[1 + \left(\frac{2b}{n+1}\right) + \left(\frac{b^2}{2n+1}\right)\right] \tag{9-39}$$

where the coefficient b clearly relates the initial and final velocities

$$\frac{u_f}{u_0} = 1 + b \tag{9-40}$$

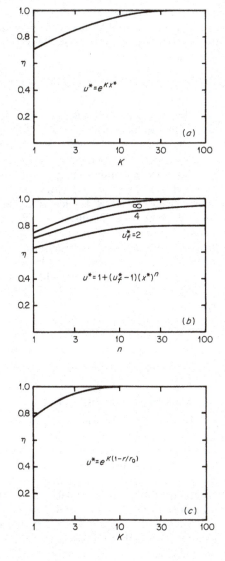

Fig. 9-8 Dynamical efficiency of one-dimensional accelerators for various sheet velocity profiles. (a) Exponential; (b) power law; (c) exponential in pinch geometry.

The efficiencies of these profiles depend both on this velocity ratio and on the value of the exponent, as displayed in Fig. 9-8b.

The effect of a variable channel cross section may be illustrated by the case of the linear pinch, where the position coordinate is now $x = r_0 - r$, and the corresponding dimensionless quantity is $x^* = 1 - r/r_0$ where r_0

refers to the initial current sheet radius. The mass-accumulation function for uniform initial density is now $m(x^*) = x^*(2 - x^*)$. Returning to the exponential velocity profile $u^* = e^{\kappa x^*}$, the evaluation of Eq. (9-36) now yields

$$\alpha = \frac{1}{2\kappa^2}(1 - e^{-2\kappa} - 2\kappa e^{-2\kappa}) \qquad (9\text{-}41)$$

with the range of efficiencies shown in Fig. 9-8c.

The results displayed in Fig. 9-8a–c show the anticipated behavior that the dynamical efficiency is favored by vigorously accelerating velocity profiles, converging channel geometries, and decreasing ambient density profiles in the direction of propagation. Clearly, these results refer only to the dynamical processes within the accelerator channel, and make no reference to other possible losses in the device, such as the end losses and switching losses discussed earlier. On the other hand, it should be noted that the dynamical conversion of energy into internal or thermal modes involved here is not necessarily irreversible, and that some fraction of this energy may be recovered before the plasma completely leaves the accelerator. This possibility adds an electrothermal overtone to pulsed plasma acceleration, much like that in the MPD arc (Sec. 8-9). The effectiveness of such electrothermal conversion at these high enthalpy levels will depend primarily on the ability of the internal collision and radiation processes to dispose of the energy on the brief time scales involved, which in turn strongly depends on the optical opacity of the hot gas to its own radiation.

9-6 INITIATION OF THE DISCHARGE

Of the three main phases of the pulsed plasma acceleration sequence—initiation, acceleration, and ejection—we have so far discussed only the second in detail. The other two phases are of equally critical importance to the effective performance of a given device, but are less amenable to analytical discussion. Regarding initiation, for example, one can impose several qualitative requirements on the breakdown event, such as proper geometric location of the discharge, adequate current density for acceleration, gas-sweeping and stability purposes, small energy expenditure compared with that imparted to the gas during the acceleration phase, negligible erosion or other deterioration of the electrode or insulator surfaces, etc., but analytical formulation of these requirements in either a continuum or particle representation is difficult. A few qualitative remarks will illustrate the problems.

The initiation geometry of discharges of this class seems to be determined mainly by transient field effects associated with the rapidly rising currents. In the linear channel configuration discussed above, for example, one can envisage the breakdown process beginning with a short-lived uniform glow discharge over the entire channel. As the current density rapidly rises, driving the discharge through an abnormal glow toward an arc, the associated distribution in the time derivative of the magnetic field $\dot{B}_z$ induces a new component of interior electric field, opposing the applied potential and preferentially discouraging current flow across the portions of channel farthest removed from the external circuit (Fig. 9-9a

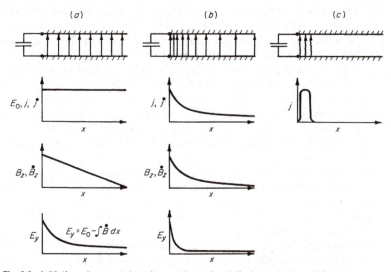

Fig. 9-9 Initiation of current sheet in one-dimensional discharge (schematic).

and b). The current thus intensifies most rapidly near the channel entrance, and a glow-to-arc transition is first accomplished there. Once this transition has occurred, the negative voltage-current characteristic of that arc zone further discriminates in favor of current intensification at this location, and a highly nonuniform current distribution is stabilized (Fig. 9-9c). The process just outlined is similar to the skin effect for high-frequency conduction in metals, and might seem amenable to a simple calculation of that type were it not for the important distinction of nonuniform electrical conductivity. In the discharge, this conductivity is strongly dependent on local electron density and temperature, and thus on the very current density it is needed to determine.

The detailed profile of current density distribution after breakdown, in particular the streamwise thickness of the current zone, depends on the rate of current rise and its ultimate magnitude and on a variety of ill-known transport processes in the gas. The rate of current rise is determined primarily by the ratio of applied voltage to initial circuit inductance. If this ratio is large, all field time derivatives will be large and the current zone will tend to be narrow and intense. Competing with this will be the tendency of migration of charged particles outward from the sheet, driven by the associated steep temperature and species gradients, again a highly nonlinear problem.

In the absence of quantitative theories for initial current sheet profiles, all available information has been experimentally determined. As a very crude generalization, current rise rates $\approx 10^{10}$ amp/(sec)(cm) of sheet length (z) typically generate bell-shaped distributions of current density several millimeters thick (x). This thickness tends to decrease with ambient gas pressure and with the current rise rate.

The desirable level of current density to be attained in the breakdown also can be discussed only in generalities. Clearly, one must achieve a self-magnetic pressure which exceeds the ambient gas pressure if the current sheet is to propagate at all. Actually, a far more intense discharge is required if this sheet is to entrain the overrun gas effectively. At the other extreme, if the current density is carried to too high a value, excessive electrode damage or radiation loss may result. Again, the available information is largely empirical, and indicates that the optimum current density levels depend strongly on ambient gas density and accelerator geometry. As a very crude index, current densities $\approx 10^3$ to 10^4 amp/cm of sheet length provide adequate sweeping and stability in the common accelerators of this class.

The requirement that breakdown initiation consume only a small fraction of the available circuit energy deals in even more obscure processes. First, a certain amount of energy must inevitably be assigned to ionization of the gas, but this need not seriously impair the overall efficiency (Prob. 9-9). A potentially more damaging loss may occur at the electrode surfaces, particularly at the cathode. In single-pulse or low-frequency intermittent operation, the cathode will not become hot enough to emit thermionically and the current-carrying electrons must be forcibly extracted from it in some less propitious manner. Although there exists considerable experimental information on spark initiation for a variety of electrode surfaces, gas types and pressures, etc., the bulk of this deals with maximum discharge currents in the milliampere range. The abrupt extraction of 10^4 amp/cm^2 or more of electron current from a metallic surface without benefit of thermionic emission has received very little study [17], but would seem to predicate extremely high local electric fields, with attendant dissipation and damage to the electrode surface. Phenomenological observation tends to confirm this suspicion. Microscopic examination of electrode surfaces in the region of discharge initiation typically shows a fine distribution of small pits, where the initial discharge presumably achieved filamentary access to the conductor's interior supply of electrons. It is also observed that electrode surfaces in this type of operation undergo an aging process over the first few hundred discharges, after which the rate of electrode erosion recedes to a lower level than that of a freshly prepared surface [18].

It is to be expected, and indeed has been observed, that a pulsed accelerator operated in rapid repetition achieves discharges with substantially different breakdown characteristics, which can be attributed to the onset of thermionic emission from the discharge-heated cathode surface. When this occurs, the discharge operates as a bona fide arc over its entire cycle of current, and cathode damage is considerably reduced. External heating of the cathode surface to enhance thermionic emission in single or slow pulse operation seems thus to suggest itself, but the power loss inherent in this procedure could be prohibitive for an efficient thruster.

9-7 EJECTION OF THE PLASMA

The process of expelling the plasma from the channel wherein it is created and accelerated is also difficult to reduce to systematic experiment or analysis. The principal concern again is the effect of the ejection phase on the overall efficiency of conversion of electrical input into ordered streaming motion of the gas, and in this regard three general questions may be posed: (1) To what extent does the exhaust orifice disturb the development of the discharge and acceleration processes within the channel? (2) To what extent does discharge current and its associated magnetic field extend out into the exhaust plume, and does their interaction there further accelerate the gas, or does it impede detachment of the plasma from the channel? (3) To what extent may internal energy, particularly random thermal energy, deposited in the plasma during breakdown and acceleration, be recovered in directed motion upon ejection from the channel?

Detailed response to these queries would seem to depend on the geometry of the specific accelerator and on its external circuit characteristics, but a few basically oriented experiments have provided some general results which bear on each of the three points [19]. Specifically, it has been shown (1) that relatively large exhaust apertures need not significantly disturb the discharge initiation and acceleration processes within the channel; (2) that sizable current densities and associated magnetic fields can extend far out into the exhaust plume, and there can aid in a further acceleration of the gas; and (3) that it is possible to recover a significant fraction of gas enthalpy in expansion through the exhaust pattern. Each of these results has favorable implications for the pulsed plasma acceleration concept, and serves to sustain the initial interest in this mode of thrust production.

9-8 COAXIAL GUNS AND PINCH ACCELERATORS

The parallel-plate accelerator which has been used as a model for much of the preceding discussion has a geometric simplicity which commends it for basic diagnostic studies [20], but at the same time presents a difficulty which reduces its effectiveness as a thruster. Namely, the gap between the edges of the finite width electrodes allows the magnetic field pattern to spill out past the edges of the current sheet, where it serves no useful purpose. Indeed, the magnetic field lines must circle outside the conductors to close on themselves (Prob. 9-10). Even within the gap, the magnetic field will be distorted somewhat near the edges, yielding undesired $j \times B$ components which may deform the current sheet in those regions. Similarly, gas accumulated by the accelerating sheet may tend to spill out this aperture, unless it is closed by a dielectric wall. This flux and gas leakage may be reduced by decreasing the ratio of channel width to breadth, h/d, but this will tend to increase the relative wall losses during acceleration. It seems better to consider other electrode-channel geometries which will allow the current sheet and self-magnetic field to close on themselves entirely within the channel. The two obvious configurations are the coaxial channel (Fig. 9-10a) and the linear pinch (Fig. 9-10b),

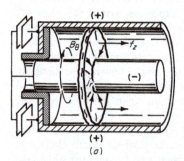 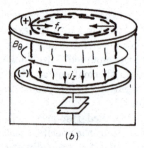

Fig. 9-10 Closed-geometry accelerators. (a) Coaxial gun; (b) dynamical pinch.

which may be generated by revolution of the linear rail geometry about a displaced axis parallel and perpendicular to the stream, respectively. In the former, the plasma sheet conducts current radially and is accelerated axially by an azimuthal self-field; in the latter, discharge current flows axially, the self-field is azimuthal, and the acceleration is radial.

Each of these configurations has been studied extensively, and each has displayed certain advantages and disadvantages. The coaxial accelerator retains the advantage of the linear rail geometry in accelerating the gas directly toward the channel exit in the final exhaust direction, while the pinch devices must somehow deflect the radially accelerated plasma to eject it. On the other hand, the pinch geometry provides a current sheet of uniform density and a self-field which is also uniform over the entire sheet surface, yielding a uniform $j_z B_\theta$ force density for acceleration. In the coaxial geometry, the current density in the discharge and the self-magnetic field both vary as $1/r$, and the body-force density thus as $1/r^2$. For this reason, and possibly from other causes, the current sheets in coaxial accelerators tend to be more susceptible to canting and to development of spoke instabilities than those in pinch devices. Finally, the dynamical efficiency analysis of Sec. 9-5 reveals an intrinsic advantage of the pinch geometry from the standpoint of providing heavier mass loading during the early portion of the acceleration profile. Thus something may be gained from coordinated study of both of these configurations.

The pulsed coaxial accelerator concept did not originate in the propulsion domain, but was first developed as a means of preparing high enthalpy plasma specimens in connection with the controlled thermonuclear fusion program. Commonly called *coaxial guns*, or *Marshall guns*, after one of their inventors [1], these devices were developed mainly as plasma injectors for large machines, and emphasis was placed on the terminal properties of the emergent plasma, rather than on the efficiency of conversion, the massiveness of the system, or the lifetime of the device under repetitive operation. Much of the experience acquired in this application, particularly the circuitry and diagnostic techniques, was useful in the early applications of coaxial guns to propulsion, but the stringent requirements on efficiency and lifetime gradually forced substantial changes in the geometrical details and modes of operation of these devices until they only slightly resembled their fusion ancestors. In particular, the barrel lengths became shorter, the central cathode became both shorter and smaller in diameter, the operating voltage was reduced, and the location, time, and speed of gas injection became critical. Figures 9-11 and 9-12 display two typical coaxial accelerators developed specifically for propulsion purposes.

Considerable performance data have been accumulated on the operation of such devices in large vacuum tanks, but systematic correlation of this performance with the controllable input parameters and details of the accelerator design have proved elusive, perhaps emphasizing the complexity of the participating physical processes [21–23]. The all-important property of accelerator efficiency becomes particularly involved here, and requires sophistication in its determination. For rapid repetitive opera-

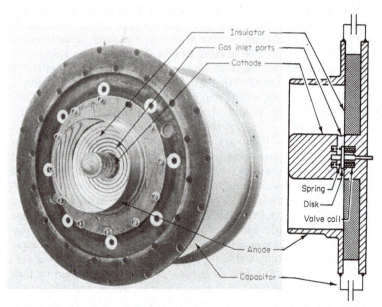

Fig. 9-11 Photograph and schematic of pulsed coaxial accelerator. *(General Dynamics Corporation, San Diego, Calif.)*

tion we may define an overall thrust efficiency as the ratio of mean thrust power in the exhaust beam to input electric power. In terms of readily measurable quantities this may be written (Prob. 9-11)

$$\eta = \frac{T^2}{\dot{m}\nu C V_0^2} \tag{9-42}$$

where T = effective thrust
 $\dot{m}$ = average mass flow rate
 ν = repetition rate
 C = capacitance
 V_0 = charging voltage

Contributing to this thrust efficiency are a sequence of interrelated factors: the efficiency with which the capacitors deliver their stored energy to the accelerator electrodes, η_c; the fraction of that energy which is imparted to the gas in any form, η_g; the fraction of gas energy in streaming motion, η_κ; and the correction of the latter for velocity distribution and exhaust-beam divergence losses, η_p:

$$\eta = \eta_c \eta_g \eta_\kappa \eta_p \tag{9-43}$$

Empirical performance measurements with thrust balances, gas flowmeters, calorimeters, and ballistic pendula indicate no simple optimization of (9-43); rather, changes in some input parameters or design features which improve some of the factors may reduce others. For example, shortening the barrel length may improve the *calorimetric efficiency* $\eta_c \eta_g$, but yield a more divergent exhaust beam with less thermal energy recovery.

More basic studies of coaxial accelerators have included measurements of the time profiles of terminal voltage and current, determination of magnetic field and current density distributions with magnetic probes, sampling of exhaust-beam characteristics with various ion-collecting devices, and photographic study of the luminous patterns in the channel and exhaust [24–26]. Figure 9-13 shows a typical sequence of magnetic field (B_θ) and current density (j_r) distributions found in one such accelerator. Figure 9-14 presents a sequence of schlieren photographs obtained through the slotted outer barrel of a coaxial gun built specifically for this purpose. Note the departures from ideal radial sheet behavior, particularly near the electrode surfaces.

Coordinated programs of such basic studies and empirical performance measurements have gradually driven coaxial accelerator design and

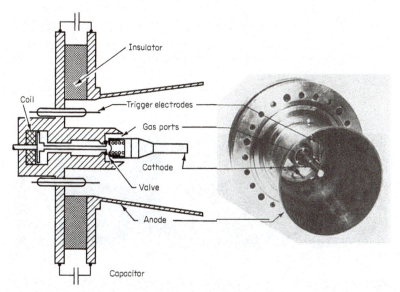

Fig. 9-12 Photograph and schematic of pulsed plasma accelerator. *(General Electric Company, Space Sciences Laboratory, Missile and Space Division, King of Prussia, Pa.)*

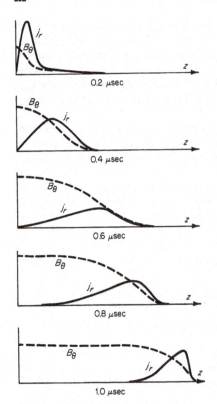

Fig. 9-13 Typical magnetic field and current density distributions in a coaxial accelerator. *(From T. J. Gooding, B. R. Hayworth, A. V. Larson, and D. E. T. F. Ashby, Development of a Coaxial Plasma Gun for Space Propulsion, General Dynamics Corp. Astronaut. Rept. GDA-DBE 64-052, Contract NAS-35759, San Diego, Calif., August, 1964.)*

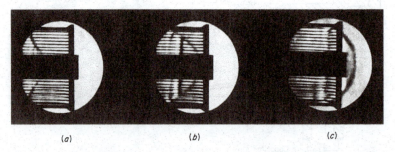

Fig. 9-14 Schlieren photographs of discharge in slotted coaxial accelerator; propellant, hydrogen at 500 μ. (a) 0.5 μsec, (b) 0.7 μsec, (c) 1.0 μsec after breakdown. *(Courtesy of R. H. Lovberg, Dept. of Physics, University of California at San Diego, La Jolla, Calif.)*

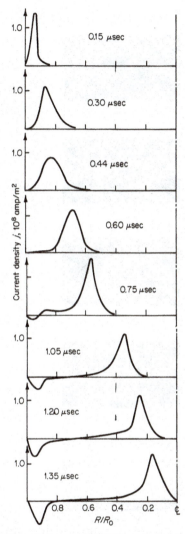

Fig. 9-15 Development of current density distribution in a large-radius pinch discharge. *(From R. L. Burton and R. G. Jahn, Structure of the Current Sheet in a Pinch Discharge, Princeton Univ. Aeron. Rept. 783, Princeton, N.J., September, 1966.)*

operation into ranges of utility for space propulsion. Detailed reviews of this technology are available in the literature, and will not be pursued further here [6].

Pulsed accelerators of the linear pinch geometry lend themselves more readily to detailed diagnostic study, but require some provision for conversion of radial motion to axial streaming if they are to function as thrusters. Figure 9-15 shows a series of current density distributions

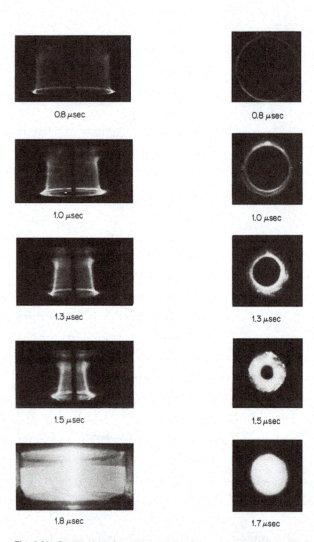

Fig. 9-16 Radial view Kerr-cell photographs of a large-radius pinch discharge in argon. (From R. L. Burton and R. G. Jahn, Electric and Magnetic Field Distributions in a Propagating Current Sheet, AIAA Paper 66-200, 5th Electric Propulsion Conference, San Diego, Calif., March 7–9, 1966.)

Fig. 9-17 Axial view Kerr-cell photographs of a large-radius pinch discharge in argon. (From R. G. Jahn and W. von Jaskowsky, Current Distributions in Large-radius Pinch Discharges, AIAA J., vol. 10, no. 2, pp. 1749–1753, October, 1964.)

obtained by magnetic probe mapping of the interior of a large-radius pinch discharge in argon. Note the initiation and acceleration of the primary current sheet and of a secondary crowbar sheet arising from current reversal in the external circuit. Figure 9-16 shows a series of radial-view Kerr-cell photographs of the development of a similar discharge. Here, too, certain departures from the ideal behavior are evident, such as the slight tilt of the luminous front with respect to the axis, and the diffuse "foot" on the anode, and to a lesser extent on the cathode. Figure 9-17 shows a sequence of axial photographs of a similar discharge which display incipient azimuthal perturbations in the luminous zone, but these seem not to develop noticeably over the available time interval of the pinch.

In an effort to retain the stability attributes of the pinch discharge while yet providing an axial exhaust jet, a hybrid configuration involving a smooth transition from pinch to coaxial geometry has been developed [27]. This device, called a *pinch engine*, is shown schematically in Fig. 9-18 and in photograph in Fig. 9-19. In the early days of pulsed thruster development, the pinch engine and pure coaxial accelerator were cast in competition with one another, but the distinctions between these two geometries have tended to diminish in the face of better understanding of the participating physical processes. For example, the plasma emerging from a pure coaxial accelerator tends to pinch itself radially in the ejection

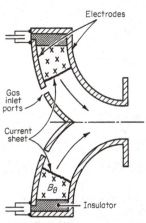

Fig. 9-18 Conversion of radial to axial current sheet motion in a pinch engine.

Fig. 9-19 Photograph of pinch engine on thrust stand. *(Republic Aviation Corporation, New York.)*

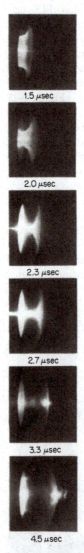

Fig. 9-20 Kerr-cell photographs of the ejection of plasma from a pinch discharge through a large orifice in one electrode. *(From R. G. Jahn et al., Ejection of a Pinched Plasma from an Axial Orifice, AIAA J., vol. 3, p. 1862, October, 1965.)*

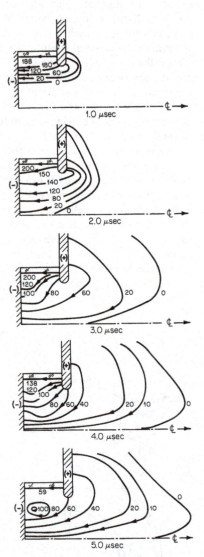

Fig. 9-21 Enclosed discharge current profiles in the exhaust pattern shown in Fig. 9-20 [$B_\theta(r,z) = (\mu/2\pi r)J_{\text{encl}}(r,z)$; numerals indicate J_{encl} in kiloamperes]. *(From R. G. Jahn et al., Ejection of a Pinched Plasma from an Axial Orifice, AIAA J., vol. 3, p. 1862, October, 1965.)*

process, while that produced in a pure pinch configuration can be readily extracted axially without a material nozzle.

The ability of a pulsed plasma to provide its own exhaust "nozzle" from its self-magnetic field has been clearly demonstrated experimentally [19]. Figure 9-20 shows a sequence of Kerr-cell photographs of the ejection of a plasma sheet from a large-radius pinch chamber in which it is formed. Figure 9-21 displays corresponding profiles of enclosed discharge current for the same process. Note that the initial radial motion is effectively converted to axial streaming in a "magnetic nozzle" provided by the self-field of the arc current, which itself billows far out into the plume. In this case the highly ionized pinch column created on the axis during the pinching phase serves as a "virtual cathode" for the ejection phase of the process, thereby closely simulating a coaxial gun configuration. The actual geometry of the anode orifice, in contrast, is found to have little primary influence on the pattern of the exhaust plume. The implication would seem to be that pulsed plasma electrode configurations will be dictated more strongly by considerations of arc initiation, stability, and efficiency of ejection than by a channeling of the flow in a classical gasdynamic sense.

9-9 ENERGY STORAGE AND SWITCHING

Except for the brief Appendix, this text does not deal in depth with the space power sources or power conditioning equipment required to drive the accelerators considered. In the realm of pulsed plasma propulsion, however, certain aspects of the power supply design are so tightly interwoven with the accelerator performance that their consideration is essential to a complete picture of the physical operation of this class of device. Each of these points has already been mentioned earlier in the discussion, namely, the necessity for energy storage units of exceedingly high performance; the importance of the arrangement of these storage units to provide optimum driving-current waveforms for the discharge; the need for an impedance match between the source circuit and the particular discharge characteristics; and the critical nature of the switching mechanism.

One may consider storage of the required electrical energy in capacitors, inductors, or batteries. The last may be immediately excluded because of their inherently high internal impedance, which would dissipate the bulk of the stored energy in this application, and in fact would completely preclude passage of the extremely high currents we require. Inductive storage has some potentiality, but adds difficult problems of effective coupling to the prime power source and of switch opening in high-current circuits, which do not arise in capacitor storage (Prob. 9-2).

The use of capacitors for storage of electrical charge is, of course, one of the most common practices of electrical engineering, but the units conventionally employed in power transmission and communication applications are totally inadequate for pulsed plasma accelerators. Beyond the obvious demands for low mass per unit energy stored, extremely long lifetime, and structural and electrical integrity in a vacuum environment, the accelerator application imposes a stringent requirement on the internal inductance of the capacitor and on the inductance of its connections to the load. We have seen that the inductance change embodied in the motion of the plasma must be comparable with the initial inductance of the entire circuit if the accelerator is to function efficiently [Eq. (9-5)] and that this inductance falls in the range of a few $\times 10^{-8}$ henry. Clearly, the capacitors are but one element of this circuit; hence they must contribute substantially less inductance than this value, lest the bulk of the stored energy be dissipated in the external circuit, rather than delivered to the plasma motion. Yet we also require total capacitance of many microfarads; hence a requirement of internal inductance $\approx 10^{-10}$ henry/µf must be imposed on the capacitor bank. Capacitors of this caliber did not exist before the interest in pulsed plasma propulsion arose, and a significant fraction of the technological effort in this field has been addressed to the improvement of this situation [28].

Assuming the availability of adequately high performance capacitors, the question of their proper disposition in the discharge circuit becomes the next crucial point. Simple parallel connection of all units to be employed, at the closest accessible position to the discharge electrodes, will provide the maximum current rise rate and peak amplitude, and is the configuration most commonly employed for such work. As we have seen, however, the resulting damped sinusoidal current waveform may produce undesirable crowbar breakdowns before the acceleration event is complete, thereby badly interfering with the electromagnetic coupling to the accelerated plasma.

An attractive alternative to this parallel bank arrangement is to interpose suitable inductance elements among the parallel capacitors, thereby forming an LC ladder network, or lumped-element transmission line (Fig. 9-22; [9]). Depending on the particular magnitudes of the

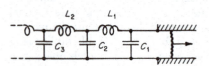

Fig. 9-22 Lumped-element transmission line to provide tailored waveform to plasma accelerator. *(From N. A. Black and R. G. Jahn, Dynamics of a Pinch Discharge Driven by a High-current Pulse-forming Network, Princeton Univ. Aeron. Rept. 778, Princeton, N.J., May, 1966.)*

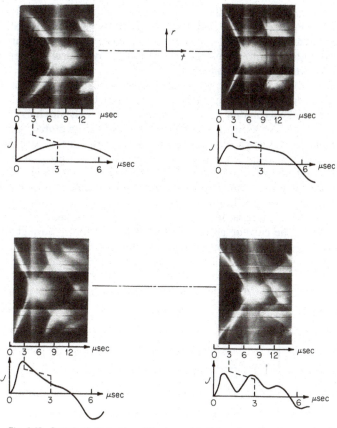

Fig. 9-23 Streak photographs of the inward radial motion of a plasma sheet in a pinch discharge driven by the indicated current waveforms. (*From R. L. Burton and R. G. Jahn, Structure of the Current Sheet in a Pinch Discharge, Princeton Univ. Aeron. Rept. 783, Princeton, N.J., September, 1966.*)

interposed inductors, the resulting current waveform delivered to the discharge may take a variety of profiles, such as a rectangular pulse, a slowly rising triangular form, a sharply rising triangular form, etc. For a given total energy storage, $\frac{1}{2}CV_0^2$, the acceleration profile and internal structure of the plasma current sheet are found to be profoundly affected by the particular shape of the driving current waveform. Figure 9-23 shows a series of streak photographs of the radial motion of a plasma sheet in a large-radius pinch discharge driven by various current waveforms.

Note that the velocity history of this current sheet and its terminal velocity as it reaches the centerline differ significantly for the cases shown. As pointed out in Sec. 9-5, for any particular accelerator geometry and ambient mass density distribution, there exists some optimum profile of driving current waveform whereby the dynamical efficiency of the acceleration event is maximized.

When employing pulse-forming networks of this sort to drive the discharge, it is important that the characteristic output impedance of the source closely match the effective impedance of the discharge. If this is achieved, the network will deliver the bulk of its stored energy to the plasma in a single pulse, and thereafter little current will flow in the circuit. If there is a substantial mismatch, only a fraction of the stored energy will be transferred to the plasma during the main pulse, and thereafter the current will continue to flow in the circuit via a series of electromagnetic wave reflections from the ends of the network. Specifically, if the characteristic impedance of the network exceeds that of the discharge, the current pattern will tend to ring down through a series of overshoots, much like an underdamped LCR circuit; if the network impedance is less than that of the discharge, the current pattern will approach zero through a series of monotonic steps, somewhat like an overdamped LCR circuit (Fig. 9-24). Clearly, secondary pulses of either

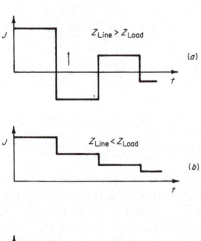

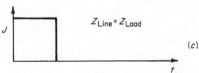

Fig. 9-24 Current waveforms delivered by a uniform distributed transmission line to a constant impedance load; characteristic impedance of line (a) greater than that of the load, (b) less than that of the load, (c) equal to that of the load.

type are undesirable, since they occur after the primary acceleration process has been completed.

Attainment of the desired impedance match may be difficult in practice. First, the typical plasma acceleration of this class embodies a dynamical impedance of only a few milliohms; to match this value in a network capable of delivering 10^5 amp requires very sophisticated capacitors, connected by minimal inductors with a minimum of connection inductance. Second, the impedance of the discharge probably will not be constant, particularly if the desired vigorous acceleration of the plasma sheet is achieved. Indeed, the problem has now become highly nonlinear, since the discharge impedance development will affect the current waveform, which determines the acceleration profile, which determines the dynamical impedance, etc. Finally, the mass and dimensions of very low impedance pulse-forming networks assembled from standard components tend to be excessive, mainly because of the large number of elements required to provide well-shaped pulses. Fortunately, techniques have recently been developed for winding distributed capacitance-inductance units which provide a given pulse profile for a specified low impedance load, in a much more compact package [28].

The remaining critical element of a pulsed plasma circuit is the switch which initiates the discharge of the capacitors through the gas. As mentioned earlier, this switch may be an external circuit element in series with the discharge, or it may be inherent in the main gap itself. External switches have the advantage of permitting precise correlation between the gas-filling cycle and the electrical cycle, but inevitably contribute some losses to the overall performance. Ideally, the switch should be capable of attaining the typical 10^5-amp level of current conduction in a fraction of a microsecond after triggering, and thereafter should display less resistance, initial inductance, and inductance change than the accelerator discharge it supplies. In all other respects its lifetime in repetitive operation should be at least as long as that of the main discharge gap. The conventional techniques for rapid switching of high currents involve either high pressure, electrically triggered spark gaps, or banks of gaseous-discharge tubes such as thyratrons and ignitrons. Each of these has a reasonable range of applicability, but both tend to complicate the external circuitry to some extent by their requirement for a triggering signal. Some simplification can be achieved in this respect by employing a low pressure gas-triggered discharge as a switching element, where the initiating gas pulse may be appropriately tapped from the main gas supply to the accelerator gap [29].

As an alternative to an external series switch, the discharge initiation may be precipitated within the accelerator gap itself. For example, one may simply employ the dependence of breakdown voltage on the product of gas density and gap spacing (Sec. 6-4) to initiate the discharge automatically at some point in the gas-filling cycle. In so coupling the electrical discharge cycle to the gas-filling cycle, however, one surrenders the ability to optimize each independently and to synchronize them most effectively. For example, for a given electrode geometry and applied voltage, the gas density level and distribution at breakdown may not be optimum for the subsequent electrical and dynamical sequence. In particular, the resistance of the discharge tends to be relatively high near the Paschen limit, and the mass loading of the accelerating sheet tends to be rather light for typical accelerator dimensions. This technique also requires that the gas inflow stop abruptly at breakdown, lest it enter the channel after the current sheet has passed and simply diffuse out the barrel orifice without acceleration.

Better separation of the gas-filling cycle from the electrical cycle can be attained by artificially triggering the accelerator gap via a small tickler discharge or by altering the potential of a judiciously placed auxiliary electrode. Unlike an external switch, these tickler discharges can operate with very small currents, and hence dissipate negligible energy. They do require an external synchronization circuit, however, and they clearly do not remove the upper Paschen limit on gas density level at breakdown.

In practice, the choice of switching mechanism is usually made as a best compromise among the factors mentioned above for the particular accelerator involved. In all cases the switch is a significant factor in the overall efficiency and lifetime of the device and, like the energy storage circuit, must be regarded as an integral element of the thruster.

9-10 QUASI-STEADY ACCELERATION

This chapter began with a discussion of the potential advantages of pulsed accelerators over steady flow devices of the same average power. In subsequent sections, various difficulties associated with the unsteady processes have been presented which tend to mitigate somewhat these inherent advantages. It may well occur that an intermediate domain of operation can be implemented that shares certain strengths of each extreme and is superior to both. In particular, the advantages of high current density operation with yet tolerable heat transfer and total power requirements could equally well be met by more widely spaced intermittent pulses of sufficient length so that effectively steady magneto-gasdynamic flow prevails over most of their duration. In this way the essentially transient processes of current sheet formation, sweeping, and uncoupling would be confined to the extremities of the pulse, while the bulk of the operation would resemble, say, the steady MPD arc (Fig.

9-25a). Various experiments on pulsed plasma accelerators driven by protracted-pulse lines have indicated that the propagating current patterns in the exhaust plume tend to revert to a more diffuse, steady configuration after a few tens of microseconds [30]. It may take somewhat longer for the discharge to heat the cathode to a satisfactory level of thermionic electron emission, but it appears that a 1-msec pulse will provide the desired *quasi-steady* operation over most of its duration. Feasibility of this concept clearly depends heavily on suitable low-impedance low-mass power sources for the required high current pro-

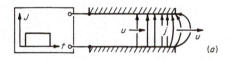

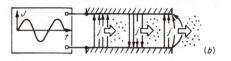

Fig. 9-25 Quasi-steady plasma acceleration concepts. *(a)* Protracted pulse; *(b)* continuous alternating wave; *(c)* self-modulated.

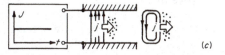

tracted pulses. The most direct approach involves some type of lumped-element transmission line, possibly followed by a pulse transformer to relieve the impedance mismatch between it and the discharge, but for the desired current levels and pulse lengths, considerable ingenuity will be required to restrain the total mass of such a system to tolerable levels.

A totally different concept of quasi-steady acceleration emerges from extrapolation of the repetitive pulse concept to the other extreme. Namely, one can imagine a steady inlet flow of gas swept along by a very rapid succession of alternating discharge pulses—so rapid that discrete cells of gas are trapped between adjacent current sheets in the channel and convected along by them (Fig. 9-25b). This type of interaction can be observed for a few cycles in pulsed accelerators driven by oscillatory ringdown of an underdamped capacitor, where a succession of secondary current sheets are generated at the voltage reversals. Extension of this mode to quasi-steady operation doubtless will introduce substantial thermionic contribution to the cathode emission process, which in turn should relieve some of the burden of high voltage from the required high-frequency power supply.

There is even some basis for conceiving such a device to be driven by a dc power source, from which discrete breakdown pulses would be initiated at the channel inlet in response to the disconnection of the previous current sheet at the exit (Fig. 9-25c). Such a self-modulated device would require no capacitors in its power supply, but total power consumption would tend to be quite high, and heat transfer could well become a serious problem again. In either concept of the device, note that, although the gas flow is quasi-steady, the electromagnetic interactions are fundamentally transient. In this sense the concept resembles the traveling-wave accelerator, to be discussed in Sec. 9-11.

9-11 PULSED INDUCTIVE ACCELERATION

In any of the pulsed plasma accelerators described above, a certain amount of electrode damage is inevitable, either from the high voltage breakdown initiation process, the high current electron emission, the ion bombardment flux, or from other forms of transient heating by the intense adjacent plasma. In all probability, the total electrode erosion rate from such effects will be the limiting factor in the useful lifetime of these devices as space thrusters. To remove this limitation, it would seem reasonable to consider coupling the electrical energy pulse from the storage circuit into the gas inductively, thereby removing the electrodes from physical contact with the plasma discharge.

A pulsed inductive accelerator must execute a slightly different sequence of physical events. The energy storage unit is first switched through an external primary circuit. The current pulse thus established sets up a rapidly growing magnetic field which permeates the gas within the accelerator. In accordance with Maxwell's induction equation, a corresponding electric field arises there which must be sufficiently strong to break down the gas to a state of high conductivity. The attempt of the primary magnetic field to penetrate this conducting zone induces a large current in it, opposite to the external current but in phase with it. Interaction of this "secondary" current with the **B** field then provides a **j** × **B** body force to accelerate the plasma away from the external circuit. It should be noted that it is the current induced by the magnetic field that interacts profitably with it. Any resistive current driven by the induced electric field is 90° out of phase with the magnetic field; hence its **j** × **B** force component reverses sign every half cycle.

UNSTEADY ELECTROMAGNETIC ACCELERATION

The simplest practical geometry which displays such a sequence of events is embodied in the *theta pinch*, a device which again substantially antedates plasma propulsion (Fig. 9-26a). Here the external circuit and induced current are concentric cylinders, and the $\mathbf{j} \times \mathbf{B}$ force is primarily radial. This particular configuration thus shares the difficulty of the

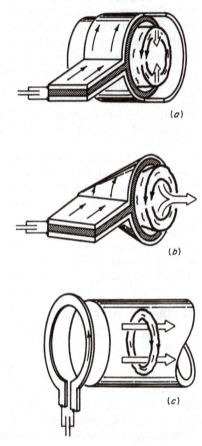

Fig. 9-26 Pulsed inductive accelerators. (a) Theta pinch; (b) conical theta pinch; (c) loop inductor.

linear pinch discharge in deflecting its plasma into an axial stream, but slight geometric modifications such as the *conical theta pinch* (Fig. 9-26b) or the *loop inductor* (Fig. 9-26c) directly impart a net streamwise acceleration.

Unfortunately, the basic attraction of this type of acceleration is somewhat illusory. Although electrode erosion is indeed removed from the problem, erosion of the dielectric channel surface near the breakdown region remains, and for the high energy discharges of pulsed propulsion interest, this can be equally serious. In addition, inductive discharges embody two inherent electrodynamic disadvantages to conversion efficiency which detract from their propulsion effectiveness. First, any delay in breakdown of the gas after application of the primary field pulse results in energy being dissipated in the external circuit, which, unlike that of the direct electrode devices, is complete without the gas loop. In the theta pinch, for example, a typical response to a ringing-capacitor input is for the neutral gas to break down only slightly on the first half cycle and to become fully conducting on the second half-cycle pulse, where it has the benefit of the residual ionization from the first pulse. This difficulty might be relieved by providing a separate preionization mechanism or by operation at a sufficiently rapid repetition rate, but it is indicative of an inherent inefficiency in coupling of the external circuit to the plasma.

Equally troublesome is the need to accomplish all the energy input to the gas before much motion of it has occurred. The current induced in the gas-loop "secondary" depends on its mutual inductance with the external primary, and thus is a strong function of the physical separation of these two current paths. As they separate under the acceleration, the coupling rapidly becomes weaker. Attempts may be made to contain the magnetic flux between the two conductors somewhat by the use of ferrite insulator materials,[1] but in general, single-pulse inductive accelerators still tend to display rather low conversion efficiencies.

9-12 TRAVELING-WAVE ACCELERATION

If the magnetic field in a pulsed inductive discharge, rather than falling off rapidly with distance from the primary conductor, can somehow be caused to propagate immediately behind the induced current sheet at undiminished strength, the acceleration may proceed more effectively. To accomplish this, a more elaborate external circuit is required, such as a succession of discharge coils, suitably programmed to establish a magnetic pulse propagating down a dielectric duct at the desired speed. In view of the aforementioned initiation inefficiency, steady or quasi-steady operation of such a device seems more attractive than isolated pulse operation. That is, one may use the same coils to establish a continuous propagating electromagnetic wave train, each half wavelength of which should create, agglomerate, and convect a blob of ionized gas down the duct.

The concept just presented is that of a traveling-wave accelerator, many forms of which have been proposed and tested. The propagating

[1] C. L. Dailey, private communication.

electromagnetic field may be produced by a series of coils driven by an rf polyphase circuit or by a transmission delay line; it may entrain the plasma blobs in "mirror" or "cusp" magnetic configurations; it may propagate at constant speed or be programmed to accelerate along the duct; it may operate in a low density particle orbit regime or a high density continuum domain; and the channel may be circular, annular, or rectangular in cross section, uniform or tapered along its length, etc. In all cases, however, this class of accelerator relies on the interaction of induced transverse currents with the crossed transverse component of the propagating magnetic field to provide the axial streamwise body force. The effectiveness of a given device, however, may hinge heavily on the role of other components of $\mathbf{j} \times \mathbf{B}$ in containing and ejecting the accelerated plasma.

Theoretical analysis of any practical traveling-wave accelerator tends to considerable complexity because of the essential three-dimensional character of the fields and interactions, the tensor conductivity, the effect of induced fields, and the difficult boundary conditions on the fields and currents at the channel walls and exit [31]. In the experimentally attractive cylindrical or conical geometries, for example, the applied magnetic field pattern has time-dependent axial and radial components which must be described by products of special mathematical functions, such as hyperbolic sines and cosines and various Bessel functions.

To illustrate the traveling-wave acceleration concept with less mathematical encumbrance, we shall resort to a much simpler, highly idealized rectangular system, which is of doubtful propulsion interest, but has nevertheless attracted some experimental study [32]. Consider the device sketched schematically in Fig. 9-27, where the polyphase coils

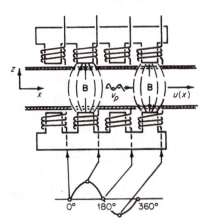

Fig. 9-27 Idealized rectangular traveling-wave accelerator. *(From L. Heflinger et al., Transverse Traveling Wave Plasma Engine, AIAA J., vol. 3, p. 1029, 1965.)*

provide a nearly transverse magnetic field propagating at uniform velocity down the duct at phase velocity v_p. Assume that the gas has uniform finite scalar conductivity σ, low enough so that induced magnetic fields are negligible, and that all properties are constant along the $\mathbf{v}_p \times \mathbf{B}$ dimension y. In the notation of Fig. 9-27, then, the magnetic field has only two components:

$$\mathbf{B} = B_x(x,z,t)\hat{\mathbf{x}} + B_z(x,z,t)\hat{\mathbf{z}} \qquad (9\text{-}44)$$

which must satisfy $\nabla \times \mathbf{B} = 0$ and $\nabla \cdot \mathbf{B} = 0$ in the form

$$\frac{\partial B_x}{\partial z} = \frac{\partial B_z}{\partial x} \qquad (9\text{-}45)$$

$$\frac{\partial B_x}{\partial x} = -\frac{\partial B_z}{\partial z} \qquad (9\text{-}46)$$

We have one boundary condition from symmetry at the centerline:

$$B_x(x,0,t) = 0 \qquad (9\text{-}47)$$

and a second from the assumption of a sinusoidal wave train:

$$B_z(x,0,t) = B_0 \cos \frac{2\pi}{\lambda}(x - v_p t) \qquad (9\text{-}48)$$

Subsequent algebra is simplified by now transforming to a coordinate system convecting with the accelerating gas. Relations (9-45) to (9-48) remain valid here, provided v_p is replaced by a slip velocity,

$$v_s = v_p - u(x) \qquad (9\text{-}49)$$

where $u(x)$ is the streaming velocity of the gas. The solution to (9-45) to (9-48) is then

$$B_z = B_0 \cosh \frac{2\pi z}{\lambda} \cos \frac{2\pi}{\lambda}(x - v_s t) \qquad (9\text{-}50)$$

$$B_x = -B_0 \sinh \frac{2\pi z}{\lambda} \sin \frac{2\pi}{\lambda}(x - v_s t) \qquad (9\text{-}51)$$

The induced electric field in this system now follows from $\nabla \times \mathbf{E} = -\partial \mathbf{B}/\partial t$:

$$E_x = E_z = 0 \qquad (9\text{-}52)$$

$$E_y = B_0 v_s \cosh \frac{2\pi z}{\lambda} \cos \frac{2\pi}{\lambda}(x - v_s t) \qquad (9\text{-}53)$$

Since $\mathbf{j} = \sigma\mathbf{E}$ in this frame, the body force density components follow directly from $\mathbf{j} \times \mathbf{B}$.

$$f_x = \sigma E_y B_z = \sigma B_0^2 v_s \cosh^2 \frac{2\pi z}{\lambda} \cos^2\left[\frac{2\pi}{\lambda}(x - v_s t)\right] \tag{9-54}$$

$$f_z = -\sigma E_y B_x = \tfrac{1}{4}\sigma B_0^2 v_s \sinh \frac{4\pi z}{\lambda} \sin\left[\frac{4\pi}{\lambda}(x - v_s t)\right] \tag{9-55}$$

The development of the acceleration along the channel can be traced from the axial force equation (9-54). At inlet, $u \ll v_p$; hence $v_s \approx v_p$ and the wave train washes over the gas at velocity v_p, exerting pulses of accelerating force at twice the field frequency, but always of the same sign. This is to be expected, since both the induced current and the field reverse every half cycle. As the gas accelerates, the apparent frequency of the imposed wave, v_s/λ, decreases, and with it the frequency of the force density pulses, whose amplitude also decays as $1/v_s$. As the gas velocity approaches the phase velocity, $v_s \to 0$, the axial body force diminishes to zero, and acceleration is complete.

The transverse force [Eq. (9-55)] is seen to be antisymmetric about the centerline and to reverse its direction inward and outward, at twice the apparent frequency. The relative importance of this force depends on the channel size, and it is not obvious without solving the gasdynamical problem whether it helps or hinders the acceleration. In addition to its dynamical contribution, this force component may substantially affect heat transfer to the channel walls, and thereby the attainable efficiency. If it acts in the mean to confine the streaming gas away from the walls, heat transfer losses will be reduced; if it acts on the average to drive the gas toward the walls, or if it excites a transverse instability of the plasma flow, it clearly is an undesirable interaction.

At this point it is perhaps worth emphasizing a basic distinction between this mode of acceleration and the isolated-pulse inductive accelerator described earlier. There the current was induced in the plasma by the attempt of the magnetic field to penetrate an excellent conductor, and hence was in phase with the B field automatically. Here, consistent with the assumption of negligible induced magnetic fields, the current is driven only by the induced electric field, and hence is in phase with that. Phase coherence with the magnetic field is provided in this case by the interplay of spatial and temporal derivatives in the wave functions describing the propagation. Note, for example, that Eq. (9-50) for the major component of $\mathbf{B}$ and (9-53) for $\mathbf{E}$ are in phase. The idealization invoked here is equivalent to allowing the traveling-wave pattern complete freedom to permeate the working fluid at will, and thus demands a modest conductivity. If this becomes too high, the magnetic field will be largely excluded from the plasma, and the earlier mode of interaction will dominate.

Returning to the question of efficiency, there is again an evident upper bound set by the dynamics of the process. Consider any convenient time segment of quasi-steady operation of this kind of device, such as one period of the applied field τ. Over this interval a certain total impulse has been imparted to the gas which may be equivalently represented by the time integral of the mean axial body force exerted on the gas throughout the channel, $F(t)$, or by the change in momentum of the total mass entering or leaving the channel during this period:

$$I = \int_0^\tau F(t)\,dt = \dot{m}\tau(u_f - u_i) \tag{9-56}$$

During this same period, the propagating field pattern performs work on the gas in amount

$$W = \int_0^\tau F(t)v_p\,dt = v_p \dot{m}\tau(u_f - u_i) \tag{9-57}$$

The dynamical efficiency may be defined as the ratio of the actual kinetic energy change of the gas to this work performed on it:

$$\eta_d = \frac{\frac{1}{2}\dot{m}\tau(u_f^2 - u_i^2)}{\dot{m}\tau v_p(u_f - u_i)} = \frac{1 + u_i/u_f}{2v_p/u_f} \tag{9-58}$$

Thus, starting from negligible inlet velocity u_i, the highest attainable efficiency is 50 percent. The remainder of the input work must appear as Joule heating, and arises fundamentally from the serious mismatch in gas and field velocities in the early portion of the acceleration.

To relieve this limitation, one may program the coils to produce an accelerating magnetic field which will better match the gas velocity profile along the channel. For example, it can be shown that if a profile of constant slip velocity, rather than constant phase velocity, is maintained, the efficiency becomes (Prob. 9-12)

$$\eta = \frac{1 + u_i/u_f}{1 + 2v_s/u_f + u_i/u_f} \tag{9-59}$$

Thus, matching of the gas and field velocities to, say, 10 percent of u_f throughout the channel raises the limiting efficiency to about 80 percent.

The crude model discussed above neglects a myriad of effects which complicate real traveling-wave accelerators, and even then describes only one possible domain of operation, but the major aspects discussed above are found empirically to remain relevant to many laboratory devices of this class. The importance of slip velocity, magnetic containment of the flow, and strong coupling between the primary coils and induced plasma currents permeates virtually all experimental attempts to drive the efficiency of this class of accelerator up into a regime of propulsion interest. As mentioned earlier, experimental work on traveling-wave acceleration has been highly diversified in its approach, and a brief review here can serve only to give some flavor of this diversification.

As a first example, we might select the device which most closely conforms to the simple rectangular model discussed above [32]. This device, sketched in Fig. 9-27, involves four pairs of equally spaced coils, driven in 90° phase succession by a transient megawatt oscillator at 480 kc, yielding a constant phase velocity of 5.4×10^4 m/sec. The duct has 4.6 cm internal dimension and 9.5 cm length. Several propellants have been tried, all at relatively high particle densities and mass flow rates ($n \approx 10^{20}$ to 10^{21} m^{-3}; $\dot{m} \approx 10^{-4}$ to 10^{-5} kg/sec), and the best performance has been obtained with helium. Thrusts of the order of a newton, at efficiencies approaching 40 percent, are indicated, the most serious problems arising from interaction of the plasma stream with the walls.

With this exception, the bulk of the experiments have been performed with cylindrically symmetric magnetic field patterns driving circular or annular ducts. In one case, a conical duct tapering from 6 to 20 cm diameter over 120 cm length is driven by 48 and 54 coil arrangements, supplied with 12-phase 115-kc rf power (Fig. 9-28a; [33]). This is a low density device ($n \approx 10^{17}$ to 10^{18} m^{-3}, $\dot{m} \approx 5 \times 10^{-6}$ kg/sec) with ion mean free path and gyro radius larger than duct dimensions, so that large axial space-charge fields develop to transmit the accelerating force to the ions. Efficiencies approaching 30 percent in argon are reported, but serious heat transfer problems arise near the duct exit, where the plasma attempts to extract itself from the propagating field pattern.

In contrast, a shorter constant-area system has been constructed using four coaxial coils driven at 90° phase separation by 150-kc 25-kw continuous rf power (Fig. 9-28b; [34]). This duct is 7.6 cm in diameter and some 30 cm long before flaring into a 15-cm-diameter exhaust nozzle. Efficiency of 23 percent is obtained in xenon at 4×10^4 m/sec exhaust speed and 4×10^{-7} kg/sec mass flow.

Traveling-wave accelerators driven by transmission lines, rather than polyphase coils, have been attempted in a few laboratories [35], but no propulsion performance data have been reported. Typically, a cylindrical duct is wound with double helices, and capacitors are distributed along the turns (Fig. 9-28c). The development of the propagating wave is similar to that in traveling-wave electronic devices, and has been shown to be effective in accelerating injected plasma. Again,

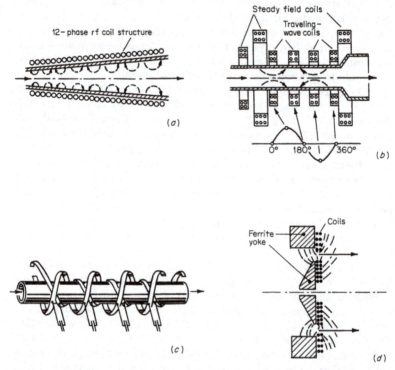

Fig. 9-28 Various inductive plasma accelerators. (a) Conical duct coil-driven traveling wave; (b) cylindrical duct coil-driven traveling wave; (c) cylindrical duct helical transmission line; (d) radio-frequency fringe-field accelerator. [*From H. Smotrich et al., Experimental Studies of a Magnetohydrodynamic C.W. Traveling Wave Accelerator, Proc. 4th Symp. Eng. Aspects Magnetohydrodynamics, p. 73 April, 1963; R. W. Palmer et al., Analytical Investigations of Coil-system Design Parameters for a Constant-velocity Traveling Magnetic Plasma Engine, NASA Tech. Note D-2278, 1964; E. B. Mayfield et al., Plasma Propulsion by Means of a Helical Transmission Line, p. 543 in E. Stuhlinger (ed.), "Electric Propulsion Development," vol. 9 of "Progress in Astronautics and Aeronautics," Academic Press Inc., New York, 1963; and A. S. Penfold, Recent Advances in the Development of Electromagnetic Thrustors for Space Propulsion, Proc. 4th Symp. Advanced Propulsion Concepts, Palo Alto, Calif., April, 1965.*]

wall losses are serious, and a certain preferential slippage along the axis has been observed.

Finally, accelerators have been proposed and constructed which, although inductively coupled in the same sense as those above, invoke little or no actual propagation of the rf wave train. One example is the *fringe-field accelerator* shown in Fig. 9-28d [36]. Two 10-turn coils, supplemented by a ferrite yoke construction, are driven as part of a 250-kc resonant circuit. The associated oscillating magnetic field pattern generates an azimuthal electric field and current density which interact with the magnetic field to provide axial acceleration and some radial containment. This device is clearly hybrid in several respects, most notably in its ability to provide electrothermal as well as electromagnetic interactions. For example, if the density is low enough to provide an electron Hall parameter greater than unity, the plasma reacts on the resonant circuit mainly as an inductive element; i.e., the current in the gas is out of phase with the coil voltage. This clearly puts it in phase with the coil current, hence with magnetic field, and a time average $\mathbf{j} \times \mathbf{B}$ body force prevails. On the other hand, if density is high enough to reduce the Hall parameter below unity, the discharge reacts resistively on the circuit, and the current is 90° out of phase with respect to the magnetic field. The time-average body force is then zero, and the discharge serves mainly to heat the plasma in the magnetic container. Presumably, some fraction of this enthalpy may be recovered by judicious expansion of the exhaust flow, but the former, direct-acceleration mode seems more attractive.

Before leaving this topic we are once again obliged to note that the eventual applicability of any of these sophisticated traveling-wave accelerators to propulsion will inevitably be prejudiced by the complexity of the space power supply needed to drive them. As in the electrodeless electrothermal accelerators (Sec. 6-7), any device requiring substantial amounts of rf input pays a penalty for the relative inefficiency with which such high-frequency power can be extracted from a prime source and delivered to the desired load. If, in addition, there are requirements for multiphasing and programming of the supplied waveform, additional mass and complexity will unavoidably appear in the power plant. It thus seems that traveling-wave accelerators will need to demonstrate high performance, indeed, before their overall system complexity can be justified for propulsion.

PROBLEMS

9-1. Derive the radial distribution of current in a uniform cylindrical conductor of radius 0.10 m, length 0.10 m, and conductivity 10^5 mhos/m for an applied sinusoidal voltage of 1-Mc frequency and 10,000-volt amplitude.

9-2. Discuss the relative merits of storing electrical energy in inductors, rather than capacitors, for propulsion purposes.

9-3. Show that if the inductance of an RCL series circuit increases monotonically, and R and C are finite constants, the circuit current is bounded by a value J_0, equal to the peak current of a circuit with the same capacitance, zero resistance, and constant inductance L_0.

9-4. Discuss whether it is possible to damp critically an LC circuit of zero resistance by a sufficiently large value of dynamical impedance $\dot{L}$.

9-5. Nondimensionalize Eqs. (9-11) and (9-12) using $t^* = t/RC$, rather than $t^* = t/\sqrt{LC}$ as done in the text. What are the scaling parameters in this case?

9-6. If the current sheet is not an ideally thin layer but has a thickness that is a significant fraction of the channel length, the concept of circuit inductance becomes difficult to apply. Show that if the current density is uniform over the total channel length, the inductance defined by flux linkage,

$$L_\phi = \frac{\phi}{J}$$

differs from that defined by field-energy considerations,

$$L_W = \frac{2W}{J^2}$$

where J = total current
ϕ = magnetic flux linked by the circuit
W = total field energy

9-7. If the ambient gas particles were ionized *before* entering the current sheet, how would their trajectories differ from those derived in the section on gas-kinetic models?

9-8. In the particle-orbit analysis of the section on gas-kinetic models, why does the electron-ion separation in the transverse (y) direction not predicate an additional component of interior electric field in that direction?

9-9. What is the minimum fraction of input power consumed by ionization of the propellant in a pulsed argon plasma accelerator operating at 5,000 sec specific impulse? What is a more realistic estimate of this loss, considering other inelastic processes which accompany the ionization events in the discharge?

9-10. Sketch the complete patterns of magnetic field lines behind the current sheet in (a) a parallel-plate accelerator; (b) a coaxial gun; (c) a linear pinch discharge.

9-11. Define a thrust efficiency for a single-pulse accelerator in terms of readily measurable quantities. Show that this is consistent with Eq. (9-42) for repetitive pulse operation.

9-12. Derive the limiting efficiency of a traveling-wave accelerator operating at constant slip speed v_s [Eq. (9-59)].

REFERENCES

1. Marshall, J.: Performance of a Hydromagnetic Plasma Gun, *Phys. Fluids*, vol. 3, p. 134, 1960.
2. McIntosh, V. G., and W. H. Bostick: A Button Source of Plasma, *Univ. Calif. Lawrence Rad. Lab. Rept.* UCRL-4688, Livermore, Calif., April, 1956.
3. McLean, E. A., et al.: Spectroscopic Study of Helium Plasmas Produced by Magnetically Driven Shock Waves, *Phys. Fluids*, vol. 3, p. 843, 1960.
4. Griem, H. R.: "Plasma Spectroscopy," McGraw-Hill Book Company, New York, 1964.
5. Ginzton, E. L., et al.: A Linear Electron Accelerator, *Rev. Sci. Instr.*, vol. 19, p. 89, 1948; Chodorow, M., et al.: Stanford High-energy Linear Electron Accelerator, *ibid.*, vol. 26, p. 134, 1955.

6. Lovberg, R. H.: Nonsteady M.H.D. Accelerators, in J. Grey (ed.), "Plasma Technology," Prentice-Hall, Inc., Englewood Cliffs, N.J., 1968.
7. Rosebrock, T. L., et al.: Pulsed Electromagnetic Acceleration of Metal Plasmas, *AIAA J.*, vol. 2, p. 328, 1964.
8. Millsaps, K., and K. Pohlhausen: The Linear Acceleration of Large Masses by Electrical Means, *Air Force Missile Defense Center Tech. Rept.* 60-11, June, 1960.
9. Black, N. A.: Dynamics of a Pinch Discharge Driven by a High-current Pulse-forming Network, *Princeton Univ. Aeron. Rept.* 778, Princeton, N.J., May, 1966.
10. Rosenbluth, M.: Infinite Conductivity Theory of the Pinch, *Los Alamos Sci. Lab. Tech. Rept.* LA-1850, Los Alamos, N.Mex., 1954.
11. Liepmann, H. W., and A. Roshko: "Elements of Gasdynamics," John Wiley & Sons, Inc., New York, 1957.
12. Rowell, G. A.: Cylindrical Shock Model of the Plasma Pinch, *Princeton Univ. Aeron. Rept.* 742, Princeton, N.J., February, 1966.
13. Düchs, D.: Numerical Calculations and Comparison with Measurements Investigating the Influence of the Neutral Gas in θ-pinch Discharges, *Phys. Letters*, vol. 5, p. 121, 1963.
14. Lovberg, R. H.: Investigation of Current Sheet Microstructure, *AIAA J.*, vol. 4, pp. 1215–1222, July, 1966.
15. Burton, R. L.: Structure of the Current Sheet in a Pinch Discharge, *Princeton Univ. Aeron. Rept.* 783, Princeton, N.J., September, 1966.
16. Black, N. A., and R. G. Jahn: Dynamic Efficiency of Pulsed Plasma Accelerators, *AIAA J.*, vol. 3, pp. 1209–1210, June, 1965.
17. Westberg, R. G.: Nature and Role of Ionizing Potential Space Waves in Glow to Arc Transitrons, *Phys. Rev.*, vol. 114, p. 1, 1959.
18. Guman, W. J., and W. Truglio: Surface Effects in a Pulsed Plasma Accelerator, *AIAA J.*, vol. 2, p. 1342, 1964.
19. Jahn, R. G., et al.: Ejection of a Pinched Plasma from an Axial Orifice, *AIAA J.*, vol. 3, p. 1862, October, 1965.
20. Lovberg, R. H.: The Measurement of Plasma Density in a Rail Accelerator by Means of Schlieren Photography, *IEEE Trans. Nucl. Sci.*, vol. NS-11, p. 187, January, 1964.
21. Larson, A. V., et al.: An Energy Inventory in a Coaxial Plasma Accelerator Driven by a Pulse Line Energy Source, *AIAA J.*, vol. 3, p. 977, 1965.
22. Gloersen, P., et al.: Energy Efficiency Trends in a Coaxial Gun Plasma Engine System, *AIAA J.*, vol. 4, p. 436, 1966.
23. Michels, C. J., et al.: Analytical and Experimental Performance of Capacitor Powered Coaxial Plasma Guns, *AIAA J.*, vol. 4, p. 823, 1966.
24. Ashby, D. E. T. F., et al.: Exhaust Measurements on the Plasma from a Pulsed Coaxial Gun, *AIAA J.*, vol. 3, p. 1140, 1965.
25. Gorowitz, B., et al.: Performance of an Electrically Triggered Repetitively Pulsed Coaxial Plasma Engine, *AIAA J.*, vol. 4, p. 1027, 1966; Karras, T., et al.: Neutral Mass Density Measurements in a Repetitively Pulsed Coaxial Plasma Accelerator, *AIAA J.*, vol. 4, p. 1366, August, 1966.
26. Lovberg, R. H.: Schlieren Photography of a Coaxial Accelerator Discharge, *Phys. Fluids*, vol. 8, p. 177, 1965.
27. Aronowitz, L., and D. P. Duclos: Characteristics of the Pinch Discharge in a Pulsed Plasma Accelerator, p. 513 in E. Stuhlinger (ed.), "Electric Propulsion Development," vol. 9 of "Progress in Astronautics and Aeronautics," Academic Press Inc., New York, 1963.

28. Hayworth, B. R., et al.: Energy Storage Capacitors for Pulsed Plasma Thrustors, *AIAA J.*, vol. 4, p. 1534, 1966; Hayworth, B. R.: Energy Storage Capacitors: An Experimental and Theoretical Investigation of Their Properties, MSE thesis, San Diego State College, San Diego, Calif., June, 1964.
29. Jahn, R. G., et al.: Gas-triggered Pinch Discharge Switch, *Rev. Sci. Instr.*, vol. 36, p. 101, January, 1965; Gas-triggered Inverse Pinch Switch, *Rev. Sci. Instr.*, vol. 34, p. 1439, December, 1963.
30. Ashby, D. E. T. F., et al.: Quasi-steady-state Pulsed Plasma Thrusters, *AIAA J.*, vol. 4, p. 831, 1966.
31. Neuringer, J. L., and E. Migotsky: Skin Effect in Magneto-fluid Dynamic Traveling Wave Devices, *Phys. Fluids*, vol. 6, p. 1164, 1963; Neuringer, J. L.: Induced Forces in Annular Magneto-fluid Dynamic Traveling Wave Devices, *AIAA J.*, vol. 2, p. 267, 1964.
32. Heflinger, L., et al.: Transverse Traveling Wave Plasma Engine, *AIAA J.*, vol. 3, p. 1029, 1965.
33. Smotrich, H., et al.: Experimental Studies of a Magnetohydrodynamic C. W. Traveling Wave Accelerator, *Proc. 4th Symp. Eng. Aspects Magnetohydrodynamics*, University of California, Berkeley, April 10–11, 1963, p. 73.
34. Palmer, R. W., et al.: Analytical Investigations of Coil-system Design Parameters for a Constant-velocity Traveling Magnetic Plasma Engine, *NASA Tech. Note* D-2278, 1964.
35. Mayfield, E. B., et al.: Plasma Propulsion by Means of a Helical Transmission Line, p. 543 in E. Stuhlinger (ed.), "Electric Propulsion Development," vol. 9 of "Progress in Astronautics and Aeronautics," Academic Press Inc., New York, 1963.
36. Penfold, A. S.: Recent Advances in the Development of Electromagnetic Thrustors for Space Propulsion, *Proc. 4th Symp. Advanced Propulsion Concepts*, Palo Alto, Calif., April, 1965.

Appendix
Space Power Supplies and Low Thrust Mission Analysis

This text on electric propulsion has dealt almost exclusively with the physical problems of gas acceleration by electrical means. In this sense its title is too general, for if electric propulsion is to achieve technological maturity, progress in this area must be matched by understanding and accomplishment in two other equally fundamental areas to which we have alluded only briefly: the conversion of basic energy sources into usable electric power in space and the mechanics of low thrust space flight. Proper presentation of the elements of either of these sciences would require another volume of at least this size, but a few remarks on each may provide an appropriate closing frame for our present effort. Like the engineering status of electric thrusters, the design and operation of low specific mass power plants and the computational analysis of low thrust space missions are presently in their infancy. Existing devices, analysis and operation procedures, and performance results are incomplete and transitory, and the student and researcher again would benefit from study of fundamental topics in each area.

SPACE POWER SUPPLIES

The development of space power sources must proceed under many of the same constraints imposed upon electric thrusters by the environment of their operation. Namely, they must be capable of unattended reliable operation in space for periods of several years; they must be compatible with other engineering equipment and payload characteristics, including man; and a premium must be placed on low specific mass and high conversion efficiency. The possibilities for meeting these requirements may be crudely divided into systems which derive the primary energy from an external source and those which are totally self-contained.

Of those systems deriving the primary energy from an external source (exogenous systems), only solar-energy converters have survived serious consideration.[1] By far the simplest and most highly developed solar power system is the solar cell panel, consisting of an extensive array of photovoltaic cells of suitable semiconductor crystals, such as silicon-germanium or lead telluride, or of thin polycrystalline films appropriately mounted, interconnected, and deployed around the power-consuming spacecraft. Solar panels have been employed on spacecraft since the earliest probes and satellites, to provide modest amounts of power for instrumentation and communications. Originally felt to be inherently too massive and extensive for propulsion-power applications, these devices were later vaulted into this arena by a rather rapid succession of technological improvements in semiconductor performance and panel fabrication, combined with some revised analyses of the missions for solar-electric-propelled spacecraft. Whereas the early solar panel systems were burdened with specific masses of hundreds of kilograms per kilowatt, application of more sophisticated materials, design, and analysis, of radiation concentration techniques, and of assembly and deployment procedures now justifies extrapolation to an eventual range of 10 kg/kw, well within the useful propulsion regime.

The physical simplicity of the solar panel systems is vitiated somewhat by certain inherent disadvantages of this class of converter. In addition to their relatively low power yield per unit surface area, necessitating the deployment of very large arrays of panels to garner useful amounts of power for propulsion, they must be protected from overheating and radiation damage, which degrade their performance in time; they must be kept properly oriented toward the sun; they must be supplemented by batteries or other storage units if the spacecraft passes into a planetary shadow; and their output varies strongly with distance from the sun. Primarily because of their inherently large surface area per unit power, it is doubtful that solar cell technology will advance to the point that the multimegawatt supplies required for manned interplanetary missions will be feasible. Nevertheless, the demonstrated

[1] The possibility of transmitting large amounts of electromagnetic radiation to a distant spacecraft from the earth's surface by a high-gain antenna network or a laser beam is, however, an interesting speculative exercise.

APPENDIX

simplicity and reliability of this class of source should keep it in the forefront for the more modest near-term applications of electric propulsion. As a first demonstration of this capability, the SERT II mission, planned for 1969, will test the space operation of an electron bombardment ion thruster from an array of solar panels.

Photovoltaic cells are by no means the only possibilities for conversion of solar energy to electric power. Solar powered thermoelectric generators, thermionic converters, and turboelectric systems, among others, may be considered. Each has received some early study, on the basis of which only the first is evidently noncompetitive with solar cells for ultimate propulsion application. These three conversion techniques are also applicable to nuclear power sources, and many of the materials and fabrications problems are common to both applications. Progress in the nuclear systems should thus benefit the development of these more sophisticated solar devices.

Of the spectrum of self-contained (endogenous) power sources, only those of nuclear origin are seriously considered. The same limitation imposed by the inherent energy of a chemical reaction on the attainable specific impulse of a rocket precludes the applicability of chemical power sources, such as fuel cells, to the generation of large quantities of electrical energy in space. In the nuclear domain, one may distinguish among sources involving controlled nuclear fission reactions, natural or artificial radioisotopes, and controlled fusion reactions. The last seem far from engineering realization, and may be fundamentally inapplicable to space systems because of the massive auxiliary equipment required. Radioisotope energy sources are attractively simple heaters for various thermal conversion processes, but the scarcity and expense of these substances would seem to preclude their general application to high-level space power requirements.

By all odds the most serviceable endogenous energy sources are the controlled nuclear fission reactors. These devices offer attractively high power densities and low specific masses, long-term reliability, ease of control, and operating temperature ranges compatible with the space environment. Their physics and engineering are well understood, and they are adaptable to a variety of processes for converting their heat source into electrical power. They have a characteristic group of disadvantages as well, centered on the damaging nuclear radiation they exude if not properly shielded from other components of the power system and spacecraft, a difficulty which intensifies when manned missions are involved. This same difficulty tends to encumber research and development efforts on all types of reactor-driven converters and to increase the expense, thereby reducing the number of parallel development efforts that can reasonably be sustained.

The choice of a particular conversion process for a nuclear space power plant must be a judicious compromise among physical sophistication, mechanical simplicity, and proven reliability of operation that leads to optimum specific mass, compatibility, and lifetime in the given application. A categorical choice is not yet justified, since at least three distinctly different candidates show substantial preliminary promise: turboelectric cycles of the Rankine or Brayton type, direct thermionic conversion by a plasma diode, and magnetogasdynamic generation in a closed channel flow.

Steam and mercury driven turbomachinery has long been the standard equipment for ground based thermal power stations and has attained a level of refinement which argues well for a similar space nuclear power system. Such extrapolation is complicated, however, by several factors: the aforementioned nuclear radiation environment, the persistent demands for low specific mass and unattended long-term reliability, and the need to reject heat at much higher temperatures to minimize radiator dimensions. The upshot of such requirements is to transform the familiar turbogenerator into a considerably more exotic system involving refractory solid elements and unusual working fluids at uncommonly high temperatures, and thus to invalidate much of the background of technological experience.

As one example, consider the large power level nuclear Rankine cycle system commonly known as SNAP-50 (Systems for Nuclear Auxiliary Power) or SPUR (Space Power Unit Reactor). The goal here is a unit providing several hundred kilowatts of power in the specific mass range of 10 kg/kw, unshielded, with a lifetime of at least 10,000 hr. In one version the concept employs a fast reactor fueled with refractory uranium carbide or nitride in a columbium alloy structure. Heat is extracted from the reactor core at about 1500°K by a liquid lithium flow loop, which in turn vaporizes a potassium flow in an intermediate heat exchanger. This potassium vapor in a power loop drives the turbogenerator, and is in turn condensed by heat transfer to a liquid sodium-potassium mixture (NaK), which disposes of the heat in a segmented radiator at about 1000°K. Subcooled potassium is also used to lubricate some bearings and to cool critical turbogenerator elements via a separate radiating loop. Clearly, the superposition of the high temperature conditions, the corrosivity of the working fluids, and the nuclear radiation environment raise severe materials problems that will determine the success of such a system.

The technological complexity of a space turboelectric system encourages parallel examination of mechanically simpler, albeit less familiar, direct conversion processes. Both thermoelectric and thermionic mechanisms are potentially exploitable for this purpose, but the former appears to be fundamentally incapable of providing high power levels at low specific mass. A thermionic converter, in essence, invokes the difference in Fermi levels or work functions for electrons in two dissimilar metallic electrodes, to derive a voltage difference from the passage of an electronic current between them. To activate this effect, one electrode (the cathode) is heated to liberate the electrons from its surface, and the interelectrode gap is filled with a conducting plasma, usually cesium, to eliminate space-charge saturation of the electron current to the anode, which of course must be kept cooler than the cathode.

In this device a premium is placed upon efficient transfer of heat to and from the electrodes, and in this connection the development of a very effective wicklike element known as a *heat pipe* appears to have much promise. Two possible implementations of a plasma diode–heat pipe system are considered. In one, the diode is heated within the reactor, and heat is rejected from the cooler electrode by a heat pipe to an external radiator. In the other, heat is transferred from the reactor to an exterior diode unit in or near the radiator. The latter presumably enjoys less serious diode irradiation difficulties, but several other more subtle considerations make the ultimate choice unclear at this time.

As a final alternative for energy conversion, one may examine a magnetohydrodynamic system, using a conducting gas or liquid as a working fluid. To a first approximation, an MHD generator is simply a crossed-field accelerator like those discussed in Chap. 8, with inlet velocity greater than E/B. In this situation, the flow will be decelerated, and current will be driven through an external load by the motional electromotive force. Again, there are distinctions between scalar conduction and Hall-current generators, solid vs. segmented electrodes, pure vs. seeded gases, etc., but much of the analysis and technology of steady electromagnetic accelerators is transcribable to this application.

Some experience with ground-based MHD generators has been achieved, but these systems can hardly be regarded as proven competitors to the turbogenerators. Implementation in a space system would therefore seem to be at least another phase removed. A major attraction in continuing their study, however, is in the possibility that a spacecraft already equipped to handle seeded-gas channel flows in its propulsion system might nicely incorporate an MHD generator channel into its power plant package.

Each of the energy conversion systems outlined above has its own peculiar output characteristics. The turboelectrics typically yield low-frequency alternating current at a few hundred volts; photovoltaic cells and thermionic diodes yield low voltage direct current; etc. Some thought thus should be given to matching particular power supplies to particular thrusters, for the benefit of minimizing the amount of power-conditioning equipment that is needed to relate one to the other. Substantial transforming of voltage, rectification of alternating current, changing of frequency, etc., can be accomplished only at the expense of additional weight in the overall system, and a mismatched power-thruster combination may become less attractive than its separate elements otherwise appear. Whatever power conditioning must unavoidably be done must also conform to the same constraints of low specific mass and unattended long-term reliability as the thrusters and primary power units, so that here too a spectrum of challenging technological problems unfolds.

LOW THRUST MISSION ANALYSIS

Given an electric thruster of particular specific impulse, thrust, and efficiency characteristics and a suitably matched power supply of the corresponding capacity, one may then inquire what space missions are accessible to this system and how it compares with competitive systems, under given constraints on payload fraction and/or mission time. In most cases, the answer will require solution of a very difficult subordinate problem: what is the optimum mode of accomplishing the given mission with the given system? The difficulty stems essentially from the presence in the equation of motion of the spacecraft of a variable forcing term corresponding to the thrust of the rocket, which operates over large portions of the dynamical trajectory.

In analyzing interplanetary flight trajectories of conventional high thrust chemical rocket systems, it is normally an acceptable approximation to regard the thrusting period as infinitesimal compared with the total flight time. The low specific impulse and high thrust of chemical rockets predicate their consuming the bulk of their propellant at the mission outset, with possible brief reignitions for trajectory adjustment or braking at the close of the mission. The resulting impulse increments change the spacecraft momentum abruptly, after which simple heliocentric motion ensues, with possible perturbing influences from the launching and target planets.

Clearly, this approximation is inappropriate for electric propulsion systems applied to the same missions. On the contrary, the inherent high specific impulse and low thrust capabilities of these systems dictate that they accelerate the spacecraft over large segments of the flight trajectory. The dynamical statements thus must be cast as nonlinear differential equations, constrained to two end conditions, but otherwise possessing innumerable degrees of freedom corresponding to the limitless possibilities for programming the thrust, specific impulse, and corresponding thrust efficiency of the propulsion unit. The accepted method of approach to this class of problem is via the calculus of variations, or more specifically, via the Pontryagin maximum principle, or similar mathemati-

APPENDIX

cal methods, implemented iteratively on a digital computer. Beyond the evident task of formulating the analysis to converge to optimal trajectories from reasonable initial estimates is the more subtle problem of identifying the absolute, or global, optimum from among the myriad of local optima that typically obtain. Illustrative of this class of difficulty is the frequent experience that the best trajectory for a particular interplanetary transfer is not confined within the solar orbits of the two planets, but may involve excursion beyond the outer planet or inside the orbit of the inner one.

This process of optimum trajectory-thrust program identification is only the first step in the more general mission analysis problem, i.e., selection of the best combination of electric thruster and power plant for a particular mission. For example, to compare the merits of a solar-powered ion thruster with a nuclear thermionic-powered MPD arc, one must not only identify the optimum thrust program and trajectory for each, but also explore their sensitivity to the imposed constraints. It may be that a small relaxation of the requirement on flight time, say, will considerably improve the performance of one or the other, even to the point of reversing the original choice. Also, a variety of subtler interactions may now bear on the final selection. For example, if the optimum thrust program calls for wide excursions in specific impulse and thrust level during the course of the flight, the effect of such operation on the reliability, lifetime, and specific mass of the thruster should be examined. As we have seen, most electric thrusters operate more effectively in rather narrow ranges of exhaust speed; to require broader ranges of operation may introduce enough technical complication into the unit to vitiate the dynamical advantages. As a second example, if the thrust program calls for large angles of inclination of the thrust vector to the trajectory, one should examine the weight and performance penalties associated with gimbaling the thruster or of continuously controlling the attitude of the spacecraft to achieve this thrust vector control.

A further complication of practical interest arises from the possibility that a hybrid propulsion system, consisting of two (or more) quite dissimilar types of thruster, might accomplish a given mission more effectively than either, separately. This might consist of a low thrust high specific impulse ion engine combined with a higher thrust lower I_s plasma device, each coupled to the same power plant, or perhaps a combination of an electric thruster with a nuclear thermal rocket, sharing some elements of a common nuclear reactor. Clearly, a new spectrum of dynamical possibilities arises in these problems, associated with the choice of thrusting time of each engine and the utilization of coasting phases between their operating periods, which must be incorporated into an optimization of the overall hybrid system.

Electric propulsion in the general sense thus comprises a triad of engineering sciences: electrical acceleration of gases, generation of electric power in space, and low thrust space mission analysis. Advances in one area must influence the development of the others; none can proceed effectively in ignorance of the others. Indeed, only if each can be brought to a high level of sophistication in its own details and to optimal interaction with the others will electric propulsion develop to a useful means of space transportation.

REFERENCES

1. Snyder, N. W. (ed.): "Energy Conversion for Space Power," vol. 3 of "Progress in Astronautics and Aeronautics," Academic Press Inc., New York, 1961.
2. Snyder, N. W. (ed.): "Space Power Systems," vol. 4 of "Progress in Astronautics and Aeronautics," Academic Press Inc., New York, 1961.
3. Zipkin, M. A., and R. N. Edwards (eds.): "Power Systems for Space Flight," vol. 11 of "Progress in Astronautics and Aeronautics," Academic Press Inc., New York, 1963.
4. Szego, G. C., and J. E. Taylor (eds.): "Space Power Systems Engineering," vol. 16 of "Progress in Astronautics and Aeronautics," Academic Press Inc., New York, 1966.
5. Szego, G. C.: Space Power Systems, p. 415 in C. T. Leondes and R. W. Vance (eds.), "Lunar Missions and Explorations," John Wiley & Sons, Inc., New York, 1964.
6. Electrical Power Generation Systems for Space Applications, *NASA SP*-79, Washington, D.C., 1965.
7. Space Power Systems Advanced Technology Conference, *NASA SP*-131, Lewis Research Center, Cleveland, Ohio, Aug. 23-24, 1966.
8. Angrist, S. W.: "Direct Energy Conversion," Allyn and Bacon, Inc., Boston, 1965.
9. Chang, S. S. L.: "Energy Conversion," Prentice-Hall, Inc., Englewood Cliffs, N.J., 1963.
10. Kaye, J., and J. A. Welsh: "Direct Conversion of Heat to Electricity," John Wiley & Sons, Inc., New York, 1960.
11. Spring, K. H. (ed.): "Direct Generation of Electricity," Academic Press Inc., New York, 1965.
12. Sutton, G. W. (ed.): "Direct Energy Conversion," McGraw-Hill Book Company, New York, 1966.
13. Direct Energy Conversion Literature Abstracts, *U.S. Naval Res. Lab.* AD635,584, June, 1966.
14. Rappaport, P.: Photovoltaic Power, *J. Spacecraft and Rockets*, vol. 4, no. 7, p. 838, 1967.
15. Elliott, D. G.: Magnetohydrodynamic Power Systems, *J. Spacecraft and Rockets*, vol. 4, no. 7, p. 842, 1967.
16. Becker, R. A.: Thermionic Space Power Systems Review, *J. Spacecraft and Rockets*, vol. 4, no. 7, p. 847, 1967.
17. Zipkin, M. A.: Alkali Metal, Rankine Cycle Power Systems for Electric Propulsion, *J. Spacecraft and Rockets*, vol. 4, no. 7, p. 852, 1967.

18. Lawden, D.: "Optimal Trajectories for Space Navigation," Butterworth & Co. (Publishers), Ltd., London, 1963.
19. Leitmann, G.: "Optimal Control," McGraw-Hill Book Company, New York, 1966.
20. Leitmann, G. (ed.): "Optimization Techniques," Academic Press Inc., New York, 1962.
21. Pontryagin, L. S., E. F. Mischenko, V. G. Boltyanski, and R. V. Gamkrelidze: "The Mathematical Theory of Optimal Processes," Interscience Publishers, Inc., New York, 1962.

Index

Index

Index

Acceleration-deceleration concept, 173–177
Accelerator (*see specific* type or class)
Ammonia (NH$_3$), arcjet, 129, 131
    as electrothermal propellant, 100–102
    frozen flow efficiency, 100–102
    MPD arc, 239
    physical properties, 100
    resistojet, 106, 109
    sparking potential, 112
Ampere's law, 17
Anode fall, 114, 135
Anode jet, 237, 246, 251
Arc, electric, 113–118
    confined, 117–118
    instabilities, 116–117, 132
    magnetic constriction, 115

Arc, electric, potential profile, 114
    stabilization, 116–118, 133
Arc chamber, 118–119
Arcjet, 119–134
    alternating current, 128–131
    analysis, 122–126
    construction and materials, 129
    cooling, 123–124, 130
    core-flow, 121–124
    efficiency, 129
    heat transfer processes, 122–126
    losses, 92–93, 129–130
    magnetically diffused and focused, 132
    performance, 120–121, 129
    propellants, 131
    radiation loss, 97, 123
    resistojet effect, 124–125

Argon (A, Ar), atom-atom ionization cross section, 63
  dielectric constant, 14
  electrical conductivity, 233
  electron-atom elastic cross section, 51–52
  equilibrium ionization, 37, 43
  ion-atom elastic and charge-transfer cross sections, 66
  ionization potential, 30
  MPD arc, 239
Aspect ratio $\mathcal{R}$, 164–165
Atom-atom collisions (see Collisions, atomic)
Atomic structure, 29
Attitude control of satellites, 9
Autocathode, 155, 172

Back emf, 215
Beam (see Ion beam)
Beryllium (Be), 100–102
Boltzmann constant k, 35
Boltzmann factor, 35
Bombardment ion source (see Electron bombardment thruster; Ion sources)
Boron (B), 100–102
Boundary layers in electromagnetic accelerators, 215
Brayton cycle, 320
Breakdown profiles for gaseous discharges, 111–112
Bremsstrahlung, 58–59, 65
Button gun, 10, 259–260

Calorimetric efficiency, 291
Capacitance C, 16
Carbon (C), 100–102
Cathode contamination in electron bombardment ion source, 155–156
Cathode erosion in electron bombardment ion source, 155–156
Cathode fall, 112–114, 135, 156
Cathode jet, 237, 246, 251–252
Cesiated cathode, 156

Cesium (Cs), charge-to-mass ratio, 147
  contact ionization source, 149, 151–153, 157
  electron-atom elastic cross section, 51
  electron bombardment ionization source, 156–157
  energy level diagram, 28
  ion thruster, 157
  ionization potential, 30
  seeded gases, 232–234
Characteristic impedance, 300–301
Characteristic velocities for electromagnetic acceleration, ξ, η, ζ, 207–208
Characteristic velocity increments Δv for planetary transfer missions, 4
Charge-exchange reactions, 32
Charge-to-mass ratio, 145–147
  of cesium ions, 147
  of colloid particles, 147, 188–191
  of mercury, 153
Charge-transfer trapping, 32
Charged-particle motion, effect of collisions on, 74–79
  in Hall accelerators, 225–226
  in MPD arc, 247–252
  for various Hall parameters and gyro radii, 76–79
Chemical rockets, 4
Chemi-ionization, 32
Child's law, 146
"Choking" in electromagnetic acceleration, 208–209
Clausius-Mosotti relation, 14
Coaxial gun, 259–260, 288–293
Coaxial Hall accelerator, 224–226
Collision frequency ν, 50, 76, 80, 219
  effective, 75
Collisions, atomic, 46–67, 96
  atom-atom, 61–64
    charge-reactive, 64
    deexcitation, 64
    elastic, 61–63
    excitation, 63
    inelastic, 63
    ionization, 63

INDEX

Collisions, atomic, atom-atom,
    radiative, 64
    resonant exchange, 64
    superelastic, 64
  charge-reactive, 46
  cross section, 48
    coulomb, 56–57, 220
    differential, 48–49
    effective, 48
    energy transfer, 56
    momentum transfer, 56–57, 75
    total, 49
  elastic, 46
  electron-atom, 50–55
    elastic, 50–51
    electron attachment, 55
    excitation, 51–53
    inelastic, 51–55
    ionization, 51–54
    radiative, 55
    superelastic, 54–55
  electron-electron, 60–61
    elastic, 60–61
    radiative, 61
  electron-ion, 55–60
    elastic, 56–57
    inelastic, 58
    radiative, 58–59
    superelastic, 58
  energy exchange in, 46–47
  with impurities, 64
  inelastic, 46
  ion-atom, 64–67
    charge exchange, 65–67
    deexcitation, 65
    elastic, 65–66
    inelastic, 65
    radiative, 65
    resonant exchange, 65
    superelastic, 65
  ion-ion, 60–61
    elastic, 60–61
    inelastic, 61
    radiative, 61
    superelastic, 61
  radiative, 46
  superelastic, 46

Collisions, atomic, three body, 46
Colloid accleration, 147, 188–191
Colloid particles, 188–191
  limiting charge-to-mass ratio, 190–191
  preparation, 188–190
    electric spraying, 188–189
    ion nucleation, 189
    preformed, 190
    surface condensation, 189
    vapor condensation, 189
  size, 188
Conductivity, electrical σ, 17, 69–83
  of collisionless ionized gas, 73–74
  complex, 80
  dimensionless, 79
  high-frequency, 79
  effect of collisions on, 74, 80
  of seeded gases, 232–234
  in steady magnetic fields, 80
  tensor, 76–80, 82–83, 219–226
Conical theta pinch, 305
Constant field accelerator, 206–209
Constricted arcs, 117–118
Contact-heated resistojet, 103, 107–108
Contact ion thruster, 178, 181–183
  (*See also* Ion acceleration; Ion sources)
Contact ionization, 149–153, 157
Core-flow in arcjet, 121–124
Corona discharge, 111
Coulomb's law, 12
Cross section (*see* Collisions, atomic)
Crossed-field accelerator, 202–219, 235–236
  adiabatic, 206
  constant field, 206–209
  electrode erosion, 215–216
  elementary, 197
  field geometry, 216–218
  isothermal, 203, 205
  losses, 214–216
  optimized, 210–211
  performance limit, 212–213
  variable channel, 211–212, 216–218

Crowbar breakdowns in pulsed plasma acceleration, 268, 271, 274
Current, electric, J, 16–18
Current density j, 16, 69, 76, 79–80, 82
  distribution, in coaxial gun, 291, 292
  in crossed-field accelerator, 227
  in exhaust plume, 296–297
  in pinch accelerator, 293–295
Current sheet, 272–273, 278–281, 288–289, 291–295, 299–300
  initiation, 284–287
  luminosity, 292, 294, 299
  profiles, 285–286
  trajectories, for slug model, 272
  for snowplow model, 274
Current waveforms, for LCR circuits, 267
  for slug model, 272
  for snowplow model, 274
  from transmission line source, 30
Cyclotron-resonance accelerator, 137–138

de Broglie wavelength, 62
Debye length λ_D, 57
Decelerating electrode, 173–177
Definition of electric propulsion, 6
Detailed physical balancing, 35
Dielectric constant κ, 13–14
  for various gases, 14
Dielectronic recombination, 34
Displacement current, 22
Dissociation, 96–98
Dissociative recombination, 34
Drift velocity $\bar{v}$, 71–74
Duoplasmatron, 158
Dynamic impedance of pulsed plasma acceleration, 271
Dynamical efficiency η_d, 280–284, 289
Dynamical pinch (see Pinch accelerator)

Effective collision frequency, 75
Effective electric field, 25
Efficiency η, calorimetric, 291
Efficiency η, dynamical, 280–284, 289
  electrical conversion, 7
  frozen flow, 99–101
  ionization, 150–152, 155–158
  thrust, 290
  (See also specific type or class)
Electric charges, 12
Electric displacement field D, 15
Electric field E, 13, 15, 20, 22
  boundary conditions, 16
  induced, 25
  time-varying, 21
Electric potential V, 15
Electric propulsion, definition, 6
Electrical conductivity (see Conductivity, electrical)
Electrical conversion efficiency η, 7
Electrode erosion, in pulsed electromagnetic accelerators, 286–287, 304–306
  in steady electromagnetic accelerators, 215–216
Electrodeless discharge acceleration, 134–139
  E-type, 135
  H-type, 135
Electromagnetic acceleration, 196–313
  classification, 197–199
  crossed-field, 202–219
  Hall current, 133, 222–227, 236
  inductive, 304–313
  magnetoplasmadynamic arc, 133, 235–253, 257
  pulsed, 259–306
  self-field, 227–232
  traveling-wave, 306–312
  unsteady, 257–313
  [See also Crossed-field accelerator; Hall current accelerators; Magnetoplasmadynamic (MPD) arc; Pulsed inductive acceleration; Pulsed plasma acceleration; Traveling-wave acceleration]
Electromagnetic "blowing," 240–245, 250
Electromagnetic body forces, 23, 200

INDEX

Electromagnetic fields, definitions, 13–20
  displacement **D**, 15
  electric **E**, 13, 20
  magnetic dipole moment **m**, 19
  magnetic **H**, 19–20
  magnetic induction **B**, 18
  polarization **P**, 13
Electromagnetic force, 12–13, 17–18, 20, 23
  on moving charge, 25
Electromagnetic propulsion, definition, 6, 196
Electromagnetic "pumping," 240, 243–244, 246, 250
Electromagnetic "swirling," 240, 252
Electromagnetic theory, 12–26
Electromagnetic waves, 22–23
  angular frequency, 23
  attenuation, 23
  in conducting media, 23
  in dielectric media, 23
  equation, 22, 83
  in ionized gas, 83–85
  propagation exponent, 23, 83–85
Electromotive force (emf) $\mathcal{U}$, 20–21
Electron-atom collisions (*see* Collisions, atomic)
Electron attachment, 34
Electron bombardment ionization, 149, 153–157, 163–164
Electron bombardment thruster, 154, 178–180, 183, 185
  (*See also* Ion acceleration; Ion sources)
Electron cyclotron resonance, 137
Electron injection (*see* Neutralization of ion beam)
Electron and ion trajectories (*see* Charged-particle motion)
Electron pressure, 261
Electron volt (ev), 30
Electronic energy levels, 29
  diagrams for helium and cesium, 28
  truncation of, 42
Electronic excitation, 29, 95, 97
Electronic orbits, 29

Electrostatic acceleration, 142–191
  (*See also* Colloid acceleration; Ion acceleration)
Electrostatic force, 13, 146–147
Electrostatic propulsion, definition, 6
Electrostatics, 12–16
Electrothermal acceleration, 90–139
  efficiency, 93, 106, 129
  exhaust velocity, 91–92, 99–101, 106, 109, 120, 129
  losses, 92–93, 104–105, 122–124
  one-dimensional model, 90–93
  performance, 91–92, 106, 109, 121, 129
  propellants, 99–103, 106
  specific impulse, 91–92, 99–101, 106, 109, 120–121, 129
  (*See also* Arcjet; Electrodeless acceleration; Resistojet)
Electrothermal propulsion, definition, 6, 90
Energy transfer cross section $Q^{(\varepsilon)}$, 56, 136
Enthalpy h, 94–103
  of high-temperature gases, 93
  of various propellants, 101
Equilibrium constant, 35
  for ionization, 35–36
    of gas mixture, 39
  for molecular ionization, 38
Escape velocity, 4
Exhaust plume of pulsed plasma accelerator, 287–288, 297
Exhaust velocity, 3
  of chemical rockets, 5
  of nuclear rockets, 5
  (*See also* Exhaust plume; Specific impulse I_s; *and specific* type or class)

Faraday's law, 20–21
Ferry missions, 9
Field fringing in electromagnetic accelerators, 218
Flight mechanics, 1–4, 7–9, 322–324

Flight test, of ion thruster, 184–188
  of resistojet, 109
Focusing of ion beam, 161–165
Free-bound electronic transitions, 32, 59–60
Free-free electronic transitions, 58–59
Fringe-field accelerator, 312–313
Frozen flow, 5, 93, 98–102
  efficiency, 99–101
    of various propellants, 101
  fraction, 99

Gaseous discharges, 110–118
Generalized Ohm's law, 17
Gladstone-Dale relation, 14
Glow-arc transition, 112, 285
Glow discharge, 111–113, 134
Goddard, Robert H., 9
Grotrian diagram, 29
Gyro frequency ω_B, 70, 76, 137, 219–220
Gyro radius r_B, 70
  in magnetoplasmadynamic arc, 248–249

Hall current, 76–79, 133, 201–202, 219, 223
Hall current accelerators, 133, 222–227, 236
  classification, 223–224
  ion slip limitation, 226–227
Hall effect, 219
Hall parameter, 76–79, 219–221
  in magnetoplasmadynamic arc, 248–249
Heat pipe, 321
Heavy-particle accelerators, 188–191
Helium (He), atom-atom ionization cross section, 63
  dielectric constant, 14
  electron-atom elastic cross section, 51
  electron-atom excitation cross sections, 53

Helium (He), as electrothermal propellant, 100–102
  energy level diagram, 28
  frozen flow efficiency, 100–102
  ion-atom elastic and charge-transfer cross sections, 66
  ionization potential, 30
  physical properties, 100
High-impulse arcjet (see Magnetoplasmadynamic arc)
Hybrid accelerators, 133–134
Hybrid propulsion system, 323
Hydrazine (N_2H_4), 100–102, 132
Hydrogen (H, H_2), arcjet, 129
  dielectric constant (H_2), 14
  dissociation, 96–99
  electron-atom elastic cross section (H_2), 51
  electron-atom ionization cross section (H, H_2), 54
  as electrothermal propellant, 100–102
  enthalpy, 91, 98–101
  frozen flow efficiency, 98–102
  ionization potential (H, H_2), 30
  MPD arc, 239
  physical properties, 100
  resistojet, 106
  sparking potential, 112
  specific heat, 91, 98–101

Impulsive thrust, 3
Induction accelerators (see Pulsed inductive acceleration; Traveling-wave acceleration)
Internal degrees of freedom, 94
Internal energy of high temperature gases, 95–103
Ion acceleration, 142–188
  beam power, 181
  efficiency, 148, 153, 155–158, 182–183
  electrode design, 159–165
  exhaust velocity, 146, 175, 181–183
  limiting electric field, 146, 175
  minimum electrode spacing, 146, 175

INDEX

Ion acceleration, one-dimensional, 143–148
  performance, 181–188
  power density, 146, 175
  sources (*see* Ion sources)
  specific impulse, 146, 175, 181–183
  specific mass, 184
  thrust, 146, 175, 181
  (*See also* Ion sources)
Ion beam, focusing, 161–165
  neutralization, 157–158, 165–173, 184–187
  power, 146
  spreading, 161–165, 168
Ion conduction sheath, 77
Ion current, 76, 83
Ion-electron recombination, 98
Ion engine (*see* Contact ion thruster; Electron bombardment thruster; Ion acceleration)
Ion gun (*see* Ion sources)
Ion gyro resonance, 137
Ion Hall parameter, 227
  in magnetoplasmadynamic arc, 248–249
Ion motor (*see* Contact ion thruster; Electron bombardment thruster; Ion acceleration)
Ion slip, 201–202, 226–227
Ion sources, 148–159
  arc, 158
  cesium-tungsten contact, 149–153, 157, 163
  colloid (*see* Colloid particles)
  current density, 148, 150, 157–158, 175
  duoplasmatron, 158
  electron bombardment, 149, 153–157, 163–164
  energy expenditure per ion, 148, 153, 155–159
  neutral fraction, 148, 152, 155–159
  oscillating electron, 156–157
  porous-plug, 151–153
  radio-frequency discharge, 158
  reliability, 149, 152, 157–158

Ion thruster (*see* Contact ion thruster; Electron bombardment thruster; Ion acceleration)
Ionization, 27–43
  of air, 39
  of argon, 37
  degree of, 36
  effect on enthalpy, 96–98
  efficiency, 150–152, 155–158
  equilibrium, 34–43
  of gas mixture, 39–43
  limit, 29
  molecular, 32, 38
  nonequilibrium, 42, 234–235
  potential, 36
    lowering of, 42
    second, 39
    of various gases, 30
  of seeded gases, 39
  (*See also* Collisions, atomic)
Ionization, second, 39
Isothermal sound speed, a_T, 203

$\mathbf{j} \times \mathbf{B}$ force, 18, 20, 23, 197, 200
Jupiter round trip, 4

Kink instabilities, 117

Lennard-Jones interatomic potential, 62
Limiting velocity, 207, 209
Linear pinch (*see* Pinch accelerator)
Liquid mercury cathode, 155
Lithium (Li), 100–102, 132
Loop inductor, 305
Lorentz accelerator (*see* Crossed-field accelerator)
Lorentz force, 18, 20, 23, 197, 200
Luminous fronts (*see* Current sheet, luminosity)

Macromolecules, 188
Magnetic body force, 18, 20, 23, 197, 200

Magnetic confinement to reduce electrode erosion, 235, 248, 252–253
Magnetic dipole moment **m**, 19
Magnetic expansion thruster, 134, 261
Magnetic field **H**, 19–20, 22
Magnetic flux ϕ, 20
Magnetic induction field **B**, 18
  from distributed current density, 18
  time-varying, 20
Magnetic interaction parameter β, 205, 210, 212–213, 253
Magnetic nozzle, 133–134, 137, 252–253, 263, 297
Magnetic permeability μ, 18
Magnetic piston, 275
Magnetic pressure, 115–116, 229, 261, 286
Magnetic Reynolds number R_B, 228
Magnetic stress tensor $\mathcal{B}$, 244–246
Magnetogasdynamic channel flow, 198–214
Magnetogasdynamic equations, 200
Magnetogasdynamic model of MPD arc, 241
Magnetohydrodynamic power generator, 321
Magnetoplasmadynamic (MPD) arc, 133, 235–253, 257
  collisional model, 248–252
  electrothermal contribution, 252
  magnetogasdynamic model, 240–246
  mass flow, 246–248
  particle model, 246–248
  performance, 236, 238–239, 246–247, 253
  zero mass flow operation, 237, 246, 248
Mars round trip, 4
Marshall gun, 289
  (*See also* Coaxial gun)
Mass-to-charge ratio (*see* Charge-to-mass ratio)
Mass ratio, payload, 3
Maxwell paradox, 21
Maxwell's equations, 22
Mean free path λ, 50

Mercury (Hg), atomic weight, 153
  autocathode, 155
  electron-atom elastic cross section, 51–52
  electron bombardment ion source, 153–157
  ion thruster, 157
  (*See also* Electron bombardment thruster)
  sparking potential, 112
Mercury round trip, 4
Microthruster, 106, 109, 181
Microwave-heated accelerators, 136–138
Migration velocity $\bar{v}$, 69–83
Mission analysis, 1–4, 7–9, 322–324
Molecular ions, 34
Molecular rotation, 95–97
Molecular susceptibility p_m, 19
Molecular vibration, 95–97
Molecular weight of various electrothermal propellants, 100
Momentum-transfer cross section, $Q^{(p)}$, 56, 75, 219–220
MPD arc [*see* Magnetoplasmadynamic (MPD) arc]
Multibeam ion accelerator, 163, 168
Mutual inductance, M_{12}, 21

Negative-ion recombination, 34
Neon (Ne), 30, 66, 112
Nernst reaction isobar, 35
Neutral fraction, 150–152, 155–158
Neutralization of ion beam, 157–158, 165–173, 184–187
Nitrogen (N, N_2), dielectric constant (N_2), 14
  electron-atom elastic cross section (N_2), 51
  as electrothermal propellant, 100–101
  ionization potential (N, N_2), 30
  physical properties, 100
Nonequilibrium ionization, 234–235
Nuclear power supplies, 319
Nuclear-thermal rocket, 5, 323

INDEX

Oberth, Herman, 9
Ohm's law, 17, 25, 201, 225–227, 229, 231
Optimization, of electromagnetic acceleration profile, 210–212
  of propulsion system, 322–324
  of specific impulse, 7–9
Orbit adjustment of satellites, 9
Oscillating-electron ion source, 156–157
Oxide-coated cathode, 156
Oxygen (O, O_2), 14, 30

Parallel-plate accelerator, 260, 263–266, 288
Partition functions, 35–41
  internal f^i, 35
  total F, 35–36
  translational f^t, 35
Paschen's law, 111, 301–302
Payload mass ratio, 3
Penning discharge, 156–159
Pentaborane (B_5H_9), 100, 102
Perveance $\mathcal{P}$, 164–165, 171
Phase velocity, of electromagnetic waves c_o, 23
  in traveling-wave acceleration $\mathbf{v}_p$, 308–310
Photoelectric emission, 32
Photoionization, 32
Physical properties of various electrothermal propellants, 100
Pierce gun, 160–161, 176
Pinch accelerator, 259–260, 283–284, 288–289, 293–300
Pinch effect, 115–117
Pinch engine, 295
Planck's constant h, 30, 35
Plasma accelerator (see Electromagnetic acceleration)
Plasma bottle, 171
Plasma bridge, 172
Plasma diode, 321
Plasma frequency ω_p, 79
Plasma jet (see Arcjet)
Plasma thruster (see Electromagnetic acceleration)

Plenum chamber for arcjet, 119
Poisson's relation, 15, 145
Polarization field $\mathbf{P}$, 13–15
  in current sheet, 278–280
Pontryagin maximum principle, 322
Porous tungsten ionizers, 151–153
Positive column, 112, 114
Potassium (K), 30, 51, 232–234
Potential, electric, V, 15
Power-conditioning, 321
Power supplies, 317–322
  high-frequency, 139
  magnetohydrodynamic, 321
  mass, 7–8
  nuclear, 319–321
  solar, 318–319
Propagation exponent k, 23, 83–85
Propellant (see specific type or class)
Propellant mass fraction, 3
Pulse-forming network, 300
  (See also Transmission line)
Pulsed electromagnetic acceleration (see Pulsed plasma acceleration)
Pulsed inductive acceleration, 304–306
Pulsed plasma acceleration, 259–304
  calorimetric efficiency, 291
  capacitance criteria, 266–269, 297–298
  circuit analyses, 263
  coaxial gun, 259–260, 288–293
  current sheet (see Current sheet)
  diagnostics, 291–297
  dynamical efficiency, 280–284, 289, 300
  dynamical models, 269–284
  ejection process, 262–263, 287–288, 297
  electrical efficiency, 265
  electrode erosion and aging, 286–287, 304–306
  electrothermal effects, 284
  gas injection, 261, 302
  gas-kinetic models, 277–280
  gasdynamic models, 275–277
  impedance matching, 274–284, 300–301
  inductance criteria, 265, 286, 298

Pulsed plasma acceleration, initiation, 284–287
  mass distribution effect, 280–284
  multifluid models, 276–277
  operation cycle, 261–262
  parallel plate accelerator, 260, 263–266, 288
  pinch accelerator, 259–260, 283–284, 288, 289, 293–300
  pinch engine, 295
  quasi-steady, 302–304
  resistance criteria, 266
  secondary discharges, 268, 271, 274, 303–304
  shock wave model, 275–276, 280
  skin effect, 258, 260, 262, 268, 285
  slug model, 269–272, 280
  snowplow model, 272–275, 280–284
  switching, 262, 301–302
  thrust efficiency, 289–291
  by transmission line source, 298–301, 311–312
  variable channel, 283–284
  voltage criteria, 266
Pulsed resistojet, 103, 109

Quanta, 29
Quasi-steady pulsed plasma acceleration, 302–304

Radiation-cooling of arcjet, 123–124
Radiation energy loss, 97, 123
Radiative recombination, 32
Rail accelerator (see Parallel-plate accelerator)
Ramsauer effect, 51
Rankine cycle, 320
Recombination, electron-ion, 32–33, 58–59, 64, 97
Regenerative cooling, of arcjet, 124–125, 128, 130
  of resistojet, 103, 107
Resistojet, 90, 103–110
  chamber pressure, 106
  efficiency, 106

Resistojet, heater configurations, 103–104, 106
  insulation, 104–105
  losses, 92–93, 103–104
  materials, 105
  missions, 92
  performance, 106, 109
Rocket equation, 1
Rubidium (Rb), 232–234
Rutherford scattering relation, 57

Saha relation, 37–43, 96–99, 234
  application to nonequilibrium gases, 43
  approximate, 37
  limitations, 42
Sastrugi ion source, 163
Saturn round trip, 4
Sausage instabilities, 117
Seeded gases, 39, 232–234
Segmented electrodes, 221–222
Self-field accelerators, 227–232
Self inductance, 21
SERT I, 10, 179, 184–188
SERT II, 319
Sheet accelerator, 230–231
Shock wave model, 275–276, 280
Skin effect, 268, 285
  in pulsed accelerators, 258, 260, 262
Slip velocity, v_s, 308–310
Slug model, 199, 269–272, 280
SNAP-50, 320
Snowplow model, 258, 272–275, 280–284
Solar cell panels, 318–319
Space-charge ion flow, 143–148
Space missions, 1–4, 7–9, 322–324
Space test, of ion thruster, 184–188
  of resistojet, 109
  of solar-electric system, 319
Sparking potential, 111–112
Specific charge (see Charge-to-mass ratio)
Specific heat c_p, of high temperature gases, 93–103
  of various electrothermal propellants, 100

Specific impulse I_s, 4
  chemical rocket, 5
  nuclear rocket, 5
  optimum, 8
  variable, 174, 322–323
  (*See also performance of specific* type or class)
Specific mass of ion thruster system, 184
Specific power plant mass, 8
Spoke instability, 133
SPUR, 320
Station keeping of satellites, 9
Statistical mechanics applied to ionization, 35
Steady electromagnetic acceleration, 202–214
  (*See also* Electromagnetic acceleration)
Stuhlinger, Ernst, 9
Swarm velocity, 69–83
Switch for pulsed plasma acceleration, 301–302

T tube, 10, 259–260
Temperature profiles in arcjet block, 124
Thermionic converter, 321
Thermionic emission, 32
Thermo-ionic accelerator (*see* Magnetoplasmadynamic arc)
Theta pinch, 305–306
  conical, 305
Three-body recombination, 34
Thrust efficiency η, 290
Thrust **T**, 3
  (*See also performance of specific* type or class)

Thruster, spelling of, 6
  (*See also specific* type or class)
Total impulse I, 3
Townsend discharge, 110, 112
Transmission line, power source for pulsed plasma acceleration, 298–301, 311–312
  for traveling-wave accelerator, 311–312
Traveling-wave acceleration, 198–199, 258–259, 306–312
  cylindrical channel, 311
  efficiency, 310–311
  performance, 311–313
  rectangular channel, 307–311
  transmission line type, 311–312
Tungsten (W), 149, 151–153
Two-stream instabilities in ion beam neutralization, 169–171

Vela satellite, 109
Velocity increment $\Delta \mathbf{v}$, 3
Velocity profiles along electromagnetic accelerator channel, 204–212
Venus round trip, 4
Virtual deceleration electrode, 176–177
Voltage-current characteristic for gaseous discharge, 110–114, 285
von Ardenne arc, 158
Vortex stabilized arc, 117–119

Wave vector **k**, 83
Wavelength λ and wave number k of electromagnetic waves, 23
Work function ϕ, 149–150
  lowering by surface contamination, 150

A CATALOG OF SELECTED
DOVER BOOKS
IN SCIENCE AND MATHEMATICS

CATALOG OF DOVER BOOKS

Physics

OPTICAL RESONANCE AND TWO-LEVEL ATOMS, L. Allen and J. H. Eberly. Clear, comprehensive introduction to basic principles behind all quantum optical resonance phenomena. 53 illustrations. Preface. Index. 256pp. 5⅜ x 8½.
0-486-65533-4

QUANTUM THEORY, David Bohm. This advanced undergraduate-level text presents the quantum theory in terms of qualitative and imaginative concepts, followed by specific applications worked out in mathematical detail. Preface. Index. 655pp. 5⅜ x 8½. 0-486-65969-0

ATOMIC PHYSICS (8th EDITION), Max Born. Nobel laureate's lucid treatment of kinetic theory of gases, elementary particles, nuclear atom, wave-corpuscles, atomic structure and spectral lines, much more. Over 40 appendices, bibliography. 495pp. 5⅜ x 8½. 0-486-65984-4

A SOPHISTICATE'S PRIMER OF RELATIVITY, P. W. Bridgman. Geared toward readers already acquainted with special relativity, this book transcends the view of theory as a working tool to answer natural questions: What is a frame of reference? What is a "law of nature"? What is the role of the "observer"? Extensive treatment, written in terms accessible to those without a scientific background. 1983 ed. xlviii+172pp. 5⅜ x 8½. 0-486-42549-5

AN INTRODUCTION TO HAMILTONIAN OPTICS, H. A. Buchdahl. Detailed account of the Hamiltonian treatment of aberration theory in geometrical optics. Many classes of optical systems defined in terms of the symmetries they possess. Problems with detailed solutions. 1970 edition. xv + 360pp. 5⅜ x 8½. 0-486-67597-1

PRIMER OF QUANTUM MECHANICS, Marvin Chester. Introductory text examines the classical quantum bead on a track: its state and representations; operator eigenvalues; harmonic oscillator and bound bead in a symmetric force field; and bead in a spherical shell. Other topics include spin, matrices, and the structure of quantum mechanics; the simplest atom; indistinguishable particles; and stationary-state perturbation theory. 1992 ed. xiv+314pp. 6⅛ x 9¼. 0-486-42878-8

LECTURES ON QUANTUM MECHANICS, Paul A. M. Dirac. Four concise, brilliant lectures on mathematical methods in quantum mechanics from Nobel Prize-winning quantum pioneer build on idea of visualizing quantum theory through the use of classical mechanics. 96pp. 5⅜ x 8½. 0-486-41713-1

THIRTY YEARS THAT SHOOK PHYSICS: THE STORY OF QUANTUM THEORY, George Gamow. Lucid, accessible introduction to influential theory of energy and matter. Careful explanations of Dirac's anti-particles, Bohr's model of the atom, much more. 12 plates. Numerous drawings. 240pp. 5⅜ x 8½.
0-486-24895-X

ELECTRONIC STRUCTURE AND THE PROPERTIES OF SOLIDS: THE PHYSICS OF THE CHEMICAL BOND, Walter A. Harrison. Innovative text offers basic understanding of the electronic structure of covalent and ionic solids, simple metals, transition metals and their compounds. Problems. 1980 edition. 582pp. 6⅛ x 9¼. 0-486-66021-4

CATALOG OF DOVER BOOKS

HYDRODYNAMIC AND HYDROMAGNETIC STABILITY, S. Chandrasekhar. Lucid examination of the Rayleigh-Benard problem; clear coverage of the theory of instabilities causing convection. 704pp. 5⅜ x 8¼. 0-486-64071-X

INVESTIGATIONS ON THE THEORY OF THE BROWNIAN MOVEMENT, Albert Einstein. Five papers (1905–8) investigating dynamics of Brownian motion and evolving elementary theory. Notes by R. Fürth. 122pp. 5⅜ x 8½. 0-486-60304-0

THE PHYSICS OF WAVES, William C. Elmore and Mark A. Heald. Unique overview of classical wave theory. Acoustics, optics, electromagnetic radiation, more. Ideal as classroom text or for self-study. Problems. 477pp. 5⅜ x 8½. 0-486-64926-1

GRAVITY, George Gamow. Distinguished physicist and teacher takes reader-friendly look at three scientists whose work unlocked many of the mysteries behind the laws of physics: Galileo, Newton, and Einstein. Most of the book focuses on Newton's ideas, with a concluding chapter on post-Einsteinian speculations concerning the relationship between gravity and other physical phenomena. 160pp. 5⅜ x 8½.
0-486-42563-0

PHYSICAL PRINCIPLES OF THE QUANTUM THEORY, Werner Heisenberg. Nobel Laureate discusses quantum theory, uncertainty, wave mechanics, work of Dirac, Schroedinger, Compton, Wilson, Einstein, etc. 184pp. 5⅜ x 8½.
0-486-60113-7

ATOMIC SPECTRA AND ATOMIC STRUCTURE, Gerhard Herzberg. One of best introductions; especially for specialist in other fields. Treatment is physical rather than mathematical. 80 illustrations. 257pp. 5⅜ x 8½. 0-486-60115-3

AN INTRODUCTION TO STATISTICAL THERMODYNAMICS, Terrell L. Hill. Excellent basic text offers wide-ranging coverage of quantum statistical mechanics, systems of interacting molecules, quantum statistics, more. 523pp. 5⅜ x 8½.
0-486-65242-4

THEORETICAL PHYSICS, Georg Joos, with Ira M. Freeman. Classic overview covers essential math, mechanics, electromagnetic theory, thermodynamics, quantum mechanics, nuclear physics, other topics. First paperback edition. xxiii + 885pp. 5⅜ x 8½. 0-486-65227-0

PROBLEMS AND SOLUTIONS IN QUANTUM CHEMISTRY AND PHYSICS, Charles S. Johnson, Jr. and Lee G. Pedersen. Unusually varied problems, detailed solutions in coverage of quantum mechanics, wave mechanics, angular momentum, molecular spectroscopy, more. 280 problems plus 139 supplementary exercises. 430pp. 6½ x 9¼. 0-486-65236-X

THEORETICAL SOLID STATE PHYSICS, Vol. 1: Perfect Lattices in Equilibrium; Vol. II: Non-Equilibrium and Disorder, William Jones and Norman H. March. Monumental reference work covers fundamental theory of equilibrium properties of perfect crystalline solids, non-equilibrium properties, defects and disordered systems. Appendices. Problems. Preface. Diagrams. Index. Bibliography. Total of 1,301pp. 5⅜ x 8½. Two volumes.   Vol. I: 0-486-65015-4   Vol. II: 0-486-65016-2

WHAT IS RELATIVITY? L. D. Landau and G. B. Rumer. Written by a Nobel Prize physicist and his distinguished colleague, this compelling book explains the special theory of relativity to readers with no scientific background, using such familiar objects as trains, rulers, and clocks. 1960 ed. vi+72pp. 5⅜ x 8½. 0-486-42806-0

Mathematics

FUNCTIONAL ANALYSIS (Second Corrected Edition), George Bachman and Lawrence Narici. Excellent treatment of subject geared toward students with background in linear algebra, advanced calculus, physics and engineering. Text covers introduction to inner-product spaces, normed, metric spaces, and topological spaces; complete orthonormal sets, the Hahn-Banach Theorem and its consequences, and many other related subjects. 1966 ed. 544pp. 6⅛ x 9¼. 0-486-40251-7

ASYMPTOTIC EXPANSIONS OF INTEGRALS, Norman Bleistein & Richard A. Handelsman. Best introduction to important field with applications in a variety of scientific disciplines. New preface. Problems. Diagrams. Tables. Bibliography. Index. 448pp. 5⅜ x 8½. 0-486-65082-0

VECTOR AND TENSOR ANALYSIS WITH APPLICATIONS, A. I. Borisenko and I. E. Tarapov. Concise introduction. Worked-out problems, solutions, exercises. 257pp. 5⅝ x 8¼. 0-486-63833-2

AN INTRODUCTION TO ORDINARY DIFFERENTIAL EQUATIONS, Earl A. Coddington. A thorough and systematic first course in elementary differential equations for undergraduates in mathematics and science, with many exercises and problems (with answers). Index. 304pp. 5⅜ x 8½. 0-486-65942-9

FOURIER SERIES AND ORTHOGONAL FUNCTIONS, Harry F. Davis. An incisive text combining theory and practical example to introduce Fourier series, orthogonal functions and applications of the Fourier method to boundary-value problems. 570 exercises. Answers and notes. 416pp. 5⅜ x 8½. 0-486-65973-9

COMPUTABILITY AND UNSOLVABILITY, Martin Davis. Classic graduate-level introduction to theory of computability, usually referred to as theory of recurrent functions. New preface and appendix. 288pp. 5⅜ x 8½. 0-486-61471-9

ASYMPTOTIC METHODS IN ANALYSIS, N. G. de Bruijn. An inexpensive, comprehensive guide to asymptotic methods–the pioneering work that teaches by explaining worked examples in detail. Index. 224pp. 5⅜ x 8½ 0-486-64221-6

APPLIED COMPLEX VARIABLES, John W. Dettman. Step-by-step coverage of fundamentals of analytic function theory–plus lucid exposition of five important applications: Potential Theory; Ordinary Differential Equations; Fourier Transforms; Laplace Transforms; Asymptotic Expansions. 66 figures. Exercises at chapter ends. 512pp. 5⅜ x 8½. 0-486-64670-X

INTRODUCTION TO LINEAR ALGEBRA AND DIFFERENTIAL EQUATIONS, John W. Dettman. Excellent text covers complex numbers, determinants, orthonormal bases, Laplace transforms, much more. Exercises with solutions. Undergraduate level. 416pp. 5⅜ x 8½. 0-486-65191-6

RIEMANN'S ZETA FUNCTION, H. M. Edwards. Superb, high-level study of landmark 1859 publication entitled "On the Number of Primes Less Than a Given Magnitude" traces developments in mathematical theory that it inspired. xiv+315pp. 5⅜ x 8½. 0-486-41740-9

CATALOG OF DOVER BOOKS

TENSOR CALCULUS, J.L. Synge and A. Schild. Widely used introductory text covers spaces and tensors, basic operations in Riemannian space, non-Riemannian spaces, etc. 324pp. 5⅜ x 8¼. 0-486-63612-7

ORDINARY DIFFERENTIAL EQUATIONS, Morris Tenenbaum and Harry Pollard. Exhaustive survey of ordinary differential equations for undergraduates in mathematics, engineering, science. Thorough analysis of theorems. Diagrams. Bibliography. Index. 818pp. 5⅜ x 8½. 0-486-64940-7

INTEGRAL EQUATIONS, F. G. Tricomi. Authoritative, well-written treatment of extremely useful mathematical tool with wide applications. Volterra Equations, Fredholm Equations, much more. Advanced undergraduate to graduate level. Exercises. Bibliography. 238pp. 5⅜ x 8½. 0-486-64828-1

FOURIER SERIES, Georgi P. Tolstov. Translated by Richard A. Silverman. A valuable addition to the literature on the subject, moving clearly from subject to subject and theorem to theorem. 107 problems, answers. 336pp. 5⅜ x 8½. 0-486-63317-9

INTRODUCTION TO MATHEMATICAL THINKING, Friedrich Waismann. Examinations of arithmetic, geometry, and theory of integers; rational and natural numbers; complete induction; limit and point of accumulation; remarkable curves; complex and hypercomplex numbers, more. 1959 ed. 27 figures. xii+260pp. 5⅜ x 8½.
0-486-63317-9

POPULAR LECTURES ON MATHEMATICAL LOGIC, Hao Wang. Noted logician's lucid treatment of historical developments, set theory, model theory, recursion theory and constructivism, proof theory, more. 3 appendixes. Bibliography. 1981 edition. ix + 283pp. 5⅜ x 8½. 0-486-67632-3

CALCULUS OF VARIATIONS, Robert Weinstock. Basic introduction covering isoperimetric problems, theory of elasticity, quantum mechanics, electrostatics, etc. Exercises throughout. 326pp. 5⅜ x 8½. 0-486-63069-2

THE CONTINUUM: A CRITICAL EXAMINATION OF THE FOUNDATION OF ANALYSIS, Hermann Weyl. Classic of 20th-century foundational research deals with the conceptual problem posed by the continuum. 156pp. 5⅜ x 8½.
0-486-67982-9

CHALLENGING MATHEMATICAL PROBLEMS WITH ELEMENTARY SOLUTIONS, A. M. Yaglom and I. M. Yaglom. Over 170 challenging problems on probability theory, combinatorial analysis, points and lines, topology, convex polygons, many other topics. Solutions. Total of 445pp. 5⅜ x 8½. Two-vol. set.
Vol. I: 0-486-65536-9 Vol. II: 0-486-65537-7

Paperbound unless otherwise indicated. Available at your book dealer, online at **www.doverpublications.com**, or by writing to Dept. GI, Dover Publications, Inc., 31 East 2nd Street, Mineola, NY 11501. For current price information or for free catalogues (please indicate field of interest), write to Dover Publications or log on to **www.doverpublications.com** and see every Dover book in print. Dover publishes more than 500 books each year on science, elementary and advanced mathematics, biology, music, art, literary history, social sciences, and other areas.